Lipid Analysis in Oils and Fats

VISIT OUR FINE CHEMISTRY SITE ON THE WEB

http://www.finechemistry.com
e-mail orders: direct.orders@itps.co.uk

Lipid Analysis in Oils and Fats

Edited by

R. J. Hamilton
Professor of Organic Chemistry
School of Pharmacy and Chemistry
Liverpool John Moores University
Liverpool
UK

BLACKIE ACADEMIC & PROFESSIONAL
An Imprint of Chapman & Hall
London · Weinheim · New York · Tokyo · Melbourne · Madras

**Published by Blackie Academic & Professional, an imprint of Thomson Science,
2–6 Boundary Row, London SE1 8HN**

Thomson Science, 2–6 Boundary Row, London SE1 8HN, UK

Thomson Science, 115 Fifth Avenue, New York, NY 10003, USA

Thomson Science, Suite 750, 400 Market Street, Philadelphia, PA 19106, USA

Thomson Science, Pappelallee 3, 69469 Weinheim, Germany

First edition 1998

© 1998 Chapman & Hall

Thomson Science is a division of International Thomson Publishing I(T)P®

Typeset in 10/12 pt Times by Pure Tech India Ltd, Pondicherry

Printed in Great Britain by St Edmundsbury Press, Bury St Edmunds

ISBN 0 7514 0414 4

All rights reserved. No part of this publication may be reproduced, stored in a retrieval system or transmitted in any form or by any means, electronic, mechanical, photocopying, recording or otherwise, without the prior written permission of the publishers. Applications for permission should be addressed to the rights manager at the London address of the publisher.
 The publisher makes no representation, express or implied, with regard to the accuracy of the information contained in this book and cannot accept any legal responsibility or liability for any errors or omissions that may be made.

A catalogue record for this book is available from the British Library

Library of Congress Catalog Card Number: 97–74428

∞Printed on acid-free text paper, manufactured in accordance with ANSI/NISO Z39.48–1992 (Permanence of Paper).

Contents

List of contributors   x

Preface   xii

List of abbreviations   xiii

Introduction   xix

1 Lipid analysis using thin-layer chromatography and the Iatroscan   1
N. C. SHANTHA and G. E. NAPOLITANO

   1.1 Introduction   1
   1.2 Principles of thin-layer chromatography   2
      1.2.1 Stationary phases and their applications   2
      1.2.2 Solvent systems and developments   7
      1.2.3 Detection methods   11
      1.2.4 Quantitation   16
   1.3 Applications of thin-layer chromatography   17
   1.4 Thin-layer chromatography–flame ionization detection for lipid analysis   19
      1.4.1 Equipment   19
      1.4.2 Optimization   20
      1.4.3 Comparison of TLC–FID system with gas chromatography   22
      1.4.4 Recent applications of TLC–FID Iatroscan system   23
   1.5 Conclusions   28
   Acknowledgements   28
   References   28

2 Characterization of lipids by supercritical fluid chromatograpahy and supercritical fluid extraction   34
L. G. BLOMBERG, M. DEMIRBÜKER and
M. ANDERSSON

   2.1 Introduction   34
   2.2 Supercritical fluid chromatography   35
      2.2.1 General aspects on the properties of supercritical media   35
      2.2.2 Speed of analysis   37
      2.2.3 Packed column diameter   40
      2.2.4 Mobile phases for supercritical fluid chromatography   40
      2.2.5 Stationary phases for supercritical fluid chromatography   42
      2.2.6 Instrumental   45
      2.2.7 Gradients   47
      2.2.8 Widening the scope of supercritical fluid chromatography   49
      2.2.9 Comparison of supercritical fluid chromatography with other separation techniques for lipid characterization   49

CONTENTS

2.3		Supercritical fluid extraction	50
	2.3.1	Analytical applications	50
	2.3.2	Semi-preparative applications	53
2.4		Reactions in supercritical media	54
2.5		Conclusions	55
Acknowledgements		55
References		55

3 Static headspace gas chromatography in the analysis of oils and fats   59
F. ULBERTH

3.1		Introduction	59
3.2		Theoretical background	60
3.3		Sampling systems	63
3.4		Applications of headspace gas chromatography in lipid chemistry	66
	3.4.1	Volatile fatty acids	67
	3.4.2	Oxidative deterioration of fats and oils	68
	3.4.3	Flavour research	80
	3.4.4	Residues of contaminants	82
	3.4.5	Miscellaneous applications	83
3.5		Conclusions	84
References		84

4 Multinuclear high-resolution nuclear magnetic resonance spectroscopy   87
B. W. K. DIEHL

4.1		Introduction	87
4.2		The instrument	88
4.3		Principles	88
4.4		Spectra	88
4.5		Response	89
4.6		Reproducibility	89
4.7		Calibration	90
4.8		Applications	90
	4.8.1	^{1}H NMR spectroscopy of fatty oils	90
	4.8.2	^{13}C NMR spectroscopy	93
	4.8.3	^{31}P NMR spectroscopy of phospholipids	117
	4.8.4	Ultra-high-resolution ^{31}P NMR spectroscopy	119
	4.8.5	Plasmalogens, O-alkyl-phospholipids and O-alkenyl-phospholipids	122
4.9		Quantitative determination of phospholipids: validation of the ^{31}P NMR method	126
	4.9.1	Test A: reproducibility	129
	4.9.2	Test B: instrument precision	129
	4.9.3	Test C: repeatability	129
	4.9.4	Tests D–I: robustness	131
	4.9.5	Validation summary	133
4.10		Instrument details	133
Acknowledgements		134
References		134

5 Cyclic fatty acids: qualitative and quantitative analysis   136
G. DOBSON

5.1	Introduction	136
5.2	Naturally occurring cyclic fatty acids	139
	5.2.1 Cyclopentenyl fatty acids	139
	5.2.2 Cyclopropane fatty acids	141
	5.2.3 Cyclopropene fatty acids	147
5.3	Cyclic fatty acids formed in heated vegetable oils	152
	5.3.1 Structural analysis of monomeric cyclic fatty acids	153
	5.3.2 Quantification of monomeric cyclic fatty acids	171
	5.3.3 Incorporation of monomeric cyclic fatty acids into biological material	175
5.4	Summary	176
References		176

6 Mass spectrometry of complex lipids   181
A. KUKSIS

6.1	Introduction	181
6.2	Equipment and principles of soft ionization mass spectrometry	181
	6.2.1 Direct probe mass spectrometry	182
	6.2.2 Gas chromatography/mass spectrometry	182
	6.2.3 Liquid chromatography/mass spectrometry	183
6.3	Applications	184
	6.3.1 Fatty acids	184
	6.3.2 Sterols and steryl esters	192
	6.3.3 Neutral glycerolipids	195
	6.3.4 Neutral glycerophospholipids	206
	6.3.5 Acidic glycerophospholipids	215
	6.3.6 Oxygenated glycerophospholipids	223
	6.3.7 Glycosylated glycerophospholipids	227
	6.3.8 Sphingomyelins, sphingolipids and gangliosides	230
	6.3.9 Lipid A and other lipopolysaccharides	236
6.4	Summary and conclusions	238
Acknowledgements		239
References		239

7 Chromatography of food irradiation markers   250
J.-T. MÖRSEL

7.1	Introduction	250
7.2	Chemical changes in irradiated foods	251
	7.2.1 Lipids	251
	7.2.2 Carbohydrates	253
	7.2.3 Proteins	254
7.3	Determination of marker substances by gas chromatography	255
	7.3.1 Determination of hydrocarbons	255
	7.3.2 Determination of cyclobutanones	257
	7.3.3 Determination of dose and method limits	259
7.4	Determination of marker substances by high-pressure liquid chromatography	260
7.5	Conclusions	264
References		264

8 Development of purity criteria for edible vegetable oils 265
J. B. ROSSELL

8.1	Introduction	265
8.2	Materials and methods	266
8.3	Results and discussion	270
	8.3.1 Fatty acid analyses	270
	8.3.2 Other traditional analyses	277
	8.3.3 Problems of maize oil analysis	282
	8.3.4 Stable carbon isotope ratio analysis	283
Acknowledgements		288
References		288

9 Analysis of intact polar lipids by high-pressure liquid chromatography–mass spectrometry/tandem mass spectrometry with use of thermospray or atmospheric pressure ionization 290
A. Å. KARLSSON

9.1	Introduction	290
9.2	Theory	291
	9.2.1 Polar lipids	291
	9.2.2 Liquid chromatography	292
	9.2.3 Mass spectrometry	294
	9.2.4 Liquid chromatography–mass spectrometry	295
9.3	Applications	296
	9.3.1 Thermospray and plasma spray	296
	9.3.2 Electrospray – atmospheric pressure chemical ionization	301
9.4	Practical experiences of analysis of polar lipids by means of liquid chromatography with (tandem) mass spectrometry	311
	9.4.1 Polyetheretherketone compared with stainless-steel tubing	312
	9.4.2 Column packing	312
	9.4.3 Solvent quality	312
	9.4.4 Adduct ions	313
	9.4.5 Wasting	313
9.5	Summary	313
References		314

10 The exploitation of chemometric methods in the analysis of spectroscopic data: application to olive oils 317
A. JONES, A. D. SHAW, G. J. SALTER, G. BIANCHI and D. B. KELL

10.1	Introduction	317
10.2	Olive oil	318
	10.2.1 Economics	318
	10.2.2 Chemistry	318
	10.2.3 Health aspects	320
	10.2.4 Analysis	322
10.3	Data acquisition methods	326
	10.3.1 Nuclear magnetic resonance	326
	10.3.2 Pyrolysis mass spectrometry	328
10.4	Multivariate methods	335
	10.4.1 Principal components analysis	335

10.4.2	Predictive models	339
10.4.3	Multiple linear regression	340
10.4.4	Ridge regression	342
10.4.5	Principal components regression	342
10.4.6	Latent variables	343
10.4.7	Validation	345
10.4.8	Partial least squares regression	350
10.4.9	Artificial neural networks	353
10.4.10	Chemometrics	357
10.4.11	Variable selection	358
10.4.12	Exploitation of multivariate spectroscopies in the identification of the geographical origin of olive oils	363

10.5 Concluding remarks and future prospects — 367
Acknowledgements — 368
References — 368

Index — 377

Contributors

M. Andersson	Department of Analytical Chemistry, Stockholm University, Arrhenius Laboratories for Natural Sciences, S–106G–91, Stockholm, Sweden
G. Bianchi	Istituto Sperimentale per la Elaiotecnica, Contrada Fonte Umano, Città St. Angelo, Pescara, Italy
L. G. Blomberg	Department of Chemistry, Karlstad University, S–651 88 Karlstad, Sweden
M. Demirbüker	Astra AB, S–151 85 Södertälje, Sweden
B. W. K. Diehl	Spectral Service GmbH, Laboratorium für Auftragsanalytik, Postfach 301186, D 50781, Köln, Germany
G. Dobson	Scottish Crop Research Institute, Invergowrie, Dundee DD2 5DA, UK
R. J. Hamilton	10 Norris Way, Formby, Merseyside L37 8DB, UK
A. Jones	Institute of Biological Sciences, University of Wales, Aberystwyth SY23 3DA, UK
A. Å. Karlsson	Nycomed Innovation AB, Ideon-Malmö, S–205 12 Malmö, Sweden and Department of Chemical Ecology and Ecotoxicology, Lund University, S-223 62 Lund, Sweden
D. B. Kell	Institute of Biological Sciences, University of Wales, Aberystwyth SY23 3DA, UK
A. Kuksis	Banting and Best Department of Medical Research, University of Toronto, Charles H. Best Institute, 112 College Street, Toronto, Ontario, Canada M5G 1L6
J.-T. Mörsel	Technische Universität Berlin, Institut für Lebensmittelchemie, Gustav-Meyer Allee 25, 13355 Berlin, Germany

G. E. Napolitano	1005 Brantley Drive, Knoxville, TN 37923–1709, USA
J. B. Rossell	Leatherhead Food Research Association, Randalls Road, Leatherhead, Surrey KY22 7RY, UK
G. J. Salter	Institute of Biological Sciences, University of Wales, Aberystwyth SY23 3DA, UK
N. C. Shantha	Nestlé Research and Development Center, PO Box 4002, 809 Collins Avenue, Marysville, OH 43040–4002, USA
A. D. Shaw	Institute of Biological Sciences, University of Wales, Aberystwyth SY23 3DA, UK
F. Ulberth	Department of Dairy Research and Bacteriology, Agricultural University, Gregor Mendel-Strasse 33, A–1180 Vienna, Austria

Preface

This book has been written to ensure that it will be of benefit to industrial analysts. Most chapters explain some of the relevant theory as well as give some historical references to place the technique in its proper context. In addition the book should appeal to academic scientists who require a good source of applications and a good set of references. Since lipids have many uses the appeal of the book will extend from the food industry to the pharmaceutical industry.

R. Hamilton
Formby
June 1997

Acknowledgement

I would wish to acknowledge the considerable help and encouragement from my wife Shiela.

Abbreviations

The following are the abbreviations used within this book and do not necessarily represent convention or internationally accepted abbreviations.

AAPH	2,2′-azobis (2-aminopropane)dihydrochloride
Ac	acetyl
AchE	acetylcholinesterase
ADC	analog digital converter
AI	artificial intelligence
ALD	aldehyde
AMPL	acetone mobile polar lipids
amu	atomic mass unit
AMVN	2,2′-azobis-2,4-dimethylvaleronitrile
ANN	artificial neural network
AOCS	American Oil Chemists' Society
APCI	atmospheric pressure chemical ionization
APE	N-acyl-phosphatidylethanolamine
API	atmospheric pressure ionization
ARA	arachidonic acid
ASG	acyl-sitosterylglyceride
ASMS	American Society for Mass Spectrometry
ATP	adenosine triphosphate
BE	backward elimination
CAD	collisionally activated dissociation
CBC	cerebroside I^3 sulphate
CBO	Certified Brands of Origin (applied to Italian virgin olive oils)
CCD	charge-coupled device
CE	capillary electrophoresis
CFAM	cyclic fatty acid monomer
CI	chemical ionization
CID	collision-induced dissociation
CL	cardiolipin
Cn:m	Hydrocarbon with n carbon atoms and m double bonds
CR	continuum regression
d.c.	direct current
DCB	2-dodecylcyclobutanone
DECB	2-tetradecenylcyclobutanone
DGB	digalactosyl-diacyl-glyceride
DHA	4,7,10,13,16,19-docosahexaenoic acid

DLCL	dilysocardiolipin
DCI	direct chemical ionization
DG	diacylglycerol
DGDG	digalactosyldi(acyl)glyceride
DGPP	diacylglycerol pyrophosphate
DGTS	diacylglyceryl-N,N,N,N-trimethylhomoserine
DHET	dihydroxyeicosatrienoic acid
DMOX	dimethyloxazoline
DMSO	dimethylsulphoxide
DNPH	dinitrophenylhydrazone
DNPU	dinitrophenylurethane
DOC	*Denominazione di Origine Controllata* (Controlled Denomination of Origin, applied to Italian virgin olive oils; see also CBO)
DOP	*Denominazione di Origine Protèggetta* (Protected Denomination of Production; applied to Italian agricultural food products)
DPG	diphosphatidylglycerol
ECA	equivalent carbon atom
ECL	equivalent chain length
EDTA	ethylenediaminetetraacetate
EET	epoxyeicosatrienoic acid
EI	electron-impact ionization
ELSD	evaporative light-scattering detector
μ-ELSD	miniaturized evaporative light-scattering detector
EPA	5,8,11,14,17-eicosopentaenoic acid
ES	electrospray
FA	fatty acid
FAB	fast atom bombardment
FAC	fatty acid composition
FAME	fatty acid methyl esters
FAO	(UN) Food and Agriculture Organisation
FD	field desorption
FFA	free fatty acid
FFAP	free fatty acid phase
FID	flame ionization detection
FOSFA	Federation of Oils, Fats and Seed Associations Ltd.
FS	forward selection
FT	Fourier transform
FTID	flame thermionic ionization detector
FTIR	Fourier transform infra-red
FT-NMR	Fourier transform nuclear magnetic resonance
GalCer	galactosyl ceramide
GC	gas chromatography
GL	glycolipid

GPC	glycerophosphatidylcholine
GPE	glycerophosphoethanolamine
GPI	glycerophosphatidylinositol or glycerophosphoinositol
GPL	glycerophospholipid
GPS	glycerophosphoserine
HAc	acetic acid
HDL	high-density lipoprotein
HETE	hydroxyeicosatetraenoic acid
HODE	hydroxyoctadecadienoic acid
HPLC	high-pressure liquid chromatography (also referred to as high-performance liquid chromatography)
HPTLC	high-performance thin-layer chromatography
HS	headspace
HSGC	headspace gas chromatography
IAEA	International Atomic Energy Association
i.d.	inner diameter
IFR	Institute of Food Research
ILPS	International Lecithin and Phospholipid Society
IR	infra-red
ISF	International Society for Fat Research
IV	iodine value
LC	liquid chromatography
LCB	long-chain base
LDL	low-density lipoprotein
LPA	lyso-phosphatidic acid
LPC	lyso-phosphatidylcholine
LPE	lyso-phosphatidylethanolamine
LOD	limit of detection
LOOH	linoleic acid hydroperoxide
LOS	lipooligosaccharide
LOQ	limit of quantification
LPS	lipopolysaccharide
LSI	liquid secondary ion
LTB_4	leukotriene B_4
LTC_4	leukotriene C_4
LTD_4	leukotriene D_4
LTE_4	leukotriene E_4
MAFF	(UK) Ministry of Agriculture, Fisheries and Food
Me	methyl group
ME	methyl ester
MG	monoacylglycerol
MGDG	monogalactosyldiacylglyceride
MHE	multiple headspace extraction
MI	matrix ionization

MIKE	mass-analysed ion kinetic energy
MLA	monophosphoryl lipid A
MLCL	monolysocardiolipin
MLR	multiple linear regression
MRM	multiple reaction mode
MS	mass spectrometry
MSD	mass selective detector
MSE	mean square error of prediction
MS–MS	tandem mass spectrometry
MUFA	monounsaturated fatty acid
MW	molecular weight
m/z	mass/charge
NI	negative ion
NICI	negative ion chemical ionization
NIPALS	non-linear iterative partial least squares
OPLC	over-pressure layer chromatography
OPTLC	over-pressure thin-layer chromatography
OTMS	trimethylsilyl ether
PA	phosphatidic acid
PAF	platelet activating factor
PC	phosphatidylcholine
PCA	principal components analysis
PCR	principal components regression
PDB	Pee Dee Belemnite (ratio of ^{12}C to ^{13}C isotopes)
PE	phosphatidylethanolamine
PFB	pentafluorobenzyl
PFBO	pentafluorobenzyloxime
PG	phosphatidylglycerol
PGD_2	prostaglandin D_2
PGE_n	prostaglandin E_n ($n = 0, 1, 2, 3$)
PGF_α	prostaglandin F_α
$PGF_{1\alpha}$	prostaglandin $F_{1\alpha}$
PGF_2	prostaglandin F_2
$PGF_{2\alpha}$	prostaglandin $F_{2\alpha}$
PGH_2	prostaglandin H_2
PI	phosphatidylinositol
PIP	phosphatidylinositol monophosphate
PIP-2	phosphatidylinositol bisphosphate
PL	phospholipid
PLS	partial least squares regression
ppb	parts per billion
ppm	parts per million
PS	phosphatidylserine
PSP	plasmaspray

PTFE	polytetrafluoroethylene
PUFA	polyunsaturated fatty acid
PVA	polyvinyl alcohol
PyMS	pyrolysis mass spectrometry
Q	quadrupole
QNP	quattro nuclei probe
r.f.	radio frequency
RMSEP	root mean square error of prediction
RP	reversed phase
RP-HPLC	reversed-phase high-pressure liquid chromatography
RR	ridge regression
SCIR	stable carbon isotope ratio [$\delta(^{13}C)$]
SFC	supercritical fluid chromatography
SFS	supercritical fluid extraction
SG	sitosterylglycoside
SIM	single ion monitoring
SMLR	stepwise multiple linear regression
SPH	sphingomyelin
Suc CL	succinylated cardiolipin
t-BDMS	tert-butyldimethylsilyl
TEA	triethylamine
TG	tri(acyl)glycerol
TH	α-tocopherol
TLC	thin-layer chromatography
TMS	trimethylsilyl
TNP	trinitrobenzenesulphonic acid
TOF	time of flight
TMS	tetramethylsilane
TNT	trinitrotoluene
TPH	total petroleum hydrocarbon
TPP	triphenylphosphate
TS	thermospray
TTLC	tubular thin-layer chromatography
TXB_2	thromboxane B_2
UV	ultraviolet
VFA	volatile fatty acid
VHC	volatile hydrocarbon
VLDL	very low density lipoprotein
WHO	World Health Organization
WMP	whole milk powder

Introduction

The revolution in analysis begun with the two seminal papers by A. J. P. Martin in 1941 and 1952 had far-reaching consequences for our understanding of lipids in our everyday life. As chromatographic techniques matured, smaller and smaller quantities of lipids could be separated and detected, with the result that studies could be initiated into biological transformations and food constituents which had not been possible before 1950. Separation science and lipid analysis have been expanding in every decade since. New techniques have uncovered unexpected compounds and forced re-examinations of earlier theories. The present volume is the second in the series and concentrates on analysis of lipids and helps to explain the importance of some recent developments. The authors range from well-known experts to some newer workers in the field, all with a distinctive contribution to make in elucidating the range of techniques which are available to the lipid scientist.

In Chapter 1, Shantha of Nestlé and Napolitano of Oak Ridge National Laboratory review the position of thin layer chromatography (TLC), describing commercial apparatus and highlighting how easy it is to modify TLC. Solvent systems, detectors and stationary phases are explained and a number of applications concerning adulteration, for example of cocoa butter, are mentioned. The authors then concentrate on the so-called weakness of the technique, namely quantitation. The authors comprehensively reject this claim and show how a TLC–flame ionization detector system gives acceptable quantitative results.

One of the newer derivatives of chromatography is supercritical fluid chromatography which Blomberg, Demirbüker and Andersson describe in Chapter 2. The theory is explained and the peculiar advantages which the supercritical fluid confers on the system are designated. Open tubular and packed columns can both be used and examples are shown.

In Chapter 3 Ulberth outlines the theory and practice of static headspace gas chromatography. He then considers the analysis of volatile lipids in the study of the digestive tract and in nutritional physiology. Equally important is the contribution headspace analysis can make in the study of oxidative rancidity. Ulberth contrasts static headspace gas chromatography with dynamic headspace analysis.

In industry, most lipid analysts are familiar with wide-line nuclear magnetic resonance (NMR). In Chapter 4 Diehl describes the application of Fourier transform NMR spectroscopy and illustrates that the technique can be used in lipid chemistry both qualitatively and quantitatively. By means of correlation tables and typical spectra he shows how the composition of fat mixtures can be determined. He also outlines the value of ^{31}P NMR to detect phosphoglycerides. Although this NMR technique has not had a major impact on industrial analysis, the non-destructive nature of the technique may enable it to challenge gas chromatography (GC) in the future.

Dobson has described in Chapter 5 how the analysis of one group of lipids, namely the cyclic fatty acids, must be attacked. In this instance a combined approach using TLC, high-pressure liquid chromatography (HPLC), GC, gas chromatography–mass spectroscopy (GC–MS), NMR and Fourier transform infra-red (FTIR) spectroscopy as well as chemical modification is required. He illustrates the analysis of cyclic fatty acids from heat-treated linseed oil and from sunflower oil. The mechanism by which these acids are produced is outlined. Considerable interest has been generated in these compounds because of the large amounts which can be formed during the frying of chips and snack foods.

In a wide-reaching chapter Kuksis explains the principles and apparatus used in soft ionization mass spectroscopy (Chapter 6). The ever increasing scope of these techniques has brought the analysis of complex high molecular weight molecules such as sphingomyelins within the reach of many scientists. He has shown that quadrupole instruments can be applied to a wide range of lipids, from hydroperoxy fatty acids through oxygenated sterols and serine glycerophospholipids to lipopolysaccharides. Sodiated adducts permit levels of 5 fmol to be detected.

Food irradiation may now be used in over 40 countries. The short chapter by Mörsel describes the need to find markers which indicate that food has been irradiated (Chapter 7). The lipids which have been examined as potential markers include hydroperoxides, alkenes, alkanes, aldehydes and cyclobutanones. Derivative formation with cyclobutanones can increase the sensitivity of detection of butanones in irradiated food.

In Chapter 8 a leading expert in the detection of adulteration of oils and fats, Rossell, explains the economic advantages which tempt unscrupulous dealers to add cheap oils to high-value oils such as olive oil. He then describes a technique which he and his coworkers at Bristol have pioneered where stable carbon isotope ratio (SCIR) measurement has simplified the task of detecting adulteration. He compares sterol and fatty acid methyl ester analysis with SCIR and shows how clearly the latter detects maize oil.

In Chapter 9 Karlsson explains the importance of combined techniques, for example joint use of HPLC, flow or loop injection and MS or tandem MS in the analysis of lipids. The techniques of thermospray, electrospray and atmospheric pressure chemical ionization are described. Applications range from platelet activating factor to cardiolipin. Karlsson mentions some of the practical procedures which he adopts in his laboratory to ensure good reproducible analysis.

Finally in Chapter 10, Jones, Shaw, Salter, Bianchi and Kell, in a joint contribution from the University of Wales at Aberystwyth and the Istituto Sperimentale per la Elaiotecnica in Italy, give a thorough description of chemometric methods. They explain how multivariate analysis can be applied to NMR data and to pyrolysis mass spectrometry. They stress the relatively cheap methodology which the latter technique offers.

1 Lipid analysis using thin-layer chromatography and the Iatroscan

N. C. SHANTHA and G. E. NAPOLITANO

1.1 Introduction

Thin-layer chromatography (TLC) was originally developed by Nikolai Izmailov and his postgraduate student Maria S. Schreiber as early as 1938. It was basically used to replace the time-consuming column chromatography so as to obtain faster results; in addition, it required less sample. Berezkin (1995) speaks of the history of planar chromatography and also includes the original paper by Izmailov and Schreiber. These articles were in Russian and were not widely read. However, this technique was further investigated and popularized following an excellent review by Mangold (1961) and a handbook on TLC by Stahl (1969). Since then TLC has become one of the most popular separation techniques in the field of lipids, oils and fats. This separation science owes its popularity to its simplicity, reliability and relatively inexpensive equipment. There are many text books and reviews published on TLC and some of the more recent ones include Sherma and Fried (1996), Touchstone (1995), Fried and Sherma (1996) and Shukla (1995). Several lipid separations have been standardized, which makes it easier to achieve effective separations even by a person not very familiar to this field; however, other separations such as for alkaloids, flavones, medicine and other natural components are still being developed, and TLC of these compounds requires the operator to have more expertise and a greater understanding of the basic concept of TLC separation and analysis. Sherma (1994, 1996) has reviewed published articles on TLC every alternate year in the journal *Analytical Chemistry*.

Christie (1990), in an overview of TLC, asks the question 'Has thin-layer chromatography had its day?'. Comparison of this simple technique with the newer and more sophisticated high pressure liquid chromatography (HPLC) makes TLC look obsolete. However, as also agreed by the author, traditional TLC is still used and is likely to be around for many years to come. Between the years 1990 and 1996 TLC has found application in several fields, and publications relate to food (18%), medical, clinical and biological applications (26%), pharmaceuticals (25%), environmental studies (11%), detection (14%) and chemical synthesis monitoring (6%) (Weins and Hauck, 1995). It is estimated that there were between 500–900 articles per year on TLC in the period 1992–94 based on *Analytical and Chemical Abstracts*. The availability of several absorbent materials including high-performing silica, bonded phases and impregnated layers have increased its versatility for numerous and quick separations.

In the field of lipids TLC is used mostly for routine separations, identification and quantitation. New developments include multidimensional and multimodal manipulations to improve separations and automations and computerization for optimum resolutions. Modern TLC includes highly automated techniques right from sample application and development to detection and quantitation. Although such automation is more expensive it has found several applications in the drug and medicinal industry. In the oils and fats industry such automated techniques are used for studying the phospholipid compositions of emulsifiers. A recent review by Muthing (1996) on glycosphingolipids emphasizes the role of TLC as a powerful analytical tool for a wide range of applications. Adaptations to traditional TLC techniques include high-pressure thin-layer chromatography (HPTLC), over-pressure thin-layer chromatography (OPTLC) and tubular thin-layer chromatography (TTLC). The perceived weakness of TLC is the quantitation aspect and this has led to the evolution of the TLC/FID (flame ionization detection) Iatroscan system, which will be discussed at length in this chapter.

1.2 Principles of thin-layer chromatography

The principles and theory behind TLC are well established. In short it is founded on the difference in the affinity of a component or components towards a stationary and a mobile phase. The four important aspects of TLC include the stationary phase, the mobile phase, detection and quantitation. The absorptive processes, the sorbent–solvent interactions and the effect of the chemical structure on adsorption have been discussed by Pomeranz and MeLoan (1994).

1.2.1 Stationary phases and their applications

There are several stationary phases that can be used for separation of various compounds including lipids, natural products, biological compounds, drugs, etc. Examples are Celite^{TM} (Supelco Inc., PA), cellulose powder, ion-exchange cellulose, starch, polyamides and Sephadex^{TM} (Supelco Inc., PA), but the most popular ones for lipid separations include silica gel, alumina and kieselguhr. These adsorbents can also be modified by impregnation with other substances so as to achieve the desired separations. Based on their characteristics these phases can be classified as normal or reversed.

Normal phases. Silica is generally the normal phase used in lipid separations. The average particle sizes in traditional TLC range between 10 μm and 50 μm, with a fairly wide size distribution. In normal phase TLC the stationary phase (such as silica) is polar and the mobile phase is relatively non-polar. Scott (1982) has extensively reviewed silica gel and its properties, the water absorption of the silica surface and its interaction with polar and non-polar solvents.

High-performance silica is used in high-performance thin-layer chromatography (HPTLC). HPTLC differs from normal TLC in that the size of the absorbent (usually silica) is only 5 μm, with a narrow distribution. This enables HPTLC to give better separations compared with TLC, which uses a standard silica, and, moreover, HPTLC requires a smaller sample size and has a lower detection limit compared with conventional TLC. HPTLC plates of varying sizes are commercially available and of late have found considerable applications in the field of lipids. Weins and Hauck (1995), in their survey of TLC, conclude that the use of HPTLC plates increased by 30% over the period 1993–95. An excellent application of HPTLC is illustrated in Fig. 1.1 for the separation of neutral and complex lipids. Yao and Rastetter (1985) have achieved separation of more than 20 lipid classes of tissue lipids on HPTLC plates using four developing solvents.

The efficiency of the available commercial silica plates allows separation within lipid classes and this has been used to advantage. Generally, longer-chain fatty acid esters have a higher mobility compared with short-chain esters. Steele and Banks (1994) have tried to determine why hydrogenated milk fat triglyceride, fractionates into three bands on TLC. The three bands from hydrogenated milk fat were resolved into individual triglycerides by means of HPLC and were analysed for fatty acid composition. From these findings the authors are of the opinion that the separation was based on differences in the fatty acid chain length of the triglycerides as well as a result of the stereospecific distribution of fatty acids within the triglycerides.

We have achieved good separation of polyunsaturated fatty acid esters from saturated, monosaturated and disaturated fatty acid esters and this has been used as a means of identifying the presence of polyunsaturates, especially fish oil, in samples (Fig. 1.2). However, if there is a considerable amount of trienoic fatty acids in the mixture, as in egg lipids, the separation is not very pronounced (Shantha and Ackman, 1991a). Nakamura, Fukuda and Tanaka (1996) have used similar TLC separation of marine lipids and were able to quantitate the amount of polyunsaturated fatty acids by using scanning densitometry followed by Coomassie blue staining.

We have used silica gel TLC to separate conjugated triglycerides and methyl esters from the non-conjugated species in the seeds of *Momordica charantia* and *Trichosanthus anguina* (unpublished results). These seeds contain between 50%–60% of their total fatty acids as α-eleostearic and punicic

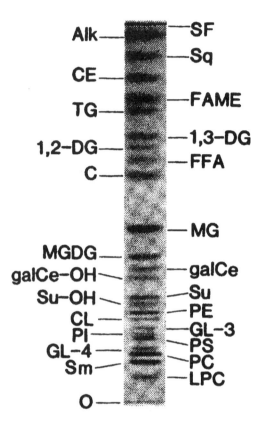

Figure 1.1 Separation of complex standard lipid mixtures on a 20 cm × 20 cm high-performance thin-layer chromatography plate using four developments. The first development was up to a distance of 5 cm above the origin in the solvent system ethyl acetate/1-propanol/chloroform/methanol/0.025% KCl, 25:25:25:10:9 vol./vol. The second development was up to 8 cm above the origin in the solvent system toluene/ether/ethanol/acetic acid, 60:40:1:0.23 vol./vol. The third development was to the full length (9 cm) in the solvent system hexane/diethyl ether, 94:6 vol./vol., followed by the last development to full length in hexane. The plates were freed of solvent between developments by blowing with hot air. Reproduced with permission from Yao, J. K. and Rastetter, G. M., Microanalysis of complex tissue lipids by high-performance thin-layer chromatography, *Analytical Biochemistry*, **150**, 111–16.

acid, respectively. The methyl esters of these oils were spotted on silica gel TLC plate and developed twice with a solvent system of hexane–diethyl ether (94:6 vol./vol.). The band corresponding to conjugated trienoic acid methyl ester resolved well from non-conjugated systems and could be detected under ultraviolet (UV). These bands were then scraped off and extracted with peroxide-free diethyl ether and were found to be 95%–99% pure as analysed by gas chromatography.

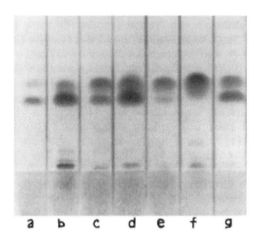

Figure 1.2 High-performance thin-layer chromatogram of methyl esters of fatty acids, showing separation based on unsaturation. The plates were developed in the solvent system hexane/diethyl ether, 92:8 vol./vol. a = a standard mixture of tetracosaenoic (24:1 vol./vol.) and docosahexaenoic (22:6 vol./vol.) fatty acid methyl esters; b = sea scallop lipids; c = dogfish liver; d = menhaden; e = redfish; f = rapeseed; g = cod liver. Reproduced with permission from Shantha, N. C. and Ackman, R. G., Silica gel thin-layer chromatographic method for concentration of longer-chain polyunsaturated fatty acids from food and marine lipids, *Canadian Institute of Food Science and Technology Journal*, **24**, 156-60, 1991.

Reversed phases. This phase is so called because of the reversed polarity of the two phases compared with the normal TLC. In the reversed phase the stationary phase is non-polar (generally hydrocarbon) and the mobile phase is polar (polar solvents including water). The solvents generally used in reversed phase TLC are alcohols, water, acetonitrile, and tetrahydrofuran, dimethylformamide and dimethylsulphoxide (DMSO). The stationary phase consists of silica gel wherein the polar hydroxyl group of silanol has been chemically modified by bonding with hydrocarbons C2, C8, C12 and C18. Layers with the same bonded group can behave differently depending on the extent to which the surface of the silica has been modified. For a given extent of surface loading the hydrophobicity increases with increase in length of carbon chain. Thus based on the extent and type of loading on silica gel the phase can exhibit either hydrophilic or hydrophobic character and can thus be manipulated to work as a normal phase, able to tolerate a more polar solvent, or as a reversed phase in TLC applications. In lipid analysis the reversed phase is used to effect separations of triglycerides and fatty acids in terms of carbon numbers. With the advent of the more sophisticated reversed-phase high-pressure liquid chromatography (RP-HPLC) reversed phase TLC is rarely used for such separations.

Modification of the stationary phase. Over the years researchers have tried several different impregnations of the silica or alumina phase to give the desired separation. The most popular among them include silver nitrate impregnations and boric acid impregnations. Silver nitrate impregnations are used to separate fatty acids, such as methyl esters, triglycerides and sterols, based on their degree of unsaturation, and these have been reviewed by Dobson, Christie and Nikolova-Damyanova (1995). The silver ion forms a reversible complex with the π electrons of the double bond of unsaturated fatty acids, thereby decreasing their mobility. Silver nitrate plates can be made by incorporating the desired amount of silver nitrate in the silica gel while preparing the plates; alternatively, ready made silica plates can be immersed in a solution of silver nitrate in acetonitrile or alcohol, or even a spray of these solutions can be used.

Silver ion TLC has also been used for separation of a variety of substituted unsaturated fatty acids such as epoxy, hydroxy and halohydroxy fatty acids, and these have been reviewed by Morris and Nichols (1972). Wax esters and steryl esters exhibit similar polarities and do not separate on column chromatography or normal silica TLC. Kiosseoglou and Boskou (1990) have separated the wax esters from steryl esters by using argentation TLC. They have used silica gel plates impregnated with 10% silver nitrate, with developing solvent hexane/chloroform (7:3 vol./vol.).

Boric acid impregnations have been used to separate the 1,2- (2,3-) and the 1,3- diglycerides and phospholipids. Boric acid impregnated plates can be prepared in a similar way to silver nitrate plates. Boric acid complexes with vicinal hydroxyl groups and leads to slower migration of these compounds. Boric acid thus complexes with phosphatidylinositol and effects its resolution. Apart from its complexation with vicinal diols, boric acid has also been found to effect the protonation equilibrium of the phosphatidic acid group and its migration (Traitler and Janchen, 1993). Other less frequently used impregnations include EDTA (ethylenediamine tetra-acetic acid) impregnated silica to improve the separation of phospholipids, and more particularly the acidic phospholipids (Allan and Cockcroft, 1982). The impregnation of silica with ammonium sulphate has been shown to improve the resolution between phosphatidylserine and phosphatidylinositol. Wang and Gustafson (1992) have used silica gel H containing 0.4% ammonium sulphate and acetone in the solvent system chloroform/methanol/acetic acid/acetone/water in the ratios 40:25:7:4:2 vol./vol. to separate phospholipids of tissue extract. The addition of acetone is said to eliminate the tailing of phosphatidylserine. However, the disadvantage with this method is that phosphatidylethanolamine partially degrades to its lyso analogue. Wang and Gustafson believed that the sulphate in the plate was responsible for the cleavage of the ester bond in phosphatidylethanolamine (PE), although it is not clear why this degradation is more selective to PE compared with phosphatidylinositol and phosphatidylserine.

Phospholipids of garlic oil have been separated into individual components by using silica gel G made in a slurry of 0.01 M sodium carbonate solution and the solvent system chloroform/methanol/water in the ratios 65:26:04 vol./vol. The alkaline sodium carbonate is believed to give a better resolution among the phospholipids (Huq et al., 1991).

1.2.2 Solvent systems and developments

Solvent systems. Unless mentioned otherwise, most of the solvent systems discussed refer to normal and high-performing silica phases. Solvents with low boiling point, low viscosity and low toxicity are suitable for TLC applications. A low boiling point helps in the quick evaporation of the solvent from the surface layer, and low viscosity helps in faster movement of the solvent during development. The selection of a suitable solvent is very important if one is to get a good separation of the lipid classes. Low-polarity solvents such as hexane and diethyl ether in varying ratios are generally suitable for separation of neutral lipids, whereas solvent systems of relatively high polarity such as chloroform, methanol and water or acetic acid are generally used for separation of complex lipids. Addition of a minute amount of acetic acid in the solvent to separate neutral lipids usually eliminates the tailing of free fatty acids. Demixing of multicomponent solvent systems should be avoided.

Table 1.1 lists with references the commonly used solvent systems for the separation of neutral and polar lipids. This table can be used as a guide for the selection of suitable solvent systems for specific separations. The humidity in the laboratory, the amount of moisture in the silica plates, chamber saturation with solvent, etc., also play an important role in effective separations and occasionally a slight modification to the solvent system mentioned may be required to get an ideal separation.

Development techniques. There are several development techniques available depending on the separation that is desired. Developments can be unidimensional or two-dimensional. Moreover, developments can be made either once or several times depending on the separations desired.

Single development. A single development is generally sufficient to separate between simple lipid classes or between polar lipids (phospholipids). The solvents for such separations are mentioned in Table 1.1, and examples of separations of neutral and complex lipids can be found in any basic text on lipid analysis (Christie, 1982). Fried (1996) has tabulated the retention data of different neutral and complex lipid classes using different solvent systems, and Aloisi, Sherma and Fried (1990) have compared 24 different solvent systems for separation of lipids by single development.

Multiple (unidimensional) development. This term is generally used to signify two or more development cycles. The development could be carried

Table 1.1 Selected solvent systems (ratios in vol./vol.) for the separation of lipids on silica gel thin-layer chromatography

Solvent system (ratios in v/v)	Reference
Neutral lipids (unidimensional development):	
petroleum ether/diethyl ether/acetic acid (90:10:1)	Mangold and Malins, 1960
hexane/diethyl ether/formic acid (80:20:2)	Storry and Tuckley, 1967
petroleum ether/diethyl ether/acetic acid (80:20:1)	Stahl, 1969
heptane/isopropyl ether/acetic acid (60:40:4)	Breckenridge and Kuksis, 1968
toluene/diethyl ether/ethyl acetate/acetic acid (80:10:10:0.2)	Storry and Tuckley, 1967
first, toluene (100); second, hexane/chloroform/methanol (30:18:2)[a]	Higgs, Sherma and Fried, 1990
Phospholipids (unidimensional development)	
chloroform/methanol/water (25:10:1)	Skipski, Peterson and Barclay, 1964
chloroform/methanol/water (65:25:4)	Wagner, Horhammer and Wolffe, 1961
chloroform/petroleum ether/methanol/acetic acid (50:3:1.6:1)	Pappas, Mullins and Gadsden, 1982
chloroform/methanol/acetic acid/water (25:15:4:2)	Skipski, Peterson and Barclay, 1964
chloroform/ethanol/triethylamine/water (30:34:30:8)	Touchstone *et al.*, 1983
chloroform/methanol/2-propanol/0.25% KCl/ethyl acetate (30:9:25:6:18)	Bradova *et al.*, 1990
Phospholipids (two-dimensional development):[a]	
First, chloroform/methanol/water (65:25:4); second, hexane/diethyl ether/acetic acid (80:20:1)	Omogbai, 1990
First, chloroform/methanol/water (65:25:4); second, n-butanol/acetic acid/water (60:20:20)	Rouser, Kritchevski and Yamamota, 1967
First, chloroform/methanol/28% aqueous ammonia (65:35:5); second, chloroform/acetone/methanol/acetic acid/water (10:4:2:2:1)	Rouser, Kritchevski and Yamamota, 1967
First, chloroform/methanol/7N ammonium hydroxide (65:30:4); second, chloroform/methanol/acetic acid/water (170:25:25:6)	Nichols, 1964
Neutral lipids and phospholipids (unidimensional development):	
First, chloroform/methanol/water (65:25:4); second, hexane/diethyl ether (4:1)[a]	Johnston, 1971

[a] The words *first* and *second* refer to development.

out twice, in which case it is referred to as double development. The two developments could be made with the same solvent, in which case a better separation is produced between the lipid classes compared with a single development. Three or four developments have also been used for intricate separations. Care has to be taken to remove completely the solvents from the plates between developments. An efficient HPTLC separation of complex tissue lipids into 20 different lipid subclasses has been achieved by Yao and Rastetter (1985) with use of four developments (Fig. 1.1). Lipids were spotted on a 20 cm × 20 cm HPTLC plate and developed with the first solvent system (methyl acetate/1-propanol/chloroform/methanol/0.25% KCl, 25:25:25:10:9 vol./vol.) to a distance of 5 cm above the origin. The second development was up to 8 cm above the origin in the solvent system benzene/diethyl ether/ethanol/acetic acid, 60:40:1:0.23 vol./vol. The third development was to the full length (9 cm) in the solvent system hexane–diethyl ether, 94:6 vol./vol., followed by the last development to full length in hexane. The plates were freed of solvent between developments by blowing with hot air.

Two-dimensional TLC. This mode of development is generally used to achieve separations between complex lipid classes and to check the purity of a standard. The sample is spotted on one corner of the TLC plate and developed in one direction. Following this the TLC plate is made free of solvent, turned by an angle of 90° and developed in a second solvent. Examples of solvent systems used for two-dimensional separation can be found in Table 1.1. Zakaria, Gonnord and Guiochon (1983) have reviewed the applications of two-dimensional TLC for analysis of several compounds, including lipids. Use of the two-dimensional technique for separation of neutral lipids is not very common; an exception would be to effect the separation of non-polar compounds of similar polarities such as hydrocarbons, steryl esters and methyl esters which migrate very closely in one-dimensional TLC. Thompson (1987) has used a two-dimensional system to effect separation of such lipid classes found in the digestive glands of snails.

Gradient development. Gradient developments have been used to achieve separation of complex mixtures such as plant extracts, dyes, etc. However, their application to lipid separation is not very common. Golkiewicz (1996) discusses in detail stationary phase gradients and mobile phase gradients, the theory behind solvent selection, automated techniques and applications of gradient elution in TLC.

Forced-flow development. Forced-flow planar chromatography is a development technique wherein pressure is used to aid the mobility of the developing solvent. Examples of this are over-pressure layer chromatography (OPLC) and over-pressure thin-layer chromatography (OPTLC). In the latter a forced-flow technique is used to decrease the development time and thus speed up the separations. A pump controls the speed of the mobile phase. Theoretically, this method is faster than when movement of the solvent is due to capillary action alone (normal TLC) and can be used to advantage if slow-moving viscous solvents are involved as developing solvents.

The main advantages of OPLC include smaller solvent volumes (< 25 ml), shorter development times (2–5 min compared with 30–90 min in normal TLC) and the reproducibility of R_f values (5% or less standard deviation). Ackman and Ratnayake (1989) give more detail of this technique as well as examples of its application to lipid separations.

Circular and anticircular development methods. This technique produces a radial chromatogram which requires special scanners and is generally not used for lipid separation. The principles of these development methods and the names of companies manufacturing chambers for use in such methods are reviewed by Cserhati and Forgacs (1996).

Multidimensional and multimodal developments. Such developments reflect the recent advances in TLC and are used to achieve maximum separation. This refers to unidimensional multiple developments as well as to two-dimensional development. The sorbent layers and mobile phase can also be manipulated to achieve better separations. It is not uncommon to use two phases at a time to effect better separations. Two sorbents may be slurried together and spread homogeneously on the plate, or portions of the plate could have different phases. An example of such an application includes the use of a plate having one strip of normal silica with the rest of the plate being impregnated with silver nitrate. Neutral lipid sample is spotted in the corner of the silica strip and developed in a solvent which can resolve the triglycerides. The plate is freed of solvent and turned by an angle of 90° and developed in the second direction into the silver-nitrate-impregnated layers by using an appropriate solvent to separate the triglycerides further, in terms of unsaturation. Thus with a single plate it is possible to achieve separation of triglycerides from other lipid classes as well as separation among the triglycerides based on their unsaturation. Similar techniques can be used to separate between the molecular species by using part of the TLC plate as a reversed phase.

Pchelkin and Vereshchagin (1992) have used a combination of silver ion TLC and reversed phase silver ion TLC to effect separation of 15 species of diglycerides and have also identified these by using reference standards. They used a two-dimensional TLC approach to effect these separations. The first development was on silver-nitrate-impregnated silica plate. The plate was freed of solvent and impregnated by spraying with *n*-tetradecane to make it a silver-nitrate-impregnated reverse phase. The plate was then turned by an angle of 90° and the second development was carried out on silver ion reversed phase. Figure 1.3 gives a schematic illustration of this separation. This represents a very good example wherein the flexibility of planar TLC makes it possible to manipulate the technique to give the desired separations. Pchelkin and Vereshchagin are of the opinion that the separation achieved using this technique is superior to separations using reversed phase HPLC and compares with gas chromatography (GC), although derivatization is a necessary step in GC. The multidimensional method for planar chromatography and theoretical considerations for optimum separation have been reviewed by Poole and Poole (1995).

TLC and developments on TLC have now become so modernized that one can optimize the mobile phase composition by means of a computer-

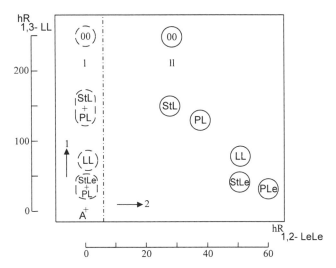

Figure 1.3 A two-dimensional thin-layer chromatogram of a mixture of six rac-1,2-diglyceride species. A = starting point; I and II = regions of the plate non-impregnated and impregnated with n-tetradecane, respectively; 1 = direction of mobile phase [silver-nitrate-saturated chloroform-isopropanol (99:1) vol./vol.)] migration during silver ion thin-layer chromatography (TLC); 2 = direction of mobile phase (5% wt/vol. boric acid solution in methanol, saturated with silver nitrate and n-tetradecane) migration during silver ion reversed phase TLC (RP-TLC). Zone limits of silver ion TLC of diglycerides in region I and after silver ion RP-TLC in region II are indicated by broken-line and solid-line enclosures, respectively. Diglycerides are named based on their fatty acids: St = stearic; P = palmitic; L = linoleic; Le = linolenic; O = oleic. Reprinted from *Journal of Chromatography*, **603**, Pchelkin, V. P. and Vereshchagin, A. G., Reversed phase thin-layer chromatography of diacylglycerols in the presence of silver ions, 213–22, 1992, with kind permission of Elsevier Science – NL, Sara Burgerhartstraat 25, 1055 KV, Amsterdam, The Netherlands.

assisted program by using the retention, R_f, as the criterion. Similar computer-assisted mobile phase optimization methods have also been developed for stepwise gradient and multiple development TLC. These computer-assisted optimization methods have been evaluated by Cavalli *et al.* (1993). Muthing (1996) describes a development chamber, TLC-MAT, which allows automatic developments without supervision. The development of solvent and distance is dictated by a sensor which senses the differences in the light-reflecting properties of dry and wet areas of the TLC plates.

1.2.3 Detection methods

There are several reagents that can be used for qualitative and quantitative assay of the separated lipids. Apart from these there are specific spray reagents available to characterize individual lipid groups. Detection methods can be classified mainly into two categories, the non-destructive methods and the destructive methods, and these can be further classified into specific and non-specific detection methods. It is obvious that if lipid purification is involved a non-destructive method needs to be used.

Non-destructive methods: non-specific spray reagents

Iodine. The most convenient and commonly used technique is exposure to iodine vapours. Iodine crystals are put in one of the covered tanks and placed under the fume hood, vapours of iodine being toxic. After it is freed from the solvent, the TLC plate is placed in the iodine tank and the spots are visualized. This is a qualitative technique, although some researchers use it for quantitation by densitometry. The theory behind iodine exposure is that the iodine is physically absorbed into the lipid spot. The disadvantage with this technique is that the saturated lipid samples do not take up iodine easily and, moreover, when unsaturated molecules are left in iodine for a sufficiently long time they tend to react with it chemically. Iodine is added at the double bonds and produces artifacts, and this has been proved by GC analysis of material extracted from plates after exposure to iodine (Vioque and Holman, 1962).

$2'7'$-Dichlorofluorescein. A 0.2% solution of $2'7'$-dichlorofluorescein in water or 95% alcohol solution is generally sprayed on the plate. This renders all classes of lipids visible (yellow) when viewed under UV light. Lipids can be eluted from the silica plate by extraction with a suitable solvent. The dye generally carries into the sample and can be separated either by washing with water if the eluted solvent is immiscible in water or by passing the organic layer through a bed of glass wool, which absorbs the dye. Silica plates with dye already incorporated are now available and are convenient to use.

Rhodamine 6G. This is similar to $2'7'$-dichlorofluorescein except that the lipids appear pink under a UV lamp. With the use of these dyes fluorimetry can be used for quantitation.

Water. Occasionally we have sprayed our plates with water when purifying large amounts of lipids, for example triglycerides, from oil samples. The hydrophobicity of triglycerides makes them appear as white spots against a translucent background when held up to the light.

Lipids containing chromophores. Some lipids contain chromophores and can be visualized directly under UV or visible light without any staining. We have separated triglycerides containing conjugated trienoic acid, as well as their methyl esters, by visualization under UV light.

Self-staining. Martinez-Lorenzo *et al.* (1994) observed that polyunsaturated fatty acid methyl esters, when separated by silver nitrate TLC using the solvent system toluene/acetonitrile (97:3 vol./vol.), darkened on standing. As such darkening was not apparent with other solvent systems, such as hexane/diethyl ether and diethyl ether/methanol, Martinez-Lorenzo et al. (1994) felt that the darkening was due to toluene-facilitated electron transfer from the olefinic bonds to the silver. The penta- and hexa- unsaturated fatty acid methyl ester darkened within 24–48 h, the trienoic esters took 7–12 days to darken, whereas the dienoic esters did not darken even after 20 days of standing. Given the oxidative susceptibility of these polyunsaturated acids it is not practical to use self-staining in purification methods. Martinez-Lorenzo et al. thought that this method would be particularly useful for locating fatty acids by autoradiography (generally 2–10 days exposure) followed by quantitation by liquid scintillation counting.

Destructive methods.

Non-specific reagents. These methods generally include spraying with a corrosive reagent and charring the plates to make the spots appear. A 50% solution of sulphuric acid, either in water or methanol, followed by heating the plate at 120°C for 1 h shows up the lipid samples as brown-black spots. Sterols initially appear as a pink-purple spot and later turn to black. The disadvantage with this method, apart from the corrosiveness of the reagent, is that saturated lipids take longer to show colour and, moreover, they char to a lesser extent than do unsaturated analogues. If the plates are removed too soon one can miss spots of saturated species which take longer to appear.

Potassium dichromate (5%) in 40% sulphuric acid works in a similar manner to sulphuric acid spray. A 50% solution of sulphuric acid in methanol may require heating at a lower temperature (100°C) for brown-black spots of lipids to appear. A 3%–6% solution of cupric acetate in 8%–10% phosphoric acid turns lipids black when heated at 160°C for 1–2 h. Molybdophosphoric acid (5%) in ethanol is another destructive reagent that turns lipids blue to black when heated at 120°C for 1 h. Coomassie blue (0.03% in 20% methanol) turns lipids to blue spots on a white background.

Specific reagents. These are reagents that react selectively with specific functional groups in the lipid moiety and give colour. There are specific reagents for almost all lipid classes and these are tabulated in Table 1.2 along with their preparations and detection (Fried, 1996). Most of these reagents are destructive and are therefore not used for purification purposes.

Overlay techniques. These techniques are generally used in the field of biology and medicine. They involve the use of ligands which specifically bind certain biologically active compounds. The selective binding of cholera toxin to gangliosides was resolved on a TLC plate (Magnani, Smith and Ginsburg, 1980).

Table 1.2 Specific chemical detection procedures for various lipids

Compound class	Reagent	Procedure	Results
Cholesterol and cholesteryl esters	Ferric chloride	Dissolve 50 mg of $FeCl_3 \cdot 6H_2O$ in 90 ml of H_2O, along with 5 ml of acetic acid and 5 ml of sulphuric acid; spray the plate and then heat it at 90°C–100°C for 2–3 min	Cholesterol and cholesteryl esters appear as red-violet spots; the cholesterol spot appears before the ester spot
Free fatty acids	2′,7′-Dichlorofluorescein/ aluminium chloride/ferric chloride	Prepare three solutions as follows: (1) 0.1% 2′,7′-dichlorofluorescein in 95% methanol; (2) 1% aluminium chloride in ethanol; (3) 1% aqueous ferric chloride; spray the plate in turn with solutions 1, 2, and 3. Warm the plate (about 45°C) briefly between each spray	Free fatty acids give a rose colour
Lipids containing phosphorus	Molybdic oxide/molybdenum Zinzadze reagent	Prepare a 4% solution of molybdic oxide in 70% H_2SO_4 (solution 1); add 0.4 g of powdered molybdenum to 100 ml of solution 1. Add 200 ml of H_2O and filter. Final spray consists of 100 ml of the above plus 200 ml of water and 240 ml of acetic acid	Phospholipids appear as blue spots on a white background within 10 min of spraying the plate
Choline-containing phospholipids (phosphatidylcholine and lysophosphatidylcholine)		Prepare a 40% aqueous solution of potassium iodide; prepare a 1.7% solution of bismuth subnitrate in 20% acetic acid; mix 5 ml of the first solution with 20 ml of the second solution and add 75 ml of water; spray the plate and then warm it	Choline-containing lipids appear within a few minutes as orange-red spots
Free amino groups (phosphatidyl ethanolamine and phosphatidyl serine)	Ninhydrin	Prepare a 0.2% solution of ninhydrin in n-butanol and add 3 ml of acetic acid; spray the plate and then heat it in an oven at 100°C–110°C	Lipids with free amino groups show as red-violet spots
Glycolipids	α-Naphthol/sulphuric acid	Prepare a 0.5% solution of α-naphthol in 100 ml of methanol–H_2SO_4 (1:1). Prepare a solution of concentrated H_2SO_4 (95:5). Spray the plate with the α-naphthol solution; allow to air dry and then spray with the H_2SO_4 solution. Heat at 120°C until colour reaches a maximum	Glycolipids (cerebrosides, sulphatides, gangliosides, and others) appear as yellow spots; cholesterol appears as a light red spot.

Glycolipids	Orcinol/sulphuric acid	Dissolve 20 mg of orcinol in 100 ml of 75% H_2SO_4. Spray the plate lightly with the reagent and then heat it at 100°C for 15 min	Glycolipids appear as blue-purple spots against a white background
Glycolipids versus phospholipids	Iodine	Place iodine crystals in a closed tank; place developed TLC plate in tank until colour appears	Phospholipids stain distinctly and glycolipids do not
Gangliosides	Resorcinol	Prepare a 2% aqueous solution of resorcinol. Add 10 ml of this solution to 80 ml of HCl containing 10.5 ml of a 0.1 M $CuSO_4$ solution. Spray the plate with this reagent and then heat at 110°C for a few minutes.	Gangliosides appear a violet-blue colour; other glycolipids appear as yellow spots
Sphingolipids	Sodium hypochlorite/benzidine reagent	Add 5 ml of sodium hypochlorite (Clorox) to 50 ml of benzene and dilute with 5 ml of acetic acid. Prepare the benzidine reagent by dissolving 0.5 g of benzidine and one crystal of KI in 50 ml of ethanol–H_2O (1:1). Spray the plate with the Clorox reagent and let it dry; then spray with the benzidine reagent	Sphingolipids (ceramides, sphingomyelin, cerebrosides, sulphatides, gangliosides) and lipids with secondary amines produce blue spots on a white background. CAUTION: benzidine is a carcinogen.

Source: Fried, B., Lipids, in *Handbook of Thin-layer Chromatography*, pp. 704–5; published by Marcel Dekker, 1996.

1.2.4 Quantitation

As mentioned earlier (section 1.1), quantitation of the separated classes has always been one of the dark areas of TLC. There are several quantitation techniques currently being used. These include the traditional scraping followed by quantitation as well as *in situ* determinations.

Sample elution and quantification. The separated lipid classes on the silica TLC can be scraped off, extracted by means of suitable solvents and then quantitated by means of gravimetry, spectrophotometry or gas chromatography. The errors in gravimetry measurement include incomplete elution of the sample, elution of minor quantities of silica and other impurities. We have derivatized the lipid classes to fatty acid methyl esters and have used an internal standard method to quantitate neutral and phospholipids by means of gas chromatography (Decker *et al.*, 1993); details of these calculations can be found elsewhere (Christie, Noble and Moore, 1970). The disadvantage of this method is that not all classes of lipids (e.g. sterols) can be derivatized to methyl esters and these need to be quantitated separately.

The eluted phospholipids have been quantitated by determining phosphorus content by means of classical spectrophotometric methods. Sugai *et al.* (1992) have developed a microscale procedure to determine soybean phospholipid by using two-dimensional TLC combined with phosphorus assay and by using a synthetic phosphate as an internal standard. The improvement over the existing methods is that the organic phosphate is converted to inorganic phosphate by perchloric nitrate digestion and then to a phosphomolybdate complex. The complex is extracted into ethyl acetate–butyl acetate (9:1 v/v) and quantitated spectrophotometrically by analysing the peak at 310 nm.

Densitometric methods. *In situ* densitometry is an often-used technique for lipid quantitation and has been extensively reviewed by Prosek and Pukl (1996). Lipids are generally sprayed with reagent and their absorption or fluorescence can be measured under UV or visible light by means of a densitometer. The method needs to be standardized and suitable calibration curves need to be constructed to avoid errors. There are several models of densitometer available and some of them are highly automated and coupled to computer systems. Apart from these the use of CCD (charge-coupled device) cameras and colour printers have further improved the densitometric capabilities for accurate quantitations (Prosek and Pukl, 1996). A recent review by Ebel (1996) compares quantitative analysis in TLC with that in HPTLC, including factors that can effect quantitation, the need for careful calibration and errors in quantitative HPTLC analyses. Ebel is of the opinion that as both HPTLC and HPLC are based on the same absorption and fluorescence phenomena they should obtain similar results with respect to quantitation.

Flame ionization detection. Special supports have been made for stationary phases and the separated lipids can be quantitated by using a flame ionization detector. Tubular thin-layer chromatography (TTLC) involves a glass or quartz tube coated on the inside with silica gel or with silica gel impregnated with 15%–20% cupric oxide. Quantitation is through flame ionization detection (FID). The advantages of TTLC include the fact that the desired coating can be applied directly by the operator, the higher sensitivity of detection compared with TLC, and, moreover, that each rod can be used up to ten times. The application of TTLC for lipid separation is limited and this has now been mostly replaced by the TLC/FID Chromarod/Iatroscan system. More details on TTLC and its application can be found in a review by Ackman and Ratnayake (1989). The TLC/FID Iatroscan system has been used extensively for lipid separation and quantitation and is therefore discussed in a section of its own towards the end of this chapter (section 1.3).

Hyphenated TLC techniques. TLC has been coupled with other instrumental techniques to aid in the detection, qualitative identification and, occasionally, quantitation of separated samples, and these include the coupling of TLC with high-pressure liquid chromatography (HPLC/TLC), with Fourier transform infra-red (TLC/FTIR), with mass spectrometry (TLC/MS), with nuclear magnetic resonance (TLC/NMR) and with Raman spectroscopy (TLC/RS). These techniques have been extensively reviewed by Busch (1996) and by Somsen, Morden and Wilson (1995). The chemistry of oils and fats and their TLC separation has been so well established that they seldom necessitate the use of these coupling techniques for their identification, although these techniques have been used for phospholipid detection. Kushi and Handa (1985) have used TLC in combination with secondary ion mass spectrometry for the analysis of lipids. Fast atom bombardment (FAB) has been used to detect the molecular species of phosphatidylcholine on silica based on the molecular ion obtained by mass spectrometry (Busch *et al.*, 1990).

1.3 Applications of thin-layer chromatography

TLC finds application in several areas in the field of lipids.

Purification, pre-purification of samples and assessment of purity. TLC has often been used for purification purposes. It is an ideal and inexpensive tool to purify small quantities of lipid samples. In addition, TLC has also been used for pre-purification of samples before chromatographic analysis. We generally purify triglycerides of tissue samples by TLC before analysis by HPLC. Similarly, the derivatized methyl esters are purified by TLC before analysis by gas chromatography. Alternatively, the purity of a given sample, including radiochemical purity, can be assessed by using two-dimensional TLC. Silver-nitrate-impregnated TLC is particularly useful for isolating *trans* fatty acids. TLC has been used to isolate metabolites of biochemical reactions for identification purposes. Taki *et al.* (1994) have separated and purified small quantities of glycosphingolipids and phospholipids by the blotting technique.

Identification. The structure of a compound can be identified by co-chromatography with a reference standard of known structure. Of course, this needs to be further verified by other analytical techniques such as MS, IR, and NMR.

Process and reaction monitoring. We have used TLC to monitor the time required for completion of the methylation reaction (Shantha, Decker and Hennig, 1993) and to study the formation of artifacts due to methylation of phthalates present in lipid samples (Shantha and Ackman, 1991b). TLC has been used to study the extent of hydrolysis of lipase-catalysed hydrolysis of triglycerides, phospholipase-catalysed hydrolysis of phospholipids and to study the activity of the enzymes. We have used TLC to study the extent of synthesis of acylated amino acids, and TLC has also been used to study the success of radiosynthesis.

Determination of cholesterol in foods. Jian, Xuexin and Ai (1996) have determined the levels of cholesterol in foods by means of TLC. The separated lipids were sprayed with 5% phosphoric acid–alcohol solution to give coloration sterols. They found this method comparable to that of GC.

Testing solvent systems for HPLC applications. HPTLC is our preliminary choice for testing solvent systems that would be suitable for HPLC separation of lipids.

Determination of lipophilicity. Reversed phase TLC has been used to predict the lipophilicity of certain drugs. Silica plates impregnated with methyl-silicone-diethyl ether, or C18 plates with octadecylsilane groups on the silica surface have been used for these applications. Biagi *et al.* (1994) have separated steroids and their acetates on reversed phase TLC based on the solvent strength and lipophilicity of the substance. It is possible that this technique may also be used to determine the relative lipophilicity of different vegetable and egg lecithins.

Detection of adulterations and contaminations. TLC on normal silica plates has been used to detect adulteration of cocoa butter by kokum butter. Deotale, Patil and Adinarayaniah (1990) have found that kokum butter (a cocoa butter substitute) appeared as a bluish green spot under UV light and could be detected up to 5% levels when mixed with cocoa butter. Similarly contamination or adulteration of edible oils with rice bran (*Oryza sativa*) and karanja (*Pongamia glabra*) seed oils were detected by the presence of oryzanol in rice bran oil and karanjin, karanjone, pongaglabrone and pongamol in Karanja oil (Nasirullah, Krishnamurthy and Nagaraja, 1992).

The separation of these unsaponifiables on TLC silica plates has been used to detect contaminants. Given the sensitivity of this method Nasirullah *et al.* felt that karanja oil at a level as low as 0.01% in other vegetable oil could be detected effectively using this TLC method. Similarly, TLC has also been used to screen large samples of meat for drug residues (Abjean, 1993).

1.4 Thin-layer chromatography–flame ionization detection for lipid analysis

The TLC/FID system, commercialized under the trade name of Iatroscan Lipid Analyzer (Iatron Laboratories, Tokyo, Japan), combines the separation capabilities of conventional TLC with the quantitation power of the flame ionization detector. This instrument has applications in the quantitative analysis of all substances separable by conventional TLC. The TLC/FID system is widely used in many fields, including food science, biology, pharmacy, medicine and petrochemistry (Ackman, McLeod and Banerjee, 1990). When the first version of the instrument was introduced more than two decades ago there were serious reservations regarding its lack of accuracy and reproducibility (Christie, 1982; Crane *et al.*, 1983). The current Mark V model, however, represents a substantial improvement on its precursors, owing to a superior detector design and greater uniformity of the TLC component (Chromarods). In spite of these improvements and its broad uses and applications, the instrument is still not fully accepted in many analytical laboratories because of claims of quantitative unreliability. The main cause for these claims are that the instrument, although superficially simple, requires special considerations with respect to calibration and standardization of many operational parameters, most of them arising from complexities of the detection system.

This section will describe various aspects of the use of the Iatroscan system for the analysis of lipids, fats, oils and other lipophilic compounds. Recent applications of TLC/FID in biology, food and environmental sciences will be presented. An exhaustive description of the instrument and discussion of the earlier literature can be found in previous reviews (Ackman, 1981; Ackman, McLeod and Banerjee, 1990; Sebedio, 1995; Shantha, 1992).

1.4.1 Equipment

The Iatroscan Lipid Analyzer consists of two instrumental units, the quartz silica-coated rods, or Chromarods, which constitute the TLC component, and the FID scanner unit. The quartz rods have a diameter of 0.9 mm, a length of 152 mm (a useful length is 12 cm) and are coated with a thin layer (75 μm) of a mixture of soft glass powder and the adsorbent, either silica gel (Chromarod SIII, particle size 5 mm) or alumina (Chromarod A, 10 μm particle size).

The Iatroscan FID scanner unit consists of a hydrogen flame jet and an ion collector. Following the sample application and solvent development, a set of ten rods are scanned and burnt, the ions are captured by the collector electrode and the signal is amplified in a similar way as in gas chromatography (GC). The more recent versions of the instrument (Mark IV and V) have an improved detector system (Oshima and Ackman, 1991) in which the collector and a circular electrode are close to the rod being scanned. These improvements minimize losses of ions and increase the sensitivity of the instrument. The reproducibility and accuracy of the TLC/FID method relies strongly on the standardization of all the operational parameters. Thus it will be useful and necessary to consider the various factors that affect the response of the flame ionization detector of the Iatroscan system.

1.4.2 Optimization

A set containing ten Chromarods mounted on a metallic frame are used in a single operation. The FID responses are shown to vary between the different sets of Chromarods, and, to a lesser extent, from rod to rod within a set. It has been suggested that one should match and select ten Chromarods of similar characteristics from a larger batch in order to avoid rod-to-rod variation (Ackman, McLeod and Banerjee, 1990). The grouped set of ten rods should then be treated in a similar fashion. Owing to cost considerations, an alternative procedure would be to treat each rod as a single isolated analytical unit and construct a calibration graph for each rod. Rod-to-rod variation was a more serious problem before the introduction of the SIII generation. Experienced users of the TLC/FID system (Sebedio and Juaneda, 1991) asserted that rod-to-rod variation among the SIII versions is low and hence it is not necessary to treat each rod as a calibration unit.

The sample application technique is an important factor in quantitative analysis by TLC/FID. A concentrated solution can be applied in a single aliquot of up to 0.5 μm. Band spreading arising from the application of large-volume and diluted samples may result in a lower FID response. If multiple applications of a diluted sample are necessary, the dropwise spotting procedure should allow the evaporation of the solvent between each application. Disposable pipettes (1–10 μl) can be used for spotting; the use of a microsyringe fitted with a repeating dispenser or an automatic applicator is more suitable for multiple spotting routines.

The activity of the adsorbent on the rods is strongly influenced by the humidity of the laboratory atmosphere, affecting R_f values and peak shapes and therefore compromising the resolution and the detector response. It is recommended that the adsorbent be reactivated prior to solvent development. This can be achieved by placing the rods in a desiccator for 5 min or in a 100°C oven for 1 min before solvent development (Ackman, McLeod and Banerjee, 1990). Alternatively, and depending on the resolution attained, rods can be conditioned at constant relative humidity (e.g. 30%) by placing them in a closed chamber containing saturated salt solution (Parrish, 1987).

The FID response also varies drastically with the hydrogen flow rate, although it is not as dramatically affected by the flow rate of air. Flow rates of 2000 ml min^{-1} for air and 160 ml min^{-1} for hydrogen are recommended by the manufacturer and adopted by most users. Studies have shown that a higher hydrogen flow (up to 180 ml min^{-1}) gives better response and reproducibility (Parrish, 1987). However, increasing the hydrogen flow may damage the adsorbent support, reducing the rod life and, more importantly, possibly increasing the volatilization of compounds owing to increased radiant heating as the flame approaches the sample.

The response of the FID to different lipid classes is quite selective, requiring careful and periodical calibrations (Parrish, 1987). Hydrocarbons have the highest ionization capability, whereas compounds containing oxygen, phosphorus or halogens have lower responses. Cholesterol and other steroids show a characteristic high detector response, probably because of the planar shape of these molecules. Selective responses of the FID have also been observed within the same lipid class (Fraser and Taggart, 1988). Some of these discrepancies are not strictly due to lipid class specific response of the FID but are associated with R_f values and band spreading. Therefore, adjustment of the TLC component of the system may result in considerable improvements in certain analyses. Alternatives to overcome these difficulties include the use of more polar solvents to improve peak shape (Shantha and Ackman, 1990), sample focusing (Parrish, 1987) and double development in the same or a different solvent system (Pollard *et al.*, 1992). Incidentally, not only for compound identification purposes, but also when quantitative results are required, it is advisable to use a reference standard that closely resembles the composition of the sample being analysed (Fraser and Taggart, 1988).

Recent work showed that Mark IV and Mark V Iatroscan models give a linear response in a working range from submicrogram up to 25 mg loads, with as little as 2.5 ng giving measurable peaks (Ackman, McLeod and Banerjee, 1990). However, reasonable signal-to-noise ratios are achieved at loads of about 10 ng (Iatron Laboratory, Japan). The introduction of the flame thermionic ionization detector (FTID), which can be fitted on the older models and is useful for specific detection of nitrogen and halogen compounds, is expected to broaden further the range of the instrument applications (Parrish, Zhou and Herche, 1988; Patterson, 1985). However, an exhaustive assessment of the performance of the FTID in an Iatroscan environment is still necessary.

1.4.3 Comparison of TLC–FID system with gas chromatography

Although the TLC–FID system does not compete with GC in the analytical arena, the speed of analysis and low maintenance cost of the TLC–FID system make it, in some cases, an attractive alternative. For example, GC–FID and GC/MS are the methods of choice for the analysis of hydrocarbons. Although GC offers superb resolution and reliability it can be a very expensive procedure for routine analysis in monitoring projects that involve the processing of many samples. The total petroleum hydrocarbon (TPH) content of biological or environmental matrices (a common end point used by regulatory agencies) can be determined by GC–FID. However, owing to costs, many laboratories prefer other procedures. Thus measurements of TPH frequently rely on gravimetric determinations ('oil and grease') or on IR spectrometry of total extractable materials. Although these are simple and useful methods for estimating the magnitude of contamination, they do not provide information about the nature of the contaminants. TLC–FID is an alternative for this type of analysis and is especially suited to synoptic studies or the screening of a large number of samples prior to the consideration of more detailed and costly analyses.

TLC–FID produced comparable results to those obtained by GC in the analysis of commercial mixtures of acylglycerides (Ranny *et al.*, 1983). Furthermore, owing to cost considerations, the TLC–FID method was found to be preferable for serial analysis in process control or technical quality control.

In a comparative study of the reproducibility of results obtained by the two techniques for the analysis of blood lipids Mares *et al.* (1983) concluded that the variations obtained with the TLC–FID system were much higher than those obtained by GC. However, recent improvements in the Iatroscan detection system are likely to increase significantly the reproducibility for this type of analyses. Sebedio and Ackman (1981) analysed a synthetic mixture of fatty acid methyl esters on silver-nitrate-impregnated Chromarods and compared the results with those obtained from conventional GC. Comparable results were obtained by the two methods for all methyl esters, except for methyl linolenate, the quantity of which was substantially overestimated by the TLC–FID method.

Bascoul *et al.* (1986) compared the levels of polar oxysterols as measured by TLC–FID and GC. The values obtained by TLC–FID analysis were considerably higher (341 ± 65 ppm) than those obtained by GC (124 ± 50 ppm). Bascoul *et al.* considered that the results obtained by TLC–FID were more accurate, because they agreed more closely with values of cholesterol content, corrected for the expected losses during heating, (376 ± 37 ppm).

Nevertheless, GC is a better method for the analysis of cholesterol autoxidation products, because of its superior resolution capabilities.

1.4.4 Recent applications of TLC–FID Iatroscan system

The separation capabilities of standard Chromarods SIII, combined with a battery of different solvent systems and multiple development techniques, provide the possibility of separating practically all the lipid classes present in natural samples (Ackman, McLeod and Banerjee, 1990). To increase further the capabilities of the TLC–FID system, and for a number of special applications, Chromarods can be treated with a variety of reagents. These modifications include exposure to iodine vapours (Banerjee, Ratnayake and Ackman, 1985) or impregnation with copper (II) sulphate (Kaimal and Shantha, 1984; Kramer, Thompson and Farnworth 1986; Ranny, Sedlacek and Michalec, 1991) to increase the detector response, impregnation with boric acid to separate triglycerides from their hydrolysis products (Tatara *et al.*, 1983), silver nitrate impregnation to separate and determine geometric and positional isomers of fatty acids and different species of triglycerides (Sebedio and Ackman, 1981; Sebedio, Farquharson and Ackman, 1985) and oxalic acid impregnation to improve the separation of phospholipids (Banerjee, Ratnayake and Ackman, 1985; De Schrijver and Vermeulen, 1991). The applications and performance of these rod treatments have recently been reviewed (Shantha, 1992). An alternative to rod impregnation is the treatment of the sample itself. Total catalytic hydrogenation of the lipid samples has been employed to improve the detector response and to increase the chemical stability of the sample (Shantha and Ackman, 1990).

Since the introduction of the Mark V Iatroscan model, there have been no further technical improvements to the TLC–FID system. However, the development of vapour phase detectors (such as the flame thermionic ionization detector for compounds containing nitrogen and halogens, and the flame emission photometric detector which detects compounds containing sulphur and phosphorus) should extend the range of applications of the TLC–FID system.

Since the introduction of Iatroscan more than 20 years ago, abundant work has been published on instrumental aspects and applications, most of them in the field of lipids. Excellent separations of most neutral and polar lipids, comparable to those achieved by conventional TLC, were reported during the early years of the TLC–FID. Recent reports of applications of TLC–FID technology for lipid and related substances can be divided into three major areas: food science; biology and ecology; and environmental sciences.

Food sciences. Three-step solvent developments, coupled with multiple scans of the rods, is the preferred procedure for the separation of lipid classes from fish and fish products. This procedure was used to separate the main neutral and polar lipid classes and acetone mobile polar lipids (AMPL) in menhaden fishmeal (Gunnlaugsdottir and Ackman, 1993). AMPL (which consist mostly of polar pigments, glycolipids and small amounts of monoglycerides) is commonly found in the lipid extracts of plants (Napolitano, 1994), and natural waters (Napolitano and Richmond, 1995; Parrish, 1987) but is much less common in fish products. In a previous analysis of the lipid class composition of herring silage by TLC–FID, Indrasena *et al.* (1990) reported that pigments and monoglycerides constituted a small proportion of the total lipids. Muscle and storage lipids in the myosepta of Atlantic salmon contained triglyceride as a major lipid component and showed only trace amounts of AMPL (Zhou, Ackman and Morrison, 1995).

A much simpler protocol consisting of two developments and a single scan was used to separate neutral lipids from phospholipids in cooked beef (St Angelo and James, 1993). This simple procedure consisted of a first development with the solvent system benzene/chloroform/formic acid (50:20:1.5 vol./vol.), followed by a short development in chloroform/methanol/29.3% ammonium hydroxide (50:50:5 vol./vol.) and a full scan. The first development separated triglycerides and cholesterol, whereas the second provided good resolution of all the major phospholipids and lysophospholipids.

In two recent reports Rosas-Romero and co-workers (Rosas-Romero, Herrera and Muccini, 1996; Rosa-Romero *et al.*, 1994) contributed to new TLC–FID procedures for the quantitative analysis of lipid classes of interest to the fat and oil industries. Concerned with the lack of linearity (particularly noticeable at high loads in the older Iatroscan systems), they suggested there is a need to establish an arithmetic transformation of the data. In order to improve quantitation of lipid classes, they proposed the data be transformed by establishing a regression of the log of the peak-area ratios against the log of the weight ratio (1996). Improvements in the design of the ion collector and the FID itself introduced for the Mark V will certainly improve the linearity of the untransformed data.

A detailed investigation of the operating conditions of the TLC–FID system for the analysis of triglycerides, and for important ingredients in the food-processing and confection industry (e.g. monoglycerides and diglycerides) has been recently carried out by Peyrou *et al.* (1996).

Ecological studies. The TLC–FID technique has enjoyed great and lasting popularity among ecologists. Early applications and operational improvements were developed for the analysis of the dissolved and particulate lipids in aquatic systems (Parrish, 1987; Parrish and Ackman, 1983). After this initial fast development the use of TLC–FID for ecological work reached a more mature stage during which no significant technical improvements were introduced. More recently, Parrish, Bodennec and Gentien (1996) described a method for the separation of algal glycolipids. Analysis of glycolipids

Table 1.3 Recent analyses of lipid classes in ecological studies by means of thin-layer Chromatography/flame ionization detection

Organism or matrix	Method[a]	Reference
Seawater, sediments, microalgae and fish oils	One-step development in hexane/diethyl ether/acetic acid (60:17:0.2)	Volkman and Nichols, 1991
Freshwater and stream surface foam	Three-step development in: hexane/diethyl ether/acetic acid (98:2:0.3); acetone 100%; chloroform/methanol/water (70:30:5)	Napolitano and Richmond, 1995
TNT contaminated soils	Two-step development	Gunderson et al., 1997
Marine diatoms (Chaetoceros gracilis)	Multiple solvent systems	Parrish and Wangersky, 1990
Freshwater diatoms (Cyclotella meneghiniana, Melosira varians, M. italica, Synedra filiformis, Stephanodiscus vinderanus)	Four-step development in: hexane/diethyl ether/formic acid (98:2:0.5); hexane/diethyl ether/formic acid (80:20:0.1); acetone 100%; dichloromethane/methanol/water (5:4:1)	Sicko-Goad and Andersen, 1991; 1993a; 1993b
Marine toxic dinoflagellate (Gyrodinium aureolum)	Multiple solvent systems	Parrish et al., 1993
Freshwater periphyton	Two-step development in: hexane/diethyl ether/acetic acid (80:20:0.3); acetone 100%	Napolitano, 1994
Estuarine shrimp (Pandalus borealis)	Two-step development in: hexane/diethyl ether/formic acid (82.2:5:0.045); hexane/diethyl ether/formic acid (55:29.7:0.075)	Ouellet, Taggart and Frank, 1992
Sea scallop (Placopecten magellanicus)	Single development in: hexane/diethyl ether/formic acid (97:3:0.1)	Napolitano and Ackman, 1990
Sea scallop (Placopecten magellanicus) and seston	Four-step development in: hexane/diethyl ether/formic acid (99:1:0.05); hexane/diethyl ether/formic acid (80:20:0.1); chloroform/methanol/water (51:40:10); chloroform/methanol/water (51:40:10)	Parrish et al., 1995
Stoneroller minnows (Campostoma anomalum)	Two-step development in: hexane/diethyl ether/acetic acid (98:2:0.3); hexane/diethyl ether/acetic acid (80:20:0.3)	Napolitano et al., 1996
Migratory shorebirds (Calidris pusilla)	Single development in: hexane/diethyl ether/formic acid (97:3:0.1)	Napolitano and Ackman, 1990

[a] Ratios of solvent system components are expressed in vol./vol.

presents a number of complications, such as interference with polar pigments and other components and the lack of commercially available standards. The two-step development procedure reported by Parrish, Bodennec and Gentien (1996) represents an improvement over previous methods.

Table 1.3 summarizes recent and important oceanographic and limnological work based on the analysis of lipid class composition by TLC–FID. A notable characteristic of this compilation of studies is the broad spectrum of matrices analysed, from lipids in environmental samples of water, sediments, soils and micro-organisms to vertebrates.

Other applications. Over recent years TLC–FID technology has been used by researchers in the petrochemical industry as well as by environmental monitoring groups. Workers in these fields realized the capability of TLC–FID for rapid sample synoptic surveys and screening routines. Most of this work was focused on the rapid characterization of crude oils, coal and petroleum products and the basic characterization of petroleum-contaminated soils. A discussion of TLC–FID applications to petrochemical

Table 1.4 Recent analyses of fossil fuels and related products by thin-layer chromatography/flame ionization detection

Sample type	Method[a]	Reference
Fossil and synthetic fuels	One step-development in, n-Pentane/isopropanol (95:5)	Poirier and George, 1983
Coal-derived liquids	Three step development either in hexane; benzene/hexane (80:29); dichloromethane/methanol (40:60) an silica; or in hexane; dichloromethane; methanol/dichloromethane (40:60) on Chromarods A	Selucky 1983, 1985
Petroleum, refinery products and reservoir rocks	Three-step development in: hexane; toluene; dichloromethane/methanol (95:5)	Karlsen and Larter, 1989, 1991
Petroleum and coal-tar creosote soils and sludges	Three-step development either in: hexane; dichloromethane; dichloromethane/methanol (98:2); or in hexane; hexane/dichloromethane (90:10); dichloromethane	Pollard et al., 1992
Hydrocarbon-degrading bacterial culture	One-step development in hexane/diethyl ether/acetic acid (60:17:0.2)	Cavanagh et al., 1995
Coal-tar pitch	Three-step development in hexane; toluene; dichloromethane/methanol (95:5)	Cebolla et al., 1996

[a] Ratios of solvent system components are expressed in vol./vol.

and environmental studies is not our main objective in this chapter. However, as these applications have been excluded from previous reviews of the TLC–FID system, we will present them briefly here.

Table 1.4 summarizes recent works in fossil fuel production and the related environmental issues. In a series of interesting reports Karlsen and Larter (1989, 1991) outlined a rapid method for analysing and mapping petroleum types within individual petroleum reservoirs. As in the analyses of biogenic lipids, most procedures used Chromarods SIII and two or three development systems with solvents of increasing polarity. Hexane is normally used for the separation of aliphatics, followed by toluene for separation of the aromatics, and finally dichloromethane–methanol for the more polar constituents. Similar procedures were employed for the analysis of petroleum-contaminated and creosote-contaminated soils. Recent experiments showed that TLC–FID allows the complete quantitation of aliphatic and aromatic hydrocarbons in contaminated sites, with no interference with

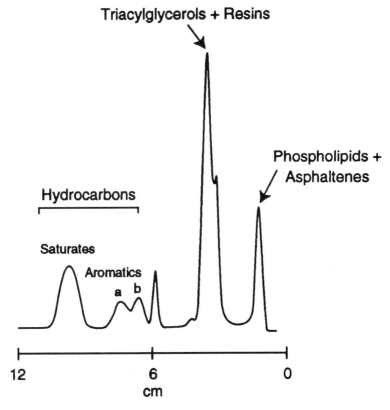

Figure 1.4 Thin-layer chromatography–flame ionization detection chromatogram showing the separation of biogenic lipids from hydrocarbons in an extract of a petroleum-contaminated soil. Solvent systems: hexane (40 min), toluene (15 min) and dichloromethane/methanol (95:5 vol./vol.) (5 min), followed by a single full scan.

background levels of biogenic lipids (e.g. triglycerides, free fatty acids, sterols) present in the soil (Fig. 1.4). These examples demonstrate the usefulness of TLC–FID for various applications.

1.5 Conclusions

The improvement and versatility of TLC enables it to be used for several applications other than traditional lipid separations. The wide array of reagents developed for detection of specific lipid classes or groups allows TLC to screen several lipid samples at a time for specific functional groups. Multidimensional and multimodal development techniques allow manipulation of TLC for specific applications. Coupling of TLC with other techniques such as HPLC, IR, MS, RS and NMR has increased its analytical power in several applications. Automation and computerization of TLC steps has increased its reproducibility and reliability. The TLC–FID Iatroscan system, which uses FID for quantitation, has now become more standardized and is being used routinely for lipid analysis in several laboratories and in several fields, including medicinal, drug, food, environmental, toxicological and ecological studies. TLC has found application in medicinal chemistry to separate enantiomers by using silica-bonded chiral phases, and it would be interesting to study such separations in lipids.

Acknowledgements

G. E. Napolitano acknowledges the support of Oak Ridge Laboratories, TN. N. Shantha wishes to thank Martha Osborn, Nestlé R&D, New Milford, CT, for her help with literature search.

References

Abjean, J. P. (1993) Planar chromatography for drug residue screening in food applications. *Chromatographia*, **36**, 359–61.

Ackman, R. G. (1981) Flame ionization detection applied to thin-layer chromatography on coated quartz rods. *Methods in Enzymology*, **72D**, 205–52.

Ackman, R. G. and Ratnayake, W. M. N. (1989) Lipid Analyses: Part 1. Properties of fats, oils and lipids: recovery and basic compositional studies with gas–liquid chromatography and thin-layer chromatography, in *Role of Fats in Human Nutrition* (eds A. J. Vergroesen and M. Crawford), Academic Press, New York, pp. 441–514.

Ackman, R. G., McLeod, C. A, and Banerjee, A. K. (1990) An overview of analyses by Chromarod-Iatroscan TLC-FID. *Journal of Planar Chromatography*, **3**, 450–90.

Allan, D. and Cockroft, S. (1982) A modified procedure for thin-layer chromatography of phospholipids. *Journal of Lipid Research*, **23**, 1373–4.

Aloisi, J. D., Sherma, J. and Fried, B. (1990) Comparison of mobile phases for separation and quantification of lipids by one-dimensional TLC on preadsorbent high performance silica gel plates. *Journal of Liquid Chromatography*, **13**, 3949–72.

Banerjee, A. K., Ratnayake, W. M. N. and Ackman, R. G. (1985) Enhanced response of more highly unsaturated lipids on thin-layer chromatography–Iatroscan flame ionization detection after exposing the developed Chromarods to iodine vapor. *Journal of Chromatography*, **319**, 215–17.
Bascoul, N., Domergue, M., Olle, M. *et al.* (1986) Autoxidation of cholesterol in tallows heated under deep frying conditions: evolution of oxysterols by GLC and TLC-FID. *Lipids*, **21**, 383–7.
Berezkin, V. G. (1995). History of planar chromatography – the discovery of thin-layer chromatography. *Journal of Planar Chromatography*, **8**, 401.
Biagi, G. J., Barbano, A. M., Japone, A. *et al.* (1994) Determination of lipophilicity by means of reversed-phase thin-layer chromatography – II. Influence of organic modifier on the slope of the thin-layer chromatographic equation. *Journal of Chromatography A*, **669**, 246–53.
Bradova, V., Simd, F., Ledvinova, J. *et al.* (1990) Improved one-dimensional thin-layer chromatography for the separation of phospholipids in biological material. *Journal of Chromatography*, **533**, 297–9.
Breckenridge, W. C. and Kuksis, A. (1968) Structure of bovine milk fat triglycerides. 1. Short and medium chain length. *Lipids*, **3**, 291–300.
Busch, K. L. (1996) Thin-layer chromatography coupled with mass spectrometry in *Handbook of Thin-layer Chromatography* (eds J. Sherma and B. Fried), Marcel Dekker, New York, pp. 241–72.
Busch, K. L., Brown, S. M., Doherty, S. J. *et al* (1990) Analysis of planar chromatograms by fast atom bombardment and laser desorption mass spectrometry. *Journal of Liquid Chromatography*, **13**, 2841–69.
Cavalli, E., Truong, T. T., Thomassin, M. *et al.* (1993) Comparison of optimization methods in planar chromatography. *Chromatographia*, **35**, 102–8.
Cavanagh, J.-A. E., Juhasz, A. L., Nichols, P. D. *et al.* (1995) Analysis of microbial hydrocarbon degradation using TLC-FID. *Journal of Microbiological Methods*, **22**, 119–30.
Cebolla, V. L., Vela, J., Membrado L. *et al.* (1996) Coal-tar pitch characterization by thin-layer chromatography with flame ionization detection. *Chromatographia*, **42**, 295–9.
Christie, W. W. (1982) *Lipid Analysis*, Pergamon, Oxford.
Christie, W. W. (1990) Has thin-layer chromatography had its day? *Lipid Technology*, **2** (1), 22–3.
Christie, W. W., Noble, R. C. and Moore, J. H. (1970) Determination of lipid classes by a gas-chromatographic procedure. *Analyst*, **95**, 940–4.
Crane, R. T., Goheen, S. C., Larking, E. C. *et al.* (1983) Complexities in lipid quantitation using thin layer chromatography for separation and flame ionization detector for detection. *Lipids*, **18**, 74–80.
Cserhati, T. and Forgacs, E. (1996) Introduction to techniques and instrumentation, in *Practical Thin-layer Chromatography: A Multidisciplinary Approach* (eds B. Fried and J. Sherma), CRC Press, New York, pp. 1–18.
Decker, E. A., Crum, A. D., Shantha, N. C. *et al.* (1993) Catalysis of lipid oxidation by iron from an insoluble fraction of beef diaphragm muscle. *Journal of Food Science*, **58** (2), 233–6, 258.
Deotale, M. Y., Patil, M. N. and Adinarayaniah, C. L. (1990) Thin layer chromatographic detection of kokum butter in cocoa-butter. *Journal of Food Science and Technology*, **27**, 230.
De Schrijver, R. and Vermeulen, D. (1991) Separation and quantitation of phospholipids in animal tissues by Iatroscan TLC/FID. *Lipids*, **26**, 74–80.
Dobson, G., Christie, W. W. and Nikolova-Damyanova, B. (1995) Silver ion chromatography of lipids and fatty acids. *Journal of Chromatography Biomedical Applications*, **671**, 197–222.
Ebel, S. (1996) Quantitative analysis in TLC and HPTLC. *Journal of Planar Chromatography*, **9**, 4–15.
Fraser, A. J. and Taggart, C. T. (1988) Thin-layer chromatography–flame ionization detection calibration with natural and synthetic standards. *Journal of Chromatography*, **439**, 404–7.
Fried, B. (1996) Lipids, in *Handbook of Thin-layer Chromatography* (eds J. Sherma, and B. Fried), Marcel Dekker, New York, pp. 704–5.
Fried, B. and Sherma, J. (eds) (1996) *Practical Thin-layer Chromatography: A Multidisciplinary Approach*, CRC Press, New York.

Golkiewicz, W. (1996) Gradient development in thin-layer chromatography, in *Handbook of Thin-layer Chromatography* (eds J. Sherma and B. Fried), Marcel Dekker, New York, pp. 149–70.

Gunderson, C. A., Kostuk, J. M., Gibbs, M. H. *et al.* (1997) Multispecies toxicity assessment of compact produced in bioremediation of explosive-contaminated soils. *Environmental Toxicology and Chemistry*, in press.

Gunnlaugsdottir, H. and Ackman, R. G. (1993) Three extraction methods for determination of lipids in fish meal: evaluation of a hexane/isopropanol method as an alternative to chloroform-based methods. *Journal of Science of Food Agriculture*, **61**, 235–40.

Higgs, M. H., Sherma, J. and Fried, B. (1990) Neutral lipids in the digestive gland–gonad complex of *Biomphalaria glabrata* snails, fed lettuce versus hen's egg yolk, determined by quantitative TLC-densitometry. *Journal of Planar Chromatography – Modern TLC*, **3**, 38–41.

Huq, F., Saha, G. C., Begum, F. and Adhikary, S. (1991) Studies on *Allium sativum linn* (Garlic) – chemical investigation on garlic oil. *Bangladesh Journal of Science and Industrial Research*, **26** (1–4), 41–51.

Indrasena, W. M., Parrish, C. C., Ackman, R. G. *et al.* (1990) Separation of lipid classes and carotenoids in Atlantic salmon feeds by thin layer chromatography with Iatroscan flame ionization detection. *Bulletin of Aquaculture Association of Canada*, **4**, 36–40.

Izmailov, N.A. and Schreiber, M.S. (1938) (in Russian). *Farmatsiya*, **3**, 1

Jian, L., Xuexin, Z. and Ai, Z. (1996) Determination of cholesterol in food by TLC. *Guangdong Yaoxueyuan Xuebao*, **12** (2), 88–90 [*CA Selects: Paper and Thin-Layer Chromatography* (1996), **19** (125), 140812r].

Johnston, P.V. (1971) *Basic Lipid Methodology*, University of Illinois Press, Urbana and Champaign, IL.

Kaimal, T. N. B. and Shantha, N. C. (1984) Quantitative analysis of lipids on copper (II) sulphate-impregnated chromarods. *Journal of Chromatography*, **288**, 177–86.

Karlsen, D. A. and Larter S. R. (1989) A rapid correlation method for petroleum population mapping within individual petroleum reservoirs: applications to petroleum reservoir description, in *Correlations in Hydrocarbon Exploration NPF* (ed. J. D. Collisson), Graham and Trotman, London, pp. 77–85.

Karlsen, D. A. and Larter, S. R. (1991) Analysis of petroleum fractions by TLC–FID: applications to petroleum reservoir description. *Organic Geochemistry*, **17**, 603–17.

Kiosseoglou, V. and Boskou, D. (1990) Separation and fatty acid composition of steryl and wax esters in hexane extracts of sunflower seed, soybeans and tomato seeds. *Lebensmittel-Wissenschaft and Technologie*, **23** (4), 340–2.

Kramer, J. K. G., Thompson, B. K. and Farnworth, T. R. (1986) Variation in the relative response factor for triglycerides on Iatroscan Chromarods with fatty acid composition and sequence of analyses. *Journal of Chromatography*, **355**, 221–8.

Kushi, Y. and Handa, S. (1985) Direct analysis of lipids on thin layer plates by matrix-assisted secondary ion mass spectrometry. *Journal of Biochemistry*, **98** (1), 265–8.

Magnani, J. L., Smith, D. F. and Ginsburg, V. (1980) Detection of gangliosides that bind cholera toxin: direct binding of 125I-labeled toxin to thin-layer chromatograms. *Analytical Biochemistry*, **109**, 399–402.

Mangold, H. K. (1961) Thin layer chromatography of lipids. *Journal of the American Oil Chemists' Society*, **38**, 708–27.

Mangold, H. K. and Malins, D. C. (1960) Fractionation of fats, oils, and waxes on thin layers of silicic acid. *Journal of the American Oil Chemists' Society*, **37**, 383–5.

Mares, P., Ranny, M., Sedlacek, J. *et al.* (1983) Chromatographic analysis of blood lipids. Comparison between gas chromatography and thin-layer chromatography with flame ionization detection. *Journal of Chromatography*, **275**, 295–305.

Martinez-Lorenzo, M. J., Marzo, I., Naval, J. *et al.* (1994) Self-staining of polyunsaturated fatty acids in argentation thin-layer chromatography. *Analytical Biochemistry*, **220** (1), 210–12.

Morris, L. J. and Nichols, B. (1972) Argentation thin-layer chromatography of lipids, in *Progress in Thin-layer Chromatography, Related Methods* (ed. A. Niederwieser), Ann Arbor–Humphrey Science Publishers, Ann Arbor, MI, pp. 74–93.

Muthing, J. (1996) High-resolution thin-layer chromatography of gangliosides, *Journal of Chromatography A*, **720**, 3–25.

Nakamura, T., Fukuda, M. and Tanaka, R. (1996) Estimation of polyunsaturated fatty acid content in lipids of aquatic organisms using thin-layer chromatography on a plain silica gel plate. *Lipids*, **31** (4), 427–32.
Napolitano, G. E. (1994) The relationship of lipids with light and chlorophyll measurements in freshwater algae and periphyton. *Journal of Phycology*, **30**, 943–50.
Napolitano, G. E. and Ackman, R. G. (1990) Anatomical distribution of lipids and their fatty acids in semipalmated sandpiper *Calidris pusilla* L. from Shepody Bay, New Brunswick, Canada. *Journal of Experimental Marine Biology and Ecology*, **144**, 113–24.
Napolitano, G. E. and Richmond, J. E. (1995) Enrichment of biogenic lipids, hydrocarbons and PCBs in stream-surface foams. *Environmental Toxicology and Chemistry*, **14**, 197–201.
Napolitano, G. E., Shantha, N. C., Hill, W. R. *et al.* (1996) Lipids and fatty acid compositions of periphyton and stoneroller minnows (*Campostoma anomalum*): trophic and environmental implications. *Archives für Hydrobiologie*, **137**, 211–25.
Nasirullah, Krishnamurthy, M. N. and Nagaraja, K. V. (1992) Methods for detection of rice-bran, mustard, karanja oils and rice-bran deoiled cake. *Fat Science and Technology*, **94** (12), 457–8.
Nichols, B. W. (1964) *New Biochemical Separations* (eds A. T. James and L. J. Morris), Van Norstrand, New York.
Omogbai, F. E. (1990) Lipid composition of tropical seeds used in the Nigerian diet. *Journal of the Science of Food and Agriculture*, **50**, 253–5.
Oshima, T. and Ackman, R. G. (1991) New developments in Chromarod/Iatroscan TLC–FID: analysis of lipid class composition. *Journal of Planar Chromatography*, **4**, 27–34.
Ouellet, P., Taggart, C. T. and Frank, K. T. (1992) Lipid condition and survival in shrimp (*Pandalus borealis*) larvae. *Canadian Journal of Fisheries and Aquatic Science*, **49**, 368–78.
Pappas, A. A., Mullins, R. E. and Gadsden, R. H. (1982) Improved one-dimensional thin-layer chromatography of phospholipids in amniotic fluid. *Clinical Chemistry*, **28** (1), 209–11.
Parrish, C. C. (1987) Separation of aquatic lipid classes by Chromarod thin-layer chromatography with measurement by Iatroscan flame ionization detection. *Canadian Journal of Fisheries and Aquatic Sciences*, **44**, 722–31.
Parrish, C. C. and Ackman, R. G. (1983) Chromarod separation for the analysis of marine lipid classes by Iatroscan thin-layer chromatography–flame ionization detection. *Journal of Chromatography*, **262**, 103–12.
Parrish, C. C. and Wangersky, P. G. (1990) Growth and lipid class composition of the marine diatom, *Chaetoceros gracilis*, in laboratory and mass culture turbidostats. *Journal of Plankton Research*, **12**, 1011–21.
Parrish, C. C., Bodennec, G. and Gentien, P. (1996) Determination of glycoglycerolipids by Chromarod thin-layer chromatography with Iatroscan flame ionization detection. *Journal of Chromatography A*, **741**, 91–7.
Parrish, C. C., Zhou, X. and Herche, L. R. (1988) Flame ionization and flame thermionic detection of carbon and nitrogen in aquatic lipids and humic-type classes with an Iatroscan Mark IV. *Journal of Chromatography*, **435**, 350–6.
Parrish, C. C., Bodennec, G., Sebedio, J.-L. *et al.* (1993) Intra- and extracellular lipids in cultures of the toxic dinoflagellate, *Gyrodinium aureolum*. *Phytochemistry*, **32**, 291–5.
Parrish, C. C., McKenzie, C. H., MacDonald, B. A. *et al.* (1995) Seasonal studies of seston lipids in relation to microplankton species composition and scallop growth in South Broad Cove, Newfoundland. *Marine Ecology Progress Series*, **129**, 151–64.
Patterson, P. L. (1985) A specific detector for nitrogen and halogen compounds in TLC on coated quartz rods. *Lipids*, **20**, 503–9.
Pchelkin, V. P. and Vereshchagin, A. G. (1992) Reversed-phase thin layer chromatography of diacylglycerols in the presence of silver ions. *Journal of Chromatography*, **603**, 213–22.
Peyrou, G., Rakotondrazafy, V., Moulonoungui, Z. *et al.* (1996) Separation and Quantitation of mono-, di- and triglycerides and free oleic acid using thin-layer chromatography with flame-ionization detection. *Lipids*, **31**, 27–32.
Pollard, S. J., Hrudey, S. E., Fuhr, B. J. *et al.* (1992) Hydrocarbon wastes at petroleum-contaminated and creosote-contaminated sites – rapid characterization of component classes by thin-layer chromatography with flame ionization detection. *Environmental Science and Technology*, **26**, 2528–34.

Pomeranz, Y. and MeLoan, C. E. (1994) Paper and thin-layer chromatography, in *Food Analysis Theory and Practice*, 3rd edn (Y. Pomeranz and C. E. MeLoan), Chapman and Hall, New York, pp. 352–65.

Poole, C. F. and Poole, S. K. (1995) Multidimensionality in planar chromatography. *Journal of Chromatography*, **703**, 573–612.

Poorier, M.-A. and George, A. E. (1983) Rapid method for the determination of malthene and asphaltene content in bitumen, heavy oils, and synthetic fuels by pyrolysis TLC. *Journal of Chromatographic Science*, **21**, 331–3.

Prosek, M. and Pukl, M. (1996) Basic principles of optical quantitation in TLC, in *Handbook of Thin-layer Chromatography* (eds J. Sherma and B. Fried), Marcel Dekker, New York, pp. 273–306.

Ranny, M., Sedlacek, J. and Michalec, C. (1991) Resolution of phospholipids on Chromarods impregnated with salts of some divalent metals. *Journal of Planar Chromatography – Modern TLC.*, **4**, 15–18.

Ranny, M., Sedlacek, J., Mares, E. *et al.* (1983) Quantitative analysis of molecularly distilled monoacyl glycerols using thin layer chromatography and the flame ionization detector (TLC–FID). *Seifen, Oele, Fette, Wachse*, **109** (8), 219–24.

Rosas-Romero, A. J., Herrera, J. C. and Muccini, M. (1996) Determination of neutral and polar lipids by thin layer chromatography with flame ionization detection. *Italian Journal of Food Science*, **1**, 33–9.

Rosas-Romero, A. J., Herrera, J. C., Martinez de Aparicio, E. *et al.* (1994) Thin-layer chromatographic determination of β-carotene, cantaxanthin, lutein, violaxanthin and neoxanthin on Chromarods. *Journal of Chromatography A*, **667**, 361–6.

Rouser, G., Kritchevski, G. and Yamamota, A. (1967) in *Lipid Chromatographic Analysis* (ed G. V. Marinetti), Edward Arnold, London, pp. 99–162.

Scott, R. P. W. (1982) The silica gel surface and its interaction with solvent and solute in liquid chromatography. *Advances in Chromatography*, **20**, 67–196.

Sebedio, J.-L. (1995) Utilization of thin-layer chromatography–flame ionization detection for lipid analyses, in *New Trends in Lipid and Lipoprotein Analyses* (eds J.-L. Sebedio and E. G. Perkins) AOCS Press, Champaign, IL, pp. 24–37.

Sebedio, J.-L. and Ackman, R. G. (1981) Chromarods-S modified with silver nitrate for the quantitation of isomeric unsaturated fatty acids. *Journal of Chromatographic Science*, **19**, 552–7.

Sebedio, J.-L. and Juaneda, P. (1991) Quantitative lipid analyses using the new Iatroscan TLC–FID system. *Journal of Planar Chromatography*, **4**, 35.

Sebedio, J.-L., Farquharson, T. E. and Ackman, R. G. (1985) Quantitative analyses of methyl esters of fatty acid geometrical isomers, and of triglycerides differing in unsaturation, by the Iatroscan TLC/FID technique using $AgNO_3$ impregnated rods. *Lipids*, **20**, 555–9.

Selucky, M. L. (1983) Quantitative analysis of coal-derived liquids by thin-layer chromatography with flame ionization detection. *Analytical Chemistry*, **55**, 141–3.

Selucky, M. L. (1985) Quantitative class separation of coal liquids using thin-layer chromatography with flame ionization detection. *Lipids*, **20**, 546–51.

Shantha, N. C. (1992) Thin-layer chromatography–flame ionization detection Iatroscan system. *Journal of Chromatography*, **624**, 21–35.

Shantha, N. C. and Ackman, R. G. (1990) Advantages of total lipid hydrogenation prior to lipid class determination on Chromarods-SIII. *Lipids*, **25**, 570–4.

Shantha, N.C. and Ackman, R.G. (1991a) Silica gel thin-layer chromatographic method for concentration of longer-chain polyunsaturated fatty acids from food and marine lipids. *Canadian Institute of Food Science and Technology Journal*, **24**, 156–60.

Shantha, N. C. and Ackman, R. G. (1991b) Behavior of common phthalate plasticizer (dioctyl phthalate) during the alkali- and/or acid-catalyzed steps in an AOCS method for the preparation of methyl esters. *Journal of Chromatography*, **587**, 263–7.

Shantha, N. C., Decker, E. A. and Hennig, B. (1993) Comparison of methylation methods for the quantitation of conjugated linoleic acid isomers. *Journal of AOAC International*, **76** (3), 644–9.

Sherma, J. (1994) Planar chromatography. *Analytical Chemistry*, **66**, 67R–83R.

Sherma, J. (1996). Planar chromatography. *Analytical Chemistry*, **68**, 1R–19R.

Sherma, J. and Fried, B. (eds) (1996) *Handbook of Thin-layer Chromatography*, Marcel Dekker, New York.

Shukla, V.K.S. (1995) Thin-layer chromatography of lipids, in *New Trends in Lipid and Lipoprotein Analyses* (eds J. L. Sebedio and E. G. Perkins), AOCS Press, Champaign, IL, pp. 17–23.
Sicko-Goad, L. and Andersen, N. A. (1991) Effect of growth and light/dark cycles on diatom lipid content and composition. *Journal of Phycology*, **27**, 710–18.
Sicko-Goad, L. and Andersen, N. A. (1993a) Effect of diatom lipid composition on the toxicity of trichlorobenzene isomers to diatoms: I. Short-term effects of 1,3,5-trichlorobenzene. *Archives of Environmental Contamination and Toxicology*, **24**, 236–42.
Sicko-Goad, L. and Andersen, N. A. (1993b) Effect of diatom lipid composition on the toxicity of trichlorobenzene: II. Long-term effects of 1,2,3-trichlorobenzene. *Archives of Environmental Contamination and Toxicology*, **24**, 243–8.
Skipski, V. P., Peterson, R. F. and Barclay, M. (1964) Quantitative analysis of phospholipids by thin-layer chromatography. *Biochemical Journal*, **90**, 374–8.
St Angelo, A. J. and James, C. Jr (1993) Analysis of lipids from cooked beef by thin-layer chromatography with flame ionization detection. *Lipids*, **70**, 1245–50.
Somsen, G. W., Morden, W. and Wilson, I. D. (1995) Planar chromatography coupled with spectroscopic techniques. *Journal of Chromatography*, **703**, 613–65.
Stahl, E. (1969) *Thin Layer Chromatography: A Laboratory Handbook*, Springer, Berlin.
Steele, W. and Banks, W. (1994) Triglyceride distribution in hydrogenated milk fat and its effects on separation by thin layer chromatography. *Milchwissenschaft*, **49** (7), 372–6.
Storry, J. E. and Tuckly, B. (1967) Thin-layer chromatography of plasma lipids by single development. *Lipids*, **2**, 501–2.
Sugai, A., Sakuma, R., Fukuda, I. *et al.* (1992) Improved method for determining soybean phospholipid composition by two-dimensional TLC-phosphorus assay. *Journal of Japanese Oil Chemists' Society (Yukagaku)*, **41** (10), 1029–34.
Taki, T., Kasama, T., Handa, S. *et al.* (1994) A simple and quantitative purification of glycosphingolipids and phospholipids by thin-layer chromatography blotting. *Analytical Biochemistry*, **223**(2), 232–8.
Tatara, T., Fuji, T., Kaware, T. *et al.* (1983) Quantitative determination of tri-, di-, monoleins and free oleic acid by thin layer chromatography–flame ionisation detection system using internal standards and boric acid impregnated chromarods. *Lipids*, **18**, 732–6.
Thompson, S. N. (1987) Effect of *Schistosoma mansoni* on the gross lipid composition of its vector *Biomphalaria glabrata*. *Comparative Biochemistry and Physiology*, **87** (2), 357–60.
Touchstone, J. C. (1995) Thin-layer chromatographic procedures for lipid separation. *Journal of Chromatography B*, **671**, 169–95.
Touchstone, J. C., Levin, S. S., Dobbins, M. F. *et al.* (1983) (3-sn-Phosphatidyl)cholines (lecithin) in amniotic fluid. *Clinical Chemistry*, **29** (11), 1951–4.
Traitler, H. and Janchen, D. E. (1993) Analysis of lipids by planar chromatography, in *CRC Handbook of Chromatography – Analysis of Lipids* (eds K. D. Mukherjee and N. Weber; editor-in-chief J. Sherma), CRC Press, Boca Raton, FL.
Vioque, E. and Holman, R. T. (1962) Quantitative estimation of esters by thin-layer chromatography. *Journal of American Oil Chemists' Society*, **39**, 63–6.
Volkman, J. K. and Nichols, P. D. (1991). Applications of thin layer chromatography–flame ionization detection to the analysis of lipids and pollutants in marine and environmental samples. *Journal of Planar Chromatography*, **4**, 19–26.
Wagner, H., Horhammer, L. and Wolffe, P. (1961) *Dünnschicht Chromatographie von Phosphatiden und Glykolipiden*. *Biochem Z*, **334**, 175–84.
Wang, W.-Q. and Gustafson, A. (1992) One-dimensional thin-layer chromatographic separation of phospholipids and lysophospholipids from tissue lipid extracts. *Journal of Chromatography*, **581**, 139–42.
Weins, C. and Hauck, H. Z. (1995) Advances and developments in thin-layer chromatography. *LC-GC*, **14** (6), 456–64.
Yao, J. K. and Rastetter, G. M. (1985) Microanalysis of complex tissue lipids by high-performance thin-layer chromatography. *Analytical Biochemistry*, **150**, 111–16.
Zakaria, M., Gonnord, M.-F. and Guiochon, G. (1983) Applications of two-dimensional thin-layer chromatography. *Journal of Chromatography*, **271**, 127–92.
Zhou, S. Y., Ackman, R. G. and Morrison, C. (1995) Storage of lipids in the myosepta of Atlantic salmon (*Salmo salar*). *Fish Physiology and Biochemistry*, **14**, 171–8.

2 Characterization of lipids by supercritical fluid chromatography and supercritical fluid extraction

L. G. BLOMBERG, M. DEMIRBÜKER and M. ANDERSSON

2.1 Introduction

Supercritical media were first applied as mobile phases for chromatography by Klesper, Corwin and Turner (1962). At this time, packed columns were employed, and for almost two decades the supercritical fluid chromatography (SFC) technique was studied and advanced only by a small group of scientists; the developments during this period have been reviewed by Sanagi and Smith (1988). It was not until 1981, with the advent of open tubular column SFC, as proposed by Novotny *et al.* (1981, 1984) that the technique received more widespread interest. Large efforts were then directed to the development of open tubular SFC, and the technique was developed to a relatively mature level, particularly by Lee and co-workers (Lee and Markides, 1989). However, in view of today's knowledge, it seems that the merits of the technique were somewhat overestimated at that time. When scientists engaged with chromatography experienced that open tubular SFC was not the miracle technique that was originally expected, the interest rapidly declined, leading to the present situation where SFC is instead underestimated. However, during recent years, the development of SFC has taken a new direction, moving back to the packed columns. Quite promising results have thus been obtained with such columns, and it seems that the potential advantages of the SFC technique finally are becoming accessible.

Supercritical fluid extraction (SFE) is a technique for which there is a high demand. This technique is thus rapidly taking off, a major driving force being the increased environmental concern regarding the use of organic solvents in Soxhlet extractions (Black, 1996). There are a number of books and reviews on SFC and SFE, some of these are dedicated to lipid methodology.*

---

* For books and reviews see Berge, 1995; Blomberg *et al.*, 1994; Charpentier and Sevenants, 1988; Dean, 1993; Jinno, 1992; Lee and Markides, 1989; Luque de Castro, Valcárcel and Tena, 1994; McHugh and Krukonis, 1994; Saito, Yamauchi and Okuyama, 1994; Smith, 1988; Taylor, 1995, 1996; Wenclawiak, 1992; Westwood, 1993; for publications dedicated to lipid morphology, see King, 1995; King and List, 1996; Laakso, 1992.

In the evaluation of a technique it is necessary to compare its performance with that of other techniques that could possibly be employed for a given type of analytical task. Currently, lipids can be separated by gas chromatography (GC), SFC, high-pressure liquid chromatography (HPLC) and capillary electrophoresis (CE). First, for separation of fatty acid methyl esters, GC is quite useful (Christie, 1992a; Duchateau, van Oosten and Vasconcellos, 1996). Also triacyglycerols (TGs) can be separated by GC, but here thermal degradation or polymerization of TG-containing polyunsaturated fatty acid moieties may occur (Hammond, 1989). This problem is not at hand in HPLC, and in fact HPLC is presently a well established technique for the analysis of TGs (Christie, 1987a). Further, for the polar lipids, HPLC is the method of choice (Christie, 1987a). The application of CE to lipid analysis is still in its initial stages (Blomberg and Andersson, 1994; Szücs *et al.*, 1996).

During recent years, SFC methods for the characterization of lipids have been developed at the Arrhenius Laboratories. The separation of complex mixtures of TGs on non-polar open tubular SFC was first attempted. However, several TG species showed overlapping in the chromatograms, and it seems that at the present level of SFC performance it is not possible to perform such separations by virtue of plate number alone. Separation of TG species was also incomplete on more polar open tubular columns (van Oosten *et al.*, 1991). Thus, separation has to be achieved by means of selectivity, and that is best performed on packed columns. Employing such columns, we have demonstrated the separation of neutral as well as polar lipids.*

In the present chapter we describe the relation between three classes of mobile phases in chromatography: gas, supercritical fluid and liquid. Further, the relation between packed and open tubular columns is discussed. Mobile and stationary phases relevant to lipid separations are reviewed as well as instrumental aspects such as injection and detection. SFE of lipids for analytical and semi-preparative purposes is discussed.

2.2 Supercritical fluid chromatography

2.2.1 General aspects on the properties of supercritical media

Carbon dioxide has a critical temperature, T_c, of 31.3°C, and a critical pressure, P_c, of 72.9 atm. When carbon dioxide is compressed at a temperature and a pressure above critical it does not liquefy but forms a dense gas (Fig. 2.1). The mobile phase in SFC is thus gaseous and solvating. Such a

---

* See Andersson, Demirbüker and Blomberg, 1997; Blomberg and Demirbüker, 1994; Blomberg, Demirbüker and Andersson, 1993; Demirbüker and Blomberg, 1990, 1991, 1992; Demirbüker, Hagglund and Blomberg, 1990, 1992; Demirbüker *et al.*, 1992; Hägglund, Demirbüker and Blomberg, 1994; Janák *et al.*, 1992a.

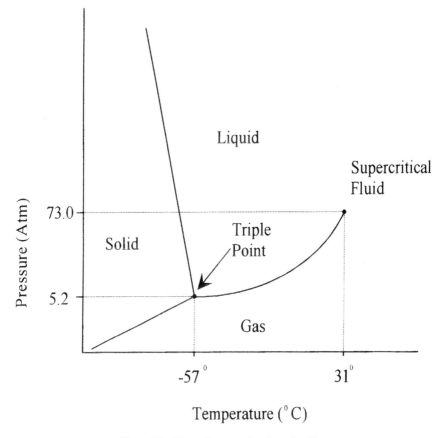

Figure 2.1 Phase diagram of carbon dioxide.

dense gas has a number of properties that makes it attractive for use as a mobile phase for chromatography.

The mobile phases may be characterized by three key factors: density, diffusivity and viscosity. Typical values of these properties regarding gases, supercritical fluids and liquids are given in Table 2.1. Let us first consider the densities. These reflect the solubilities of the analytes. The mobile phase in GC does not dissolve the analytes and as a consequence the analytes will have to pass through the column in a gaseous state, and that is the inherent limitation of GC. Nowadays there are GC columns that can withstand up to 600°C, but there are few analytes that are stable at those temperatures. In liquid chromatography (LC) the situation is different because here the mobile phase, by virtue of its density, transports the analytes through the separation column and there is no immediate need for elevated temperatures. However, there is a price to pay for the increase in analyte solubility, and that is a drastically reduced diffusivity (Table 2.1).

Table 2.1 Order of magnitude comparison of physical properties of liquids, gases and supercritical fluids.

Phase	Density (g cm^{-3})	Diffusion (cm^2 s^{-1})	Viscosity (g cm^{-1} s^{-1})
Gas (1 atm, 21°C)	10^{-3}	10^{-1}	10^{-4}
Supercritical fluid	0.3–0.8	10^{-3}–10^{-4}	10^{-4}–10^{-3}
Liquid	1	< 10^{-5}	10^{-2}

Source: van Wasen, U., Swaid, I. and Schneider, G. M., Physicochemical principles and applications of supercritical fluid chromatography, *Angew. Chem. Int. Ed. Engl.*, **19**, 575–87, 1980

As a consequence the chromatographic processes are quite slow in LC compared with those in GC. Here SFC comes in as an alternative, providing relatively high densities and higher diffusivities than observed in conventional liquids (Blomberg, 1988; Blomberg *et al.*, 1994).

The viscosities of supercritical media are relatively low (Table 2.1) and therefore the pressure drop applied over the column is much lower when supercritical media are being used as the mobile phase than when conventional liquids are employed. This is an important advantage because when a compressible mobile phase is being used there will inevitably be a mobile phase velocity gradient over the column, and that will lead to broader analyte bands. The decrease in plate number is, however, partly offset by the increase in optimal mobile phase velocity that is achieved as the mobile phase density is decreased towards the end of the column (Cramers and Rijks, 1979). There is a density gradient as well. This gradient is of particular significance in SFC; the reason for this is that the solubility of the analytes is directly related to the mobile phase density. Thus if the density becomes too low the analytes may precipitate. To avoid this problem, the column end is provided with a restriction over which the pressure drop to the atmosphere takes place. However, in cases when there is a relatively large pressure drop over the column, for example with long packed columns, analyte precipitation may occur towards the end of the column unless a sufficiently high pressure has been selected.

SFC has been performed with packed as well as with open tubular columns. In the former case, the aim is to provide a substitute for packed column HPLC and in the latter case it is to provide a substitute for open tubular GC.

2.2.2 Speed of analysis

Open tubular columns. The band broadening obtained in an open tubular column can be expressed by the Golay equation (Golay, 1958). This summarizes the contribution to band broadening from longitudinal diffusion, non-uniform local mobile phase velocities (the parabolic flow profile) and diffusion-controlled mass transfer in the stationary phase:

$$H = \frac{2D_G^o j}{\bar{u}} + \frac{r^2(1 + 6k + 11k^2)\bar{u}}{24D_G^o(1+k)^2 j} + \frac{2kd_f^2 \bar{u}}{3(1+k)^2 D_L},\quad (2.1)$$

Where
H is the height of theoretical plates;
D_G^o = solute molecular diffusion coefficient in the mobile phase at outlet pressure;
$j = \bar{u}/u_o$;
$\bar{u}$ is the average mobile phase velocity;
u_o is the mobile phase velocity at column outlet;
r is the capillary radius;
k is the capacity factor;
d_f is the stationary phase film thickness;
D_L is the solute molecular diffusion coefficient in the stationary phase.

The contribution from the last factor in equation (2.1) is often relatively small and it may be neglected for simplicity (Lee, Yang and Bartle, 1984). If one neglects the band broadening in the stationary phase, the Golay equation can be solved for the minimum height of theoretical plates:

$$H_{min} = 1.125r \left[\frac{1 + 6k + 11k^2}{3(1+k)^2} \right]^{1/2}. \quad (2.2)$$

Equation (2.2) shows that H_{min} is dependent on only the capillary radius and the capacity factor. Solution of the Golay equation regarding optimal mobile phase velocity, again neglecting band broadening in the stationary phase, gives:

$$\bar{u}_{opt} = \frac{4D_G^o j}{r} \left[\frac{3(1+k)^2}{1 + 6k + 11k^2} \right]^{1/2}. \quad (2.3)$$

Liquids provide diffusion coefficients that are 10^4 times lower than those obtained with gases (Table 2.1). According to equation (2.3), this would, for a given capillary diameter, lead to extended separation times. However, the slow diffusion can be compensated for by a decrease in capillary radius. Moreover, the decreased radius also results in a decreased plate height, thus a shorter column length is then required to get a given number of theoretical plates.

The speed of analysis obtained with a 20 m × 0.25 mm capillary column which gives 80 000 theoretical plates under GC conditions can be used as a starting point for a comparison of the different classes of mobile phases. To obtain with SFC a speed of analysis in the same range as that obtained with GC, the capillary would have to have an inner diameter (i.d.) of c. 20 μm,

Table 2.2 Order of magnitude comparison of the contribution to band broadening by non-uniform local mobile phase velocities, the flow profile (r^2/D_M) in capillary columns (the C_G term in Golay's equation)

Mobile phase	Column inner diameter (µm)	r^2/D_M (s)
Gas chromatography	250	1.6×10^{-3}
Supercritical fluid chromatography	50	15.6×10^{-3}
Supercritical fluid chromatography	20	2.5×10^{-3}
Liquid chromatography	5	6.2×10^{-3}

and in LC the capillary i.d. would have to be c. 5 µm. An additional way to increase speed of analysis is to apply above optimal mobile phase velocities. Band broadening at such velocities is basically governed by the non-uniform local mobile phase velocities, the middle term on the right-hand side of equation (2.1). Table 2.2 indicates the relative magnitude of this term for different combinations of mobile phase and capillary diameter. Evidently, the capillary i.d. for SFC will have to be c. 20 µm to be comparable with the GC capillary in this respect.

The state of the art in capillary SFC is the application of 50 µm i.d. capillaries. As shown above [equation (2.3)], a smaller i.d. would be required to compensate for the relatively slow diffusion compared with that obtained in gases. However, practical considerations presently restrict the use of such capillaries. There are three main factors involved here: difficulties with injecting with the precision necessary for quantitative analysis, difficulties with preparing the separation capillaries and difficulties with detection. A general feature of miniaturization is that, unless one has very sensitive detection available, for example fluorescence detection, the linear range will decrease when going to smaller dimensions. This is a result of overloading problems. The consequence of the difficulties with the small i.d. columns is that open tubular SFC presently cannot be utilized to its full potential.

Packed columns. The techniques for packed columns for HPLC have been extensively developed during recent decades. Thus very small particles with narrow size distributions are available. With such particles, reasonable plate counts can be obtained in short column lengths. The optimal mobile phase velocities are often in the range of 1–2 mm s^{-1} in such columns, and thus they need to be short in order to avoid excessive analysis times. Compared with the plate numbers obtained in GC, the plate numbers of HPLC are quite low. The classical remedy in chromatography in such cases is to opt for selectivity, and that is precisely what makes HPLC such a versatile technique. Selectivity profits from intensive partitioning of the analytes between the mobile phase and the stationary phase. Thus the ratio β, the volume of mobile phase, V_{mobile}, the volume of stationary phase, $V_{stationary}$, should be low. In open tubular GC, the ratio is c. 300, whereas in HPLC it is 25–10. During recent years the benefits of supercritical media as mobile phases in packed columns have been extensively demonstrated (Berger, 1995).

It is, in this context, of interest to compare the speed of analysis in packed and open tubular columns. The relation between the i.d., d_c, of an open tubular column and the particle size, d_p, in a packed column for which the speed of analysis, using the same mobile phase, would be equal was derived by Guiochon (1981). It was found that equal speed of analysis would be obtained when

$$d_c \approx 2.1 d_p. \tag{2.4}$$

Thus, comparing a packed column having 5 μm particles with an open tubular column, the latter would have to have an i.d. of c. 11 μm to give the same speed of analysis.

2.2.3 Packed column diameter

Packed columns of different diameters have been used in SFC. Large-diameter columns (4.6 mm i.d.) provide high sample capacities, and, because of the relatively large flow of mobile phase, variable end column restrictors can readily be employed. Such restrictors make it possible to programme the flow of mobile phase, keeping the pressure constant. It is possible to use a variable end column restrictor also with micropacked columns (Janák et al., 1992b; Janssen, Rijks and Cramers, 1990) but that is quite complicated and systems for this are not commercially available.

Micropacked columns offer a number of advantages over wide-bore columns. They are much cheaper to use as the consumption of mobile and stationary phases is quite low. Moreover, the evaluation of the performance of a large number of packing materials and subsequent *in situ* modifications can be made in a simple and cheap manner. Further, an *in situ* column modification can be performed very effectively with relatively small amounts of the modifier, for example for the preparation of argentation columns. In addition, the low thermal mass of micropacked columns provides excellent possibilities for temperature programming. Last, the low flow of mobile phase is an advantage *per se* when concentration-sensitive detectors are used and when the mobile phase has to be removed prior to detection, for example when using a mass spectrometer (MS) or an evaporative light-scattering detector.

2.2.4 Mobile phases for supercritical fluid chromatography

Carbon dioxide (CO_2) is by far the most commonly used mobile phase in SFC. It has a number of advantages. It has a low critical temperature and is highly inert (except when used with amines); it is non-toxic, non-explosive, and is easily purified and it can be used with the flame ionization detector, which was, in fact, one of the major issues when open tubular SFC was introduced (Chester, 1989). It was considered that this would satisfy the long desired universal detector for chromatography with a liquid mobile phase. However, since that time, bench-top liquid chromatography/mass spectrometry (LC–MS) has arrived and thereby universal detection for HPLC.

There is, however, one major problem with CO_2 as a mobile phase and that is its low polarity. Thus only relatively non-polar analytes can be dissolved in CO_2. Moreover, in columns packed with silica-based material there are always residual adsorptive sites. In reversed-phase HPLC the mobile phase deactivates these sites, but the CO_2 is not polar enough to do that. As a consequence, the more polar analytes are adsorbed and these are then eluted as severely tailing peaks or are not eluted at all. It should be mentioned here that reports on more inert packings have been published (Li, Malik and Lee, 1994). There are some supercritical mobile phases other than CO_2 that can be used, but those that are realistic to use are all non-polar. The only alternatives are the freons, of which chlorine-free freons are considered to be less harmful to the environment (Blackwell and Schallinger, 1994).

The generally employed approach to the problems encountered with a non-polar supercritical fluid is to add a polar organic modifier. This provides analyte solubility and packing deactivation. The flame ionization detector cannot be employed with use of such modifiers but there are other detectors that can be used, for example the evaporative light scattering detector (ELSD) or the mass spectrometer. A further advantage of the polar organic additives is that they provide a means of applying mobile phase composition gradients (Blomberg, Demirbüker and Andersson, 1993). Finally, it should be emphasized that the mixture of CO_2 and modifier must be homogeneous under all conditions applied, from the pump to the restrictor.

Mobile phases for argentation chromatography. It is necessary to add acetonitrile to the CO_2 to facilitate elution of TG. Acetonitrile modifies the interactions between the analytes and the stationary phase and also improves the solubility of the analytes in the mobile phase. Although the solubility of TG in supercritical CO_2 is, in general, good, the presence of a polar modifier may diminish the risk of analyte precipitation as the pressure is released in the restrictor.

In order to achieve a mobile phase that is homogeneous under all conditions applied, a minor amount of 2-propanol is added to the mobile phase. Typically, a mobile phase for the elution, on an argentation column, of TG from a vegetable oil consists of carbon dioxide–acetonitrile–2-propanol (92.8:6.5:0.7 mol%) (Demirbüker and Blomberg, 1991; Demirbüker, Hägglund and Blomberg, 1990; Demirbüker et al., 1992).

2.2.5 Stationary phases for supercritical fluid chromatography

Open tubular columns. Here, the separation, in most cases, takes place according to the volatility of the analytes by means of a non-polar stationary phase. Stationary phases, with greater polarity are also used, but owing to the high β-value (section 2.2.2), selectivity is not particularly powerful in open tubular columns.

Packed columns. In this technique separation is by virtue of selectivity. A number of different types of modified silica particles developed for HPLC are applicable. In addition, there are a number of *in situ* modified packings that are of interest in connection with lipid analysis (Demirbüker and Blomberg, 1991, 1992; Demirbüker, Hägglund and Blomberg, 1990; Hägglund, Demirbüker and Blomberg, 1994)

The physical stability of the packing is an important issue in packed column SFC. First, there should be no void volumes in the column. Often, the packed bed is compacted the first time supercritical CO_2 is applied, and proper care must be taken regarding the end frits so that band broadening is avoided. Further, if the packing is not tightly secured it may rearrange when the pressure is rapidly released after a pressure-programmed run or on back flushing (Andersson, Demirbüker and Blomberg, 1993).

Argentation chromatography or silver ion chromatography has, since its introduction in 1962 (de Vries, 1962; Morris, 1962), been extensively employed in thin-layer chromatography (TLC) and HPLC for the separation of unsaturated lipids. At first, the separations were performed on silica impregnated with silver nitrate. However, the silver ions were gradually leached out of columns packed with this material. In 1987 Christie (1987b) demonstrated that a very high column stability could be obtained when a silica-based cation exchanger was employed as a support for the silver ions. This technique was adopted by us for the separation of TG in vegetable oils (Demirbüker and Blomberg, 1990, 1991; Demirbüker, Hägglund and Blomberg, 1990; Demirbüker *et al.*, 1992) and in a fish oil (Blomberg, Demirbüker and Andersson, 1993; Demirbüker, Hägglund and Blomberg, 1990). Under the conditions employed in SFC, type of mobile phase, temperature, etc., the columns were remarkably stable; they could be used for years and they could withstand temperatures up to 115°C for prolonged lengths of time.

Initially, 5 µm particles were used in our columns, but an appreciable improvement in separation efficiency was achieved when these were replaced by 4 µm particles (Demirbüker *et al.*, 1990).

In SFC, argentation columns separate TG according to degree of chemical unsaturation, chain length [e.g. POL (P = palmitate, O = oleate, L = linoleate) is separated from SOL (S = stearate) (Fig. 2.2)], position of double bonds in a fatty acid moiety [e.g. TG containing an α-linolenic moiety is separated from a TG where this moiety is exchanged for a γ-linolenic moiety

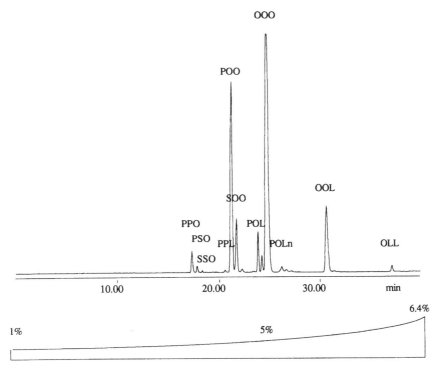

Figure 2.2 Supercritical fluid chromatogram, with use of a miniaturized evaporative light-scattering detector, of an olive oil. Column: glass-lined metal tubing, 150 mm × 0.7 mm, packed with Nucleosil™ 4SA and impregnated with silver nitrate. Conditions: temperature 100°C, pressure 300 atm. Mobile phase: gradient of carbon dioxide/(acetonitrile/isopropanol, 90:10 mol%). Restrictor: fused silica capillary tubing 130 mm × 10 μm inner diameter. Peaks: triacylglycerols. Abbreviations: P = palmitate; S = stearate; O = oleate; L = linoleate. Reproduced with permission from Blomberg, L. G., Demirbüker, M., Hägglund, I. and Andersson, P. E., Supercritical fluid chromatography: open tubular *vs.* packed columns, *Trends Anal. Chem.*, **13**, 126–37, 1994.

(Blomberg, Demirbüker and Andersson, 1993)] and separation of saturated TG according to carbon number (illustrated in Fig. 2.3 by the separation of a mixture of two oils – cohune oil, which is an oil consisting basically of saturated TG, and sunflower seed oil). Separation according to carbon number may be the result of analyte interactions with the hydrocarbon spacer situated between the silica and the cation exchanging functional group. A column that possesses such dual selectivities can find application in the separation of interesterified oils (Fig. 2.4). In addition, separation of *cis* and *trans* isomers of fatty acids has been demonstrated (Demirbüker, Hägglund and Blomberg, 1992). Finally, it should be mentioned that separation according to the position of a fatty acid on the glycerol moiety, for example, PPO from POP, could not be achieved with the present system.

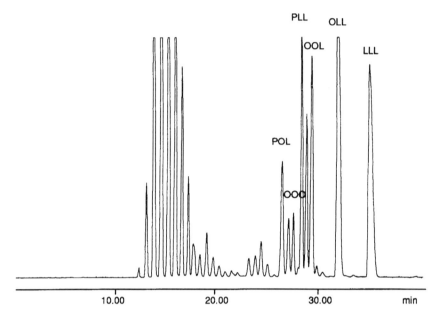

Figure 2.3 Supercritical fluid chromatogram with use of miniaturized evaporative light-scattering detector, of a mixture of cohune oil and sunflower seed oil (50:50). Column as in Fig. 2.2. Conditions: injection at 115°C and 180 atm; after 1 min, programmed at $-1°C\ min^{-1}$ to 75°C and at 2 atm min^{-1} to 210 atm, then 4 atm min^{-1} to 320 atm. Mobile phase: carbon dioxide/acetonitrile/isopropanol (92.8:6.5:0.7 mol%). Peaks and abbreviations as in Fig. 2.2.

As well as the silver-ion impregnated cation exchanger, some other types of stationary phases have been examined. Columns packed with an anion exchanger and impregnated with permanganate gave group separation according to number of double bonds (Demirbüker and Blomberg, 1992). For the separation of hydroxyl-containing TG, castor oil, an anion exchanger was impregnated with periodate (Demirbüker, Hägglund and Blomberg, 1990). Further, silica-based 8-quinolinol, with and without metal ions, was examined as a selective stationary phase for the separation of fatty acid methyl esters (Hägglund, Demirbüker and Blomberg, 1994). Further, the application of argentation chromatography in open tubular columns has been attempted (Janák et al., 1992a; Shen et al., 1995).

For the group separation of polar lipids, the packing material should be as inert as possible. Otherwise lipids such as phosphatidylethanolamine (PE) are eluted as tailing peaks. However, when there is some residual adsorptive activity from surface silanol groups on the packing these can be partially deactivated when the mobile phase contains an additive that has a deactivating ability, for example methanol. Class separation of moderately polar lipids, under subcritical conditions, was obtained on columns packed with diol-modified silica and a mobile phase modified with 19 mol% methanol (Fig. 2.5). Also, lipids with greater polarity, such as PE, could be

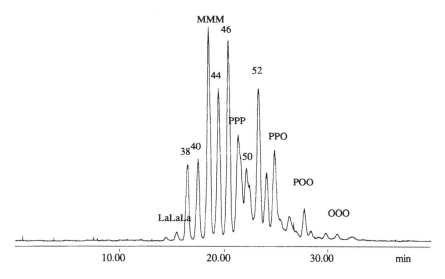

Figure 2.4 Supercritical fluid chromatogram, with use of a miniaturized evaporative light-scattering detector, of interesterified palm oil/palm kernel oil (50:50). Column: as in Fig. 2.2. Conditions: injection at 115°C and 180 atm; after 1 min programmed at $-1°C\,min^{-1}$ to 75°C and $2\,atm\,min^{-1}$ to 220 atm, and then at $4\,atm\,min^{-1}$ to 360 atm. Mobile phase: carbon dioxide with 6.6 mol% of a 9:1 vol./vol. acetonitrile/isopropanol mixture. Peaks indicated by carbon number and by triacylglycerol composition. Abbreviations: L = linoleate; La = laurate; M = myristate; O = oleate; P = palmitate; S = stearate.

eluted under the conditions applied in Fig. 2.5, but the peaks showed some tailing.

2.2.6 Instrumental

Injection. For quantitative analysis, direct injection on the column, in our view, gives the best result. For such injections, unless one has the analytes efficiently focused on the column head, the flow rate of the mobile phase should not be too low. It is our experience that the direct injection works well on packed columns having an i.d. of 0.7 mm and larger.

Detection. Detection of lipids separated by open tubular SFC has been by flame ionization or mass spectrometry. For packed columns, ultraviolet (UV) or detectors or ELSDs have been employed. ELSDs have been used extensively for detection of lipids in connection with HPLC, and it is of interest to apply this detector also to packed column SFC.

ELSDs were developed for connection to HPLC with conventional column diameters (Christie, 1992b; Upnmoor and Brunner, 1992) but they have also been used for packed column SFC with wide (Carraud *et al.*, 1987; Lafosse, Dreux and Morin-Allory, 1987; Lafosse *et al.*, 1992) and narrow bore columns (Demirbüker, Andersson and Blomberg 1993;

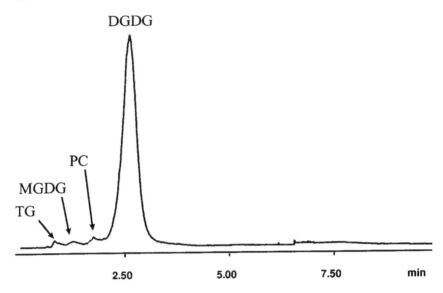

Figure 2.5 Supercritical fluid chromatogram, with use of a miniaturized evaporative light-scattering detector, of an extract obtained from oat bran. Column: 100 mm × 0.9 mm, packed with LiChrosorb Diol, 5 μm. Conditions: temperature 22°C; pressure 300 atm. Mobile phase: carbon dioxide modified with 19 mol% methanol. Restrictor: 45 mm × 9 μm at 90°C. Peaks: TG = triacylglycerols; MGDG = monogalactosyldiacylglycerols; PC = phosphatidylcholines; DGDG = digalactosyldiacylglycerols.

Hagen, Landmark and Greibrokk, 1991; Hoffmann and Greibrokk, 1989). Using conventional ELSD there is always some loss of analytes within the mobile phase evaporation compartment (the drift tube) of the detector. With large bore columns (4.6 mm i.d.) this is not a problem, but when using narrow bore packed columns the losses are, in general, too large. Therefore, a miniaturized evaporative light scattering detector (μ-ELSD), dedicated to the application with micropacked SFC, has been developed at the Arrhenius Laboratories (Demirbüker, Andersson and Blomberg, 1993). The drift tube is quite small in our detector, 25 mm × 1.5 mm i.d. Larger drift tubes are not needed here as the mobile phase is easily evaporated in SFC and, moreover, the flow rates are quite low with micropacked columns.

A modified version of the detector is shown in Fig. 2.6. In this version, the earlier used He–Ne laser is substituted for a diode laser, MDL-200-670-5 (Laser Max Inc., Rochester, NY). In order to allow a major part of the analytes to pass through the light beam, the drift tube ends as closely as possible to the beam. The diode laser produces a faint halo around the light beam, and if this halo hits the drift tube, an increased photodiode background current results. Therefore, the halo must be removed, and for this purpose a slit is included. The position of the slit is 45 mm from the end of the photodiode. In addition, a mirror has been inserted opposite the photodiode in the detector cell. With this detector, the limit of detection for

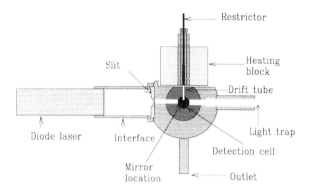

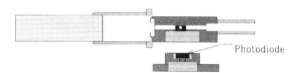

Figure 2.6 Schematic diagram of the miniaturized evaporative light-scattering detector.

trimyristin was c. 5 ng and the slope of the calibration curve for trimyristin (log–log plot) was 1.52. The response for some different TGs is shown in Fig. 2.7.

The μ-ELSD is thus an integrated part of our miniaturized chromatographic system for quantitative analysis of neutral as well as more polar lipids.

2.2.7 Gradients

A large number of gradients are possible in SFC. These include temperature, pressure and/or density, velocity, eluent composition and combinations of these (Klesper and Schmitz, 1987). There are thus many possibilities for optimizing performance.

Most commonly, gradients resulting in increased mobile phase densities are applied. Such gradients result in a decrease in analyte diffusion coefficients, which leads to impaired chromatographic performance. The optimal mobile phase velocity will thus be decreased and the slope of the high-velocity branch of the van Deemter curve will be steeper. Moreover, pressure programming, without application of constant flow regulation, results in

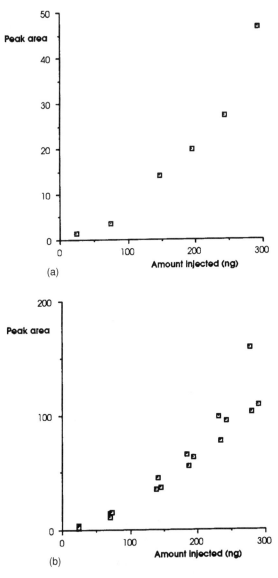

Figure 2.7 Response curves of some different triacylglycerols (TGs) on a miniaturized evaporative light-scattering detector: (a) saturated TGs CyCyCy, PPP and SSS; (b) unsaturated TG LLL. Abbreviations: Cy = caprylate; L = linoleate; P = palmitate; S = stearate.

increased mobile phase velocities, leading to decaying separation efficiencies. When applying positive pressure programming, we have thus been obliged to apply quite slow mobile phase velocities, typically in the range of 2–3 mm s^{-1} at the start of a run, to maintain the highest separation efficiency throughout the run. In addition, negative temperature programming is not really optimal in connection with argentation chromatography because the strength of the olefin–silver ion complex will be increased with decreasing temperature.

Application of moderate modifier gradients on packed column SFC is an attractive approach. With such gradients, the mobile phase elution strength will be greatly enhanced, whereas diffusion coefficients will be only moderately increased. We have constructed a chromatographic system for the application of such gradients in connection with micropacked columns in SFC (Blomberg and Demirbüker, 1994; Blomberg, Demirbüker and Andersson, 1993). Application of a mobile phase gradient is beneficial for separation of oils containing components that have a widely differing degree of unsaturation, for example fish oils, as in the applications cited above, but it also improves the chromatography of oils having a narrower range of unsaturation (Fig. 2.2). One important advantage of the use of gradients is that the analytes are focused at the column head, leading to narrow starting bands.

2.2.8 Widening the scope of supercritical fluid chromatography

The main virtue of SFC compared with HPLC is that it provides enhanced diffusivity. Enhanced diffusivity can also be obtained, however, in other ways. First, CO_2 in its liquid state provides higher diffusivity than do conventional liquids; application of this mobile phase is often called subcritical fluid chromatography. Another possibility to improve diffusivity is to use mixtures of liquid CO_2 and conventional liquids as the mobile phase. This approach has been taken by Olesik and co-workers (Cui and Olesik, 1991; Lee and Olesik, 1995; Lee, Olesik and Fields, 1995) who showed that the mixing of methanol with up to 50% liquid CO_2 resulted in increased diffusivity and decreased viscosity and the polarity was largely maintained. A third approach to HPLC with enhanced diffusivity is the application of enhanced separation temperatures as demonstrated by Trones, Iveland and Greibrokk (1995).

2.2.9 Comparison of supercritical fluid chromatography with other separation techniques for lipid characterization

As mentioned above (section 2.2.1), the intention with open tubular SFC is to act as a substitute for GC. The state of the art is to have the analysis temperature much lower in SFC than in GC; thus TG-containing polyunsaturated fatty acids pass through the column without problems. However, the separation performance is not as good as in GC, resulting in the co-elution of TGs.

The performance of packed columns should be compared with that of packed column HPLC. As the diffusion coefficients are ≥ 20 times greater, depending on the density, higher than in HPLC, SFC should in theory be more rapid and/or efficient than HPLC. The examples given here indicate that this is the case. In our view, powerful selectivities can be achieved in packed column SFC. However, to compete with HPLC one also needs to include mobile phase composition gradients. This is technically demanding, but it can be done.

2.3 Supercritical fluid extraction

2.3.1 Analytical applications

SFE has for many years been applied in the food industry for preparative purposes (Dean, 1993; Luque de Castro, Valcárcel and Tena, 1994; Rizvi,

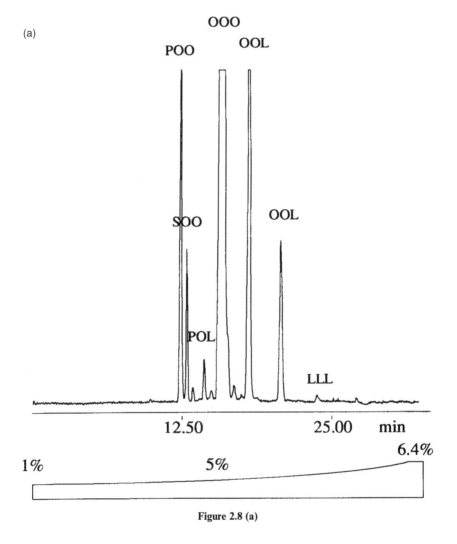

Figure 2.8 (a)

1994). In connection with the more recent interest in SFC, methods for the analytical application of extraction with supercritical fluids have been extensively developed.* At the Arrhenius Laboratories SFE has been applied to the extraction of different types of rapeseeds and other oil seeds as a part of a breeding project. SFCs of two of these extracts are shown in Fig. 2.8. Traditionally, such seeds are examined for total fat content, either gravimetrically after Soxhlet extraction or by means of elementary analysis. In addition, the fatty acid pattern is analysed by GC after hydrolysis and methylation. Soxhlet extraction can be replaced by SFE followed by a gravimetric analysis to quantify the extract. Alternatively, quantification can be performed on a detector (ELDS) that has been connected directly to

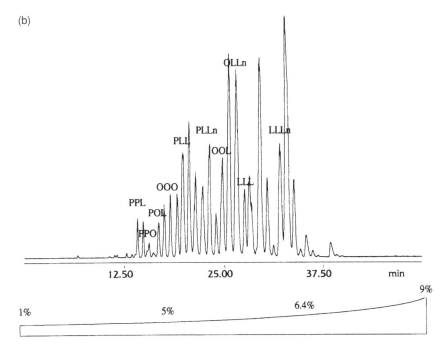

Figure 2.8 Supercritical fluid chromatography, with use of a miniaturized evaporative light-scattering detector, of seed extracts from (a) an autumn rapeseed; (b) *Camelina sativa* (triacylglycerides containing long-chain monoenes have not been identified because reference substances are not available). SFE/SFC was off-line. Column: as in Fig. 2.2. Separation conditions: temperature 140°C; pressure 160 atm, after 1 min programmed at $-10°C\,min^{-1}$ to 100°C and 25 atm min^{-1} to 260 atm, then programmed at $-1°C\,min^{-1}$ to 75°C and at 2 atm min^{-1} to 260 atm. Extraction conditions: temperature 90°C, pressure 160 atm. Abbreviations: L = linoleate; Ln = linolenate; O = oleate; P = palmitate; S = stearate; SFC = supercritical fluid chromatography; SFE = supercritical fluid extraction.

* Berger, 1995; Dean, 1993; Jinno, 1992; King, 1995; Luque de Castro, Valcárcel and Tena, 1994; McHugh and Krukonis, 1994; Saito, Yamauchi and Okuyama, 1994; Wenclawiak, 1992; Westwood, 1993.

the extraction cell (van Lenten *et al.*, 1996). Recently, fat content has also been measured by means of near infra-red (Orman and Schumann, 1991, 1992). These measurements give relatively crude information concerning the composition of the fat. Often, more detailed information is called for and this can be obtained by chromatographic analysis of extracts, for example by means of SFC. A fruitful approach in this context is the application of on-line SFE–SFC. The merits of this technique were described in a review by Greibrokk (1995).

In general, SFE–SFC has been applied to trace analysis and the extracted material has therefore been trapped prior to SFC analysis. However, such trapping is not necessary for the characterization of the TG pattern of seeds. For example, SFE may be directly (i.e. without intermediate trapping), coupled to SFC (Fig. 2.9). In this system, the extraction medium coming from the extraction cell is allowed to pass through a 60 nl internal loop injector, and this amount is subsequently injected on to the chromatograph. In order to focus the analytes at the column head, the column is held at an elevated temperature during injection; a low mobile phase density is thus employed. During the chromatographic separation step, the extraction cell can be loaded with the next sample to be extracted. The system shown in Fig. 2.9 is a one-pump system, thus the same CO_2/modifier mixture is

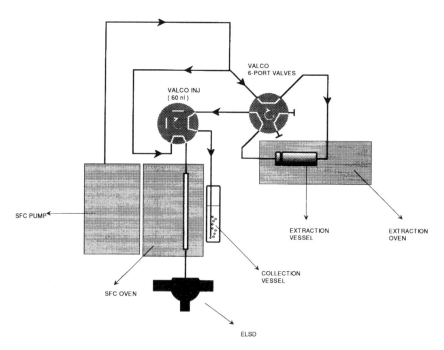

Figure 2.9 Schematic diagram of the set up for on-line SFE/SFC (supercritical fluid extraction/supercritical fluid chromatography).

used for extraction and chromatography. With two pumps this is, of course, not necessary. In addition, analyte focusing can be done by virtue of a reduced pressure during injection. In general, focusing by means of a reduced pressure is more effective than by means of an elevated temperature.

SFE can also be employed for extraction of volatile compounds. This was demonstrated in the analysis of volatiles in oxidized oils by SFE–GC (Snyder, 1995).

2.3.2 Semi-preparative applications

One of the major advantages of supercritical media is the possibility to regulate the solubility of different compounds in the medium. This can be easily accomplished with high accuracy by a change in pressure. This property makes supercritical fluids extremely suitable for selective extractions. A method for the selective extraction of lipid classes has been developed at the Arrhenius Laboratories and its performance was demonstrated by the extraction of digalactosyldiacylglycerol (DGDG) from oat bran (Andersson, Demirbüker and Blomberg, 1992).

For preparative purposes, extractions are generally performed in extraction cells in a batchwise mode. To obtain a high throughput the cells need to be relatively large. As extractions are performed at high pressures (≥ 400 atm) the cells must be thick-walled and thus they are relatively expensive. To avoid large volumes, we introduce a liquid extraction step, where we extract the oat bran with refluxing acetone. The oil resulting after acetone evaporation is then deposited in an extraction cell and purified with supercritical CO_2 in a batchwise mode. A further development is to have a continuously operating device for the purification of the oil (Fig. 2.10). Here a spray of a solution of the oil is formed, and, in the spray, the TG is dissolved in CO_2 and transported out of the cell. The DGDG is precipitated in the cell. This system is similar to that constructed by Tom and Debenedetti (1991) for the formation of fine particles from CO_2 expansion. Extraction systems, operating with supercritical CO_2, have also been devised (Eggers, 1996; Eggers and Wagner, 1993; Wagner and Eggers, 1996). However, they extracted from a viscous matrix whereas we extracted from a low viscous solution. The system shown in Fig. 2.10 is on a laboratory scale, but it can be scaled up. In its present form it can produce 1 g DGDG h^{-1}. The purity of the product is shown in Fig. 2.5. In conclusion, it seems that extraction and/or purification is an area where supercritical media may be of great importance in the future.

In SFE, *in situ* derivatization of the analytes is sometimes applied in order to facilitate the extraction of target analytes (Hawthorne *et al.*, 1992). Such a procedure leads to improved extraction selectivity. The application of enzymatic reactions is a powerful means to increase further the selectivity in

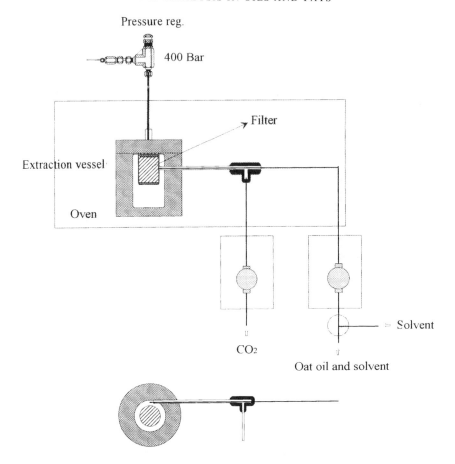

Figure 2.10 Schematic diagram of the set up for jet extraction.

supercritical extractions. This was shown, in a preliminary investigation, by Greibrokk and Berg (1993). Direct methanolysis of TGs in flowing supercritical CO_2 by an immobilized lipase has been described by Jackson and King (1996). Nakamura (1996) presented a detailed discussion of enzymatic reactions in supercritical fluids.

2.4 Reactions in supercritical media

As a consequence of the very low viscosity, reaction rates can be very high in supercritical media. This was recently demonstrated for hydrogenation of TGs in supercritical propane (Härröd and Möller, 1996). The reaction took place in a small cell, and reaction rates up to 1000 times higher than those obtained with traditional techniques were reported.

2.5 Conclusions

The main advantage of SFC in relation to GC is the relatively low analysis temperatures. In comparison with HPLC, analyte diffusion coefficients are higher in SFC which leads to faster and/or more efficient separations. On the other hand, the possibilities of obtaining selectivities via the mobile phase are not as good in SFC as in HPLC. In addition, HPLC instrumentation is currently in a more advanced state than is SFE instrumentation, and the experience among analysts concerning HPLC is much greater than that concerning SFC. Nevertheless, SFC may be a viable alternative, which we have attempted to demonstrate in this chapter.

SFE is a very promising method for sample preparation and it has already been employed extensively in connection with lipid analysis, where it can be a substitute for traditional Soxhlet extraction. In addition, there is considerable potential for further development of preparative SFC.

Acknowledgements

This work was supported by Karlshamns AB and the Swedish Natural Science Research Council. Thanks are due to Utku Sayin for his contribution to the development of the ELSD.

References

Andersson, P. E., Demirbüker, M. and Blomberg, L. G. (1993) Characterization of fuels by multi-dimensional supercritical fluid chromatography and supercritical fluid chromatography–mass spectrometry. *J. Chromatogr.*, **641**, 347–55.

Andersson, M., Demirbüker, M. and Blomberg, L.G. (1997) Selective extraction/purification of lipids by means of supercritical media, in press.

Berger, T. A. (1995) *Packed Column SFC*, Royal Society of Chemistry, Cambridge.

Black, H. (1996) Supercritical carbon dioxide: the 'greener' solvent. *Environ. Sci. Technol.*, **30** 124A–7A.

Blackwell, J. A. and Schallinger, L.E. (1994) Hydrofluorocarbon and perfluorocarbon mobile phases for capillary supercritical fluid chromatography. *J. Microcol. Sep.*, **6**, 551–6.

Blomberg, L. (1988) Comparison of gas, supercritical fluid and liquid as mobile phase for chromatography. *Chimica Oggi*, (1), 17–21.

Blomberg, L. and Andersson, P. E. (1994) Capillary electrophoresis for lipid analysis. *INFORM*, **5**, 1030–7.

Blomberg, L. G. and Demirbüker, M. (1994) Analysis of tricylglycerols by argentation supercritical fluid chromatography, in *Developments in the Analysis of Lipids* (eds J. H. P. Tyman and M. H. Gordon), Royal Society of Chemistry, Cambridge, pp. 42–58.

Blomberg, L. G., Demirbüker, M. and Andersson, P. E. (1993) Argentation supercritical fluid chromatography for quantitative analysis of triacylglycerols. *J. Am. Oil Chem. Soc.*, **70**, 939–46.

Blomberg, L. G., Demirbüker, M., Hägglund, I. and Andersson, P. E. (1994) Supercritical fluid chromatography: open tubular vs. packed columns. *Trends Anal. Chem.*, **13**, 126–37.

Carraud, P., Thiebaut, D., Caude, M. *et al.* (1987) Supercritical fluid chromatography/light-scattering detector: a promising coupling for polar compounds analysis with packed columns. *J. Chromatogr. Sci.*, **25**, 395–8.

Charpentier, B. A. and Sevenants, M. R. (eds) (1988) *Supercritical Fluid Extraction and Chromatography*, ACS Symposium Series 366, American Chemical Society, Washington, DC.

Chester, T. L. (1989) Practice and applications of supercritical fluid chromatography in the analysis of industrial samples, in *Microbore Column Chromatography* (ed. F. J. Yang) Chromatography Science Series 45, Marcel Dekker, New York, pp. 369–97.

Christie, W. W. (1987a) *High-performance Liquid Chromatography and Lipids*, Pergamon, Oxford.

Christie, W. W. (1987b) A stable silver-loaded column for the separation of lipids by high performance liquid chromatography. *J. High Resolut. Chromatogr. Chromatogr. Commun.*, **10**, 148–50.

Christie, W. W. (1992a) *Gas Chromatography and Lipids*, The Oily Press, Ayr.

Christie, W. W. (1992b) Detectors for high-performance liquid chromatography of lipids with special reference to evaporative light-scattering detection, in *Advances in Lipid Methodology – One* (ed. W. W. Christie), The Oily Press, Ayr, pp. 239–71.

Cramers, C. A. and Rijks, J. A. (1979) Micropacked columns in gas chromatography: an evaluation, in *Advances in Chromatography 17* (eds J. C. Giddings, E. Grushka, J. Cazes and P. R. Brown), Marcel Dekker, New York, pp. 101–61.

Cui, Y. and Olesik, S. V. (1991) High-performance liquid chromatography using mobile phases with enhanced fluidity. *Anal. Chem.*, **63**, 1812–19.

Dean, J. (ed.) (1993) *Applications of Supercritical Fluids in Industrial Analysis*, Blackie Academic & Professional, London.

Demirbüker, M. and Blomberg, L. G. (1990) Group separation of triacylglycerols on micropacked argentation columns using supercritical media as mobile phases. *J. Chromatogr. Sci.*, **28**, 67–72.

Demirbüker, M. and Blomberg, L. G. (1991) Separation of triacylglycerols by supercritical fluid argentation chromatography. *J. Chromatogr.*, **550**, 765–74.

Demirbüker, M. and Blomberg, L. G. (1992) Permanganate-impregnated packed capillary columns for group separation of triacylglycerols using supercritical media as mobile phases. *J. Chromatogr.*, **600**, 358–63.

Demirbüker, M., Andersson, P. E. and Blomberg, L. G. (1993) Miniaturized light scattering detector for packed capillary supercritical fluid chromatography. *J. Microcol. Sep.*, **5**, 141–7.

Demirbüker, M., Hägglund, I. and Blomberg, L. G. (1990) Separation of lipids by packed fused silica capillary SFC: selectivity of some stationary phases, in *Contemporary Lipid Analysis* (eds N. U. Olsson and B. G. Herslöf), Lipid Teknik, Stockholm, pp. 30–47.

Demirbüker, M., Hägglund, I. and Blomberg, L. G. (1992) Separation of unsaturated fatty acid methyl esters by packed capillary supercritical fluid chromatography. Comparison of different column packings. *J. Chromatogr.*, **605**, 263–7.

Demirbüker, M., Blomberg, L. G., Olsson *et al.* (1992) Characterization of the seeds of *Aquilegia vulgaris* by chromatographic and mass spectrometric methods. *Lipids*, **27**, 436–41.

de Vries, B. (1962) Quantitative separations of lipid materials by column chromatography on SiO_2 impregnated with $AgNO_3$. *Chem. Ind. (London)*, 1049–50.

Duchateau, G. S. M. J. E., van Oosten, H. J. and Vasconcellos, M. A. (1996) Analysis of *cis*- and *trans*-fatty acid isomers in hydrogenated and refined vegetable oils by capillary gas–liquid chromatography. *J. Am. Oil Chem. Soc.*, **73**, 275–82.

Eggers, R. (1996) Supercritical fluid extraction of oilseeds/lipids in natural products, in *Supercritical Fluid Technology in Oil and Lipid Chemistry* (eds J. W. King and G. R. List), AOCS Press, Champaign, IL, pp. 35–64.

Eggers, R. and Wagner, H. (1993) Extraction device for high viscous media in a high-turbulent two-phase flow with supercritical CO_2. *J. Supercrit. Fluids*, **6**, 31–7.

Golay, M. J. E. (1958) Theory of chromatography in open and coated tubular columns with round and rectangular cross-sections, in *Gas Chromatography 1958* (ed. D. H. Desty), Butterworth, London, pp. 36–55.

Greibrokk, T. (1995) Applications of supercritical fluid extraction in multidimensional systems. *J. Chromatogr. A*, **703**, 523–36.

Greibrokk, T. and Berg, B. E. (1993) Trace analysis in capillary supercritical fluid chromatography: sample introduction. *Trends Anal. Chem.*, **12**, 303–8.
Guiochon, G. (1981) Conventional packed columns *vs.* packed or open tubular microcolumns in liquid chromatography. *Anal. Chem.*, **53**, 1318–25.
Hagen, H. M., Landmark, K. E. and Greibrokk, T. (1991) Separation of oligomers of polyethylene glycols by supercritical fluid chromatography. *J. Microcol. Sep.*, **3**, 27–31.
Hägglund, I., Demirbüker, M. and Blomberg, L. G. (1994) Performance of silica-bonded quinolinol as selective stationary phase for packed capillary supercritical fluid chromatography. *J. Microcol. Sep.*, **6**, 223–8.
Hammond, E. W. (1989) Chromatographic techniques for lipid analysis. *Trends Anal. Chem.*, **8** 308–13.
Härröd, M. and Möller, P. (1996) Hydrogenation at supercritical conditions. Lecture presented at European Section of AOCS, 1st Meeting, Dijon, 19–20 September.
Hawthorne, S. B., Miller, D. J., Nivens, D. E. *et al.* (1992) Supercritical fluid extraction of polar analytes using in situ chemical derivatization. *Anal. Chem.*, **64**, 405–12.
Hoffmann, S. and Greibrokk, T. (1989) Packed capillary supercritical fluid chromatography with mixed mobile phases and light-scattering detection. *J. Microcol. Sep.*, **1**, 35–40.
Jackson, M. A. and King, J. W. (1996) Methanolysis of seed oils in flowing supercritical carbon dioxide. *J. Am. Oil Chem. Soc.*, **73**, 353–6.
Janák, K., Demirbüker, M., Hägglund, I. and Blomberg, L. G. (1992a) Modifications of poly (methyl-3-propylthiol) siloxane to give stationary phases for open tubular supercritical fluid chromatography. *Chromatographia*, **34**, 335–41.
Janák, K., Hägglund, I., Blomberg, L. *et al.* (1992b) Universal set-up for measurement of diffusion coefficients in supercritical carbon dioxide with flame ionization detection. *J. Chromatogr.*, **625**, 311–21.
Janssen, H.-G., Rijks, J. A. and Cramers, C. A. (1990) Flow rate control in pressure-programmed capillary supercritical fluid chromatography. *J. Microcol. Sep.*, **2**, 26–32.
Jinno, K. (ed.) (1992) *Hyphenated Techniques in Supercritical Fluid Chromatography and Extraction*, Journal of Chromatography Library 53, Elsevier, Amsterdam.
King, J. W. (1995) Analytical-process supercritical fluid extraction: a synergistic combination for solving analytical and laboratory scale problems. *Trends Anal. Chem.*, **14**, 474–81.
King, J. W. and List, G. R. (eds) (1996) *Supercritical Fluid Technology in Oil and Lipid Chemistry*, AOCS Press, Champaign, IL.
Klesper, E. and Schmitz, F. P. (1987) Single and multiple gradients in supercritical fluid chromatography. *J. Chromatogr.*, **402**, 1–39.
Klesper, E., Corwin, A. H. and Turner, D. A. (1962) High pressure gas chromatography above critical temperatures. *J. Org. Chem.*, **27**, 700–1.
Laakso, P. (1992) Supercritical fluid chromatography of lipids, in *Advances in Lipid Methodology – One* (ed. W. W. Christie), The Oily Press, Ayr.
Lafosse, M., Dreux, M. and Morin-Allory, L. (1987) Application fields of a new evaporative light scattering detector for high-performance liquid chromatography and supercritical fluid chromatography. *J. Chromatogr.*, **404**, 95–105.
Lafosse, M., Elfakir, C., Morin-Allory, L. *et al.* (1992) The advantages of evaporative light scattering detection in pharmaceutical analysis by high performance liquid chromatography and supercritical fluid chromatography. *J. High Resolut. Chromatogr.*, **15**, 312–18.
Lee, M. L. and Markides, K. E. (1989) *Analytical Supercritical Fluid Chromatography and Extraction*, Chromatography Conference, Provo, UT.
Lee, M. L., Yang, F. J. and Bartle, K. D. (1984) *Open Tubular Column Gas Chromatography*, John Wiley, New York.
Lee, S. T. and Olesik, S. V. (1995) Normal-phase high-performance liquid chromatography using enhanced-fluidity liquid mobile phases. *J. Chromatogr.*, **707**, 217–24.
Lee, S. T., Olesik, S. V. and Fields, S. M. (1995) Applications of reversed–phase high performance liquid chromatography using enhanced–fluidity liquid mobile phases. *J. Microcol. Sep.*, **7**, 477–83.
Li, W., Malik, A. and Lee, M. L. (1994) Fused silica packed capillary columns in supercritical fluid chromatography. *J. Microcol. Sep.*, **6**, 557–63.
Luque de Castro, M. D., Valcárcel, M. and Tena, M. T. (eds) (1994) *Analytical Supercritical Fluid Extraction*, Springer, Berlin.

McHugh, M. A. and Krukonis, V. J. (1994) *Supercritical Fluid Extraction, Principles and Practice*, Butterworth, Boston, MA.
Morris, L. J. (1962) Separation of higher fatty acid isomers and vinylogues by thin layer chromatography. *Chem. Ind. (London)* 1238–40.
Nakamura, K. (1996) Enzymatic synthesis in supercritical fluids, in *Supercritical Fluid Technology in Oil and Lipid Chemistry* (eds J. W. King and G. R. List), AOCS Press, Champaign, IL, pp. 306–20.
Novotny, M., Lee, M. L., Peaden, P. A. *et al.* (1984) US Patent 4 479 380.
Novotny, M., Springston, S. R., Peaden, P. A. *et al.* (1981) Capillary supercritical fluid chromatography. *Anal. Chem.*, **53**, 407A–14A.
Orman, B. A. and Schumann, R. A. Jr. (1991) Comparison of near-infrared spectroscopy calibration methods for the prediction of protein, oil, and starch in maize grain. *J. Agric. Food Chem.*, **39**, 883–6.
Orman, B. A. and Schumann, R. A. Jr. (1992) Non destructive single-kernel oil determination of maize by near-infrared transmission spectroscopy. *J. Am. Oil Chem. Soc.*, **69**, 1036–8.
Rizvi, S. S. H. (ed.) (1994) *Supercritical Fluid Processing of Food and Biomaterials*, Blackie Academic & Professional, London.
Saito, M., Yamauchi, Y. and Okuyama, T. (eds) (1994) *Fractionation by Packed-column SFC and SFE, Principles and Applications*, VCH, Weinheim.
Sanagi, M. M. and Smith, R. M. (1988) The emergence and instrumentation of SFC, in *Supercritical Fluid Chromatography* (ed. R. M. Smith), Royal Society of Chemistry, London, pp. 29–52.
Shen, Y., Reese, S. L., Rossiter, B. E. *et al.* (1995) Silver-complexed dicyanobiphenyl-substituted polymethylsiloxane encapsulated particles for packed capillary column supercritical fluid chromatography. *J. Microcol. Sep.*, **7**, 279–87.
Smith, R. M. (ed.) (1988) *Supercritical Fluid Chromatography*, Royal Society of Chemistry, London.
Snyder, J. M. (1995) Volatile analysis of oxidized oils by a direct supercritical fluid extraction method. *J. Food Lipids*, **2**, 25–33.
Szücs, R., Verleysen, K., Duchateau, G. S. M. J. E. *et al.* (1996) Analysis of phospholipids in lecithins. *J. Chromatogr. A.*, **738**, 25–9.
Taylor, L. T. (1995) Strategies for analytical SFE. *Anal. Chem.*, **67**, 364A–70A.
Taylor, L. T. (1996) *Supercritical Fluid Extraction*, John Wiley, New York.
Tom, J. W. and Debenedetti, P. G. (1991) Particle formation with supercritical fluids – a review. *J. Aerosol Sci.*, **22**, 555–84.
Trones, R., Iveland, A. and Greibrokk, T. (1995) High temperature liquid chromatography on packed capillary columns with nonaqueous mobile phases. *J. Microcol. Sep.*, **7**, 505–12.
Upnmoor, D. and Brunner, G. (1992) Packed column supercritical fluid chromatography with light scattering detection. Optimization of parameters with a carbon dioxide/methanol mobile phase. *Chromatographia*, **33**, 255–60.
van Lenten, F., Drews, J. M., Ivey, K. *et al.* (1996) The utilization of an ELDS/UV-Vis HPLC detector system as an alternative to gravimetric analysis in SFE or ASE for QC applications. *Pittsburg Conf.* New Orleans, LA, Abstract 1323.
van Oosten, H. J., Klooster, J. R., Vandeginste, B. G. M. and de Galan, L. (1991) Capillary supercritical fluid chromatography for the analysis of oils and fats. *Fat Sci. Technol.*, **93**, 481–7
van Wasen, U., Swaid, I. and Schneider, G. M. (1980) Physicochemical principles and applications of supercritical fluid chromatography. *Angew. Chem. Int. Ed. Engl.*, **19**, 575–87.
Wagner, H. and Eggers, R. (1996) Extraction of spray particles with supercritical fluids in a two-phase flow. *A.I.Ch.E.J.*, **42**, 1901–10.
Wenclawiak, B. (ed.) (1992) *Analysis with Supercritical Fluids: Extraction and Chromatography*, Springer, Berlin.
Westwood, S. A. (ed.) (1993) *Supercritical Fluid Extraction and its use in Chromatographic Sample Preparation*, Blackie Academic & Professional, London.

3 Static headspace gas chromatography in the analysis of oils and fats
F. ULBERTH

3.1 Introduction

In modern analytical chemistry more time is spent on sample preparation than on the actual determination process, which is mostly a technique related to chromatography, electrophoresis or spectroscopy. Major improvements have been obtained over past decades to simplify the whole chain of preparative steps a raw sample has to go through before it is ready for injection into a chromatograph or a spectrometer. Miniaturized solid phase extraction columns prepacked with a vast array of chromatographic media, affinity chromatography, extraction with supercritical fluids and highly sophisticated coupled techniques such as high-pressure liquid chromatography–mass spectrometry (HPLC–MS) or HPLC–MS–MS have increasingly substituted classical liquid–liquid or solid–liquid extraction, column or thin-layer chromatography and various forms of distillation as the principal methods used for sample preparation. Although these advanced methods have revolutionized sample preparation and thus sample throughput the ultimate goal of a simple 'one-pot reaction' as the only pretreatment before the sample is analysed is only rarely achieved. Static headspace gas chromatography (HSGC) is one of the exceptions to the rule. In its simplest form, the sample is placed in a sealed container and after an appropriate incubation time the gas in the headspace (HS) of the container is sampled with a gas-tight syringe and analysed by GC. Since in HSGC only the gas phase over the liquid or solid sample is used, the fundamental prerequisite for applying HSGC is that the target analyte(s) has (have) to be volatile. HSGC is the method of choice for the rapid determination of compounds with a comparatively high vapour pressure at sampling temperatures ranging from room temperature to $c.$ 100°C. No wonder it has become the primary analytical tool in flavour and fragrance research and related fields in food chemistry. Automation of the sampling procedure has further fostered the use of this technique in investigations concerning odorous substances. Other examples include the determination by HSGC of volatile residues or contaminants in water and sewage, the packaging industry and in food stuffs. Monitoring of environmental or occupational exposure to low boiling toxic compounds, and numerous applications in clinical chemistry or forensic medicine, for instance in the determination of ethanol or cyanide or carbon monoxide in blood, are other areas where HSGC is used to advantage.

Substances with high boiling points are, on the other hand, bad candidates for successfully employing HSGC as an analytical method. As a consequence, the majority of compounds subsumed under the term 'lipids' is not eligible for HSGC analysis. Strictly speaking, short-chain fatty acids (FAs) are the only lipid-related materials where HSGC can be directly applied. The method has, nevertheless, the potential for being used in a variety of tasks in connection with lipids or lipid-derived materials. The determination of residues in solvent extracted meals and of the aroma components of oils and fats (e.g. diacetyl in butter fat) and the measurement of the oxidative stability of edible oils via the assessment of volatile secondary lipid oxidation products (e.g. ethane, pentane and hexanal) are examples of the part HSGC plays in lipid analysis. Not only is the simplicity of sample preparation an asset, but also the high boiling, involatile materials originating from the matrix are excluded from the analysis and will therefore not contaminate the chromatographic system. Moreover, HSGC is well suited for trace analysis, as low boiling substances are greatly enriched in the HS. This effect is potentiated when the major components in the mixture (matrix) differ greatly in terms of volatility from the trace component, leading to a vastly altered ratio of the trace component to matrix components in the vapour phase. In addition, in HSGC some very low boiling components will not be obscured by a solvent peak, something which usually occurs when injecting liquid samples.

One of the weaknesses of HSGC is that quantitation is not as straightforward as in classical GC. Calibration of a HSGC system relies either on the availability of the matrix void of the analyte or on the composition of the matrix being known so that it can be simulated by individual ingredients. If neither option works, for instance if the matrix is not available and cannot be simulated, the so-called 'standard addition method' has to be used. Even in situations where a standard addition method may lead to uncertainties, the multiple headspace extraction (MHE) technique allows a reliable quantification of trace constituents.

3.2 Theoretical background

In principle two versions of HS analysis have been developed to date, one dealing with vapour in a thermodynamic equilibrium with a liquid or solid sample (static HSGC), and one in which equilibrium is not reached (dynamic HSGC). Only the first approach, where a volatile substance partitions between the liquid or solid sample and the surrounding gas phase in a closed container, will be dealt with in this review.

Fundamental principles concerning sampling and quantitation in HSGC have been covered in more detail by Hachenberg and Schmidt (1977), Kolb, Pospisil and Auer (1981), Kolb and Pospisil (1985) and Ioffe (1989).

In HSGC – and this is also the case with other GC techniques – the property directly measured is the peak area, A, or the peak height, H, which is proportional to the partial vapour pressure, p_i, of substance i in the HS. Given the substance concerned and the GC system used, this relation has the form:

$$A_i = c_i p_i, \tag{3.1}$$

where c_i is a proportionality factor specific to substance i. The partial vapour pressure is related to the concentration of substance x_i in a two-component mixture (expressed as mole fraction or any other suitable unit of concentration) to the vapour pressure of the pure substance, p_{0i} by Henry's law:

$$p_i = p_{0i} x_i \gamma_i. \tag{3.2}$$

The activity coefficient γ_i describes the intermolecular attraction forces between solutes and solvent in a mixture. It depends on the chemical nature of substance i but is also influenced by the concentration of all the other components making up the sample mixture. Moreover, the coefficient is temperature dependent and to a lesser extent pressure dependent.

The vapour pressure p_{0i} is related to temperature by an exponential function (Clausius–Clapeyron equation). Consequently, the temperature during sample equilibration has to be carefully controlled and the time to reach the equilibrium at the chosen temperature of thermostatting has to be determined experimentally.

Combination of equations (3.1) and (3.2) gives a fundamental relationship relating the measured peak area (height) to x_i:

$$x_i = \frac{A_i}{c_i p_{0i} \gamma_i}. \tag{3.3}$$

The denominator of equation (3.3) is a constant for a given separation system and has to be determined in calibration runs. Equation (3.3) is valid only when a dilute HS gas mixture at a comparatively low pressure (< 2 bar) is concerned, which holds true for most practical situations.

Calibration in HSGC is profoundly influenced by parameters which may alter either p_i or γ_i. That the equilibration temperature plays an important role has already been stressed, but γ_i is also altered if the composition of the mixture to be analysed changes. In other words, γ_i is influenced by the concentration of other substances present. This means that simple binary solutions cannot be used for calibration purposes. A good example to illustrate this point is the determination of halogenated pollutants in sea water. Calibration with use of solutions based on distilled water can give rise to errors because of the considerable effect salt and other dissolved components may exert on the value of γ_i. The matrix, either as received or in a simulated form, has to be included in the calibration procedure

The 'external standard' method is mostly employed for making quantitative measurements in HSGC:

$$x_{Un} = \frac{A_{Un} x_{St}}{A_{St}} \qquad (3.4)$$

where x_{Un} and x_{St} are the unknown and standard concentrations of x, respectively, and A_{Un} and A_{St} are the areas produced by the unknown and standard concentrations of x, respectively. When the sample matrix is available in a pure form, calibration is simply done by the addition of known amounts of the analyte and treating these standard samples in the same way as the unknowns (equilibration temperature and time, volume sampled, etc.). For some special applications (e.g. analysis of alcoholic beverages), where the matrix is composed of only a few well known major components, the matrix can be simulated by mixing the pure substances. A standard addition method is recommended when very high accuracy is required or should be carried out during method development to check the reliability of the calibration.

Use of an 'internal standard', as is commonly used in most chromatographic analysis protocols, not only accounts for differences in detector response but also for slight variations in the sampling procedure, especially minor changes in equilibration temperature and time. The internal standard compound has to be selected so that its activity coefficient is similar to that of the analyte. With dedicated HSGC instrumentation the addition of an internal standard is not normally necessary. Automatic HS samplers also offer the possibility of running an MHE analysis unattendedly. The MHE technique is especially suited to the analysis of solid samples (e.g. residual solvents in printed packaging materials). It is difficult to obtain samples of a solid material with exactly known content of volatile analyte that can be used as calibration standards. MHE is in principle a repeated HS analysis from a single vial containing the sample. By analogy to exhaustive liquid–liquid extraction, one could use gas to extract a volatile substance from the sample, analyse each extract by GC and, by totalling the peak areas for all analyses, quantitate the amount of the volatile present in the sample. This approach is, though straightforward, laborious and time-consuming. Repeated sampling of the HS gas in equilibrium with a sample reduces the concentration of a volatile according to first-order kinetics, leading to chromatograms where the peak area or height of the volatile diminishes accordingly. A plot of the logarithm of the peak areas against the number of injections (extractions) gives a straight line, and by applying regression analysis the slope of the line is determined. The total peak area can be represented by a geometrical progression which can be solved by using data from the regression analysis. This mathematical approach allows an accurate quantitation of volatile substances with only a few HS gas extractions. A detailed description of the mathematical background of MHE is found in Ettre, Jones and Todd (1984).

3.3 Sampling systems

Withdrawal of the HS gas sample with a gas-tight syringe is by far the simplest way of performing HSGC. A basic arrangement for HS sampling is shown in Fig. 3.1. The sample vial is tightly sealed by an aluminium crimp-cap and a septum, which should be lined with polytetrafluoroethylene (PTFE) to prevent adsorption of sensitive sample compounds. To avoid condensation in the barrel, the syringe should be thermostatted, preferably slightly above the equilibration temperature. It is recommended the sample vials be filled up to 10%–30% of the nominal volume of the container. This creates proper conditions for reaching rapidly the gas–liquid equilibrium.

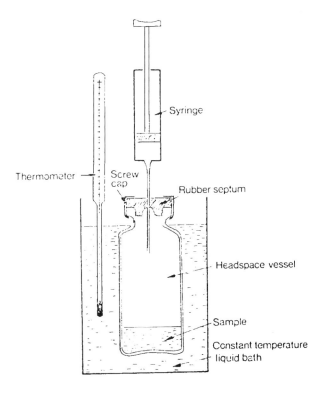

Figure 3.1 Basic set-up for headspace analysis. Reproduced from Hachenberg, H. and Schmidt, A.P., *Gas Chromatographic Headspace Analysis*; published by Heiden, 1977.

Stirring of the vial content also helps to keep equilibration time at a minimum.

Thermostatting the HS vial at an elevated temperature to increase the volatility of the analyte(s) concomitantly increases the pressure inside the vial. During sample withdrawal this pressure extends also to the syringe. As the syringe needle is not usually sealed, part of the sample may be lost to the atmosphere when the pressure inside the syringe is considerably higher than the atmospheric pressure. The fraction of sample lost depends on the internal pressure in the syringe and hence on the sample composition. When appropriate calibration standards are used these losses are compensated for, provided actual samples and standards are handled in exactly the same way, which is usually best done by an autosampler. Other sampling techniques for introducing HS gas into a gas chromatograph have been reviewed by Hachenberg and Schmidt (1977).

Principal layouts of autosamplers for sample introduction in HSGC analysis are given in Fig. 3.2. In one of the designs an electromechanically driven, thermostatted gas-tight syringe is used as the sampling device [Fig. 3.2(a)]. Another, fundamentally different, approach makes use of the build-up of an internal pressure in the HS vial (Kolb and Pospisil, 1985). At the end of the equilibration time the sample vial is pierced by a moveable needle and is pressurized with carrier gas so that the pressure in the vial equals the column-head pressure. Then the carrier gas is interrupted by a solenoid valve causing the pressure at the head of the column to drop. This pressure drop is compensated for by the expansion of the pressurized gas in the HS vial into the GC column. After a preselected time period the solenoid valve opens again, restoring carrier gas and thus immediately stopping sample transfer from the vial to the column. This technique has been termed 'balanced-pressure sampling'. It is also possible to apply a pressure to the HS vial which is higher than the column-head pressure ('high-pressure sampling'). In both cases the HS gas in the vial is diluted to some extent with carrier gas before actual sampling. The third commercially available category of HSGC autosamplers is based on the so-called 'valve and loop' sampling technique (Wylie, 1986). A six-port gas-sampling valve is used to pressurize the HS vial, and the pressurized gas is allowed to expand into a sample loop. Once filled, the loop is connected via the six-port valve to the carrier gas line thereby transferring the sampled HS gas to the injection port of the GC.

With few exceptions the split injection technique has to be employed when capillary columns are used for HSGC, unless some special precautions are taken to refocus the sample at the column head. Split injection is the easiest way to increase gas flow in the injection port in order to facilitate a rapid sample transfer to the column and thus ensure sharp starting bands. Problems associated with split injection of liquid samples, for instance discrimination effects due to differences in volatility of sample constituents, do not

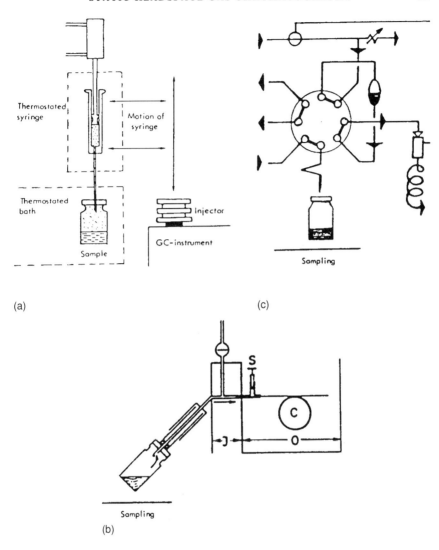

Figure 3.2 Automated headspace sampling systems. Descriptions of each system are provided in text. Reproduced from Schomburg, G., Temperature programmed sample transfer, in *Sample Introduction in Capillary Gas Chromatography* (ed. P. Sandra), pp. 55–76; published by Huethig Verlag, 1985.

come into play in HSGC because the sample injected is already in the gas phase.

Cryogenic focusing is an efficient way to reconcentrate large HS gas injection volumes into a narrow initial sample plug (Kolb, Licbhardt and Ettre, 1986) By focusing the sample, peak widths become narrower and resolution and signal-to-noise ratios improve. 'Focusing' in chromatography means the slowing down of the migration rates of solutes at the beginning of the chromatographic process by increasing their capacity factors. This can be accomplished by increasing the thickness of the liquid film of the capillary column and/or by lowering the temperature of the column. Most gas chromatographs can be fitted with cryogenic cooling units which allows the oven temperature to be reduced to $-70°C$. An alternative would be to dip just the first coil of a flexible fused-silica column into a Dewar flask filled with coolant (Takeoka and Jennings, 1984) or to jacket the first 15–25 cm of the capillary column by a suitable tubing and conducting a gas or compressed air precooled in liquid nitrogen through it (Kolb, Liebhardt and Ettre, 1986). An extra heater to heat up the cooled section of the capillary to start chromatography is unnecessary as the low thermal mass of the column follows the temperature of the oven rapidly when the flow of cooling gas is stopped. The trapping effect of the latter method is as good as that obtained when the whole oven is cooled down, but the cool-down period and the consumption of coolant is shortened considerably.

Cryofocusing allows the injection of a larger HS sample volume without sacrificing chromatographic resolution or peak shape deformations, especially for early eluting substances. This technique can improve sensitivity by a factor of 10–100. Not only can larger volumes be injected, but also it is possible to sample two or even more vials and cryotrap the subsamples at the column head. This variant has been termed 'multiple headspace injection' (Wylie, 1986).

In general, the sensitivity of static HSGC varies, depending on the properties of the analyte, in the range of parts per billion (ppb) (volatile halocarbons detected with an electron-capture detector) and parts per million (ppm) (volatiles detected with a flame-ionization detector). Bassette and Ward (1975) suggested a simple though very efficient way of enriching HS vapours from foodstuffs. Volatile materials are steam distilled and collected in a tube immersed in an ice bath. Subsequent static HSGC on a part of the distillate allows quantitation of volatiles at the ppb level.

3.4 Applications of headspace gas chromatography in lipid chemistry

The term 'lipid' is applied to a relatively diverse array of natural products which are insoluble in water and soluble in organic solvents. No general applicable definition of this class exists and one of the best interpretations was given by Christie (1989, p. 10): 'Lipids are fatty acids and their derivatives, and substances related biosynthetically or functionally to these compounds'. Most substances belonging to this class have high molecular weights and are involatile. Only a few FAs (formic acid up to decanoic acid) have vapour pressures which permit the application of HSGC for their qualitative and quantitative analysis. Nevertheless, other lipid-derived or lipid-related substances are volatile in nature and are thus traceable by HSGC.

3.4.1 Volatile fatty acids

Besides milk fat of ruminant animals and lauric fats, volatile fatty acids (VFAs) are lacking in edible oils and fats. Consequently, the main area of HSGC in the analysis of VFAs is not lipid chemistry but microbiology and related fields. Anaerobic fermentation of saccharides results in the production of VFAs. In particular, VFAs play a major role in the digestive tract and in nutritional physiology of ruminants as well as non-ruminants (Cummings and MacFarlane, 1991). In ruminants they are formed in the rumen whereas in non-ruminants they are the result of colonic bacterial degradation of unabsorbed starch and non-starch polysaccharides (fibre). Interest in VFAs has been spurred by findings that these substances may exert proliferative effects on the colonocyte and that diets low in dietary fibre, which results in low production of VFAs in the colon, may explain the high occurrence of colonic disorders seen in industrialized civilization (Scheppach, 1994). In addition, analysis of VFAs is a valuable tool for the identification of anaerobic bacteria (Anaerobe Laboratory Manual, 1977). Traditional methods for VFA analysis are based either on the direct injection of liquid samples or on diethyl ether extraction of VFA and use of packed column GC (Cottyn and Boucque, 1968). Alternatively, HSGC can be utilized for the determination of VFAs (Larson, Mardh and Odham, 1978; Taylor, 1984). Column contamination with non-volatile, high molecular weight compounds, as is inevitably the case with direct injection of fermentation broth, rumen fluid, etc., does not occur. Another advantage is that extraction with a highly flammable liquid is avoided. Packed and capillary columns have both been successfully applied to HSGC analysis of VFAs (Figs 3.3 and 3.4). Samples are acidified with strong acid and equilibrated at around 100°C for times of 30 min up to 2 h. Although VFAs are quite aggressive substances and tend to get adsorbed onto the column material, no memory effects are seen when fused-silica capillary columns coated with SP-1000 or free fatty acid phase (FFAP), preferably chemical bonded, are chosen.

VFA analysis is important in the ripening control of cheeses, particularly of hard cheeses. HSGC offers a rapid route to check for the presence of unwanted *Clostridia* species. These bacteria ferment lactic acid and are capable of forming large amounts of gas and butyric acid, which spoils the cheese. An elegant way of sample preparation for final HSGC analysis was devised by Osl (1988). A cheese slurry is made; it is then refrigerated in order to crystallize the fat, filtered to remove the solidified fat and diluted with ethanol. To this ethanolic solution concentrated sulphuric acid is added as well as valeric acid as an internal standard; this solution is then thermostatted for 1 h at 80°C, and the ethyl esters formed are sampled from the HS.

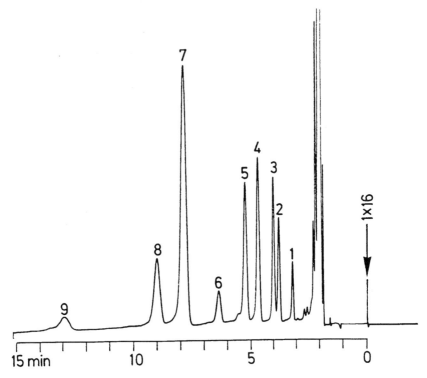

Figure 3.3 Headspace analysis of volatile fatty acids from *Peptostreptococcus anaerobius*. Chromatographic conditions: 6′ × 1/8″ stainless steel column packed with 8% SP–1000™ + 1% H_3PO_4 on Chromosorb™ W/AW, 80/100 mesh, 150°C isothermal. Sample thermostatted at 90°C. Peaks: 1 = acetic acid; 2 = propionic acid; 3 = isobutyric acid; 4 = butyric acid; 5 = isovaleric acid; 6 = valeric acid; 7 = isocaproic acid; 8 = caproic acid; 9 = heptanoic acid. Reproduced from Kolb, B., Beyaert, G., Pospisil, P. *et al.*, Applications of gas chromatographic head space analysis, application note 26; published by Bodenseewerk Perkin Elmer, 1980.

Chromatograms of a high-quality and a poor-quality Dutch cheese are given in Fig. 3.5.

3.4.2 Oxidative deterioration of fats and oils

Edible oils in contact with air deteriorate eventually owing to oxygen uptake and the formation of primary and secondary oxidation products of unsaturated FAs. The kinetics of the formation and the identity of these oxidation products have been the subject of several monographs and reviews (Chan, 1987; Hamilton, 1989).

The first intermediates in the chain reaction leading to volatile secondary oxidation products are hydroperoxides. They react further in a very complex way to give a diverse spectrum of short-chain secondary reaction products pertinent to oxidative rancidity. Grosch (1987) isolated

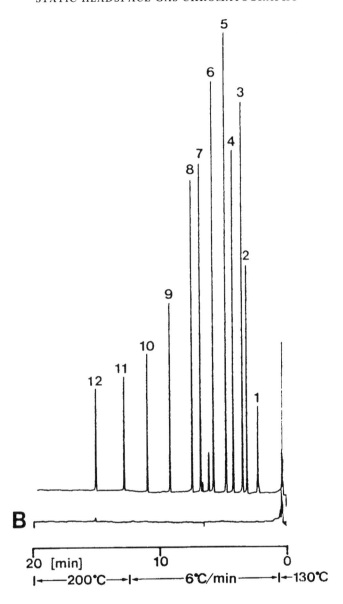

Figure 3.4 Headspace analysis of volatile fatty acids from aqueous solution. Chromatographic conditions: 15 m × 0.23 mm fused silica column coated with free fatty acid phase, 130°C (1 min), then increased by 6°C min^{-1} to 200°C. Sample thermostatted at 120°C. Trace B is a blank. Peaks: 1 = acetic acid; 2 = propanoic acid; 3 = 2-methyl propanoic acid; 4 = butanoic acid; 5 = 3-methyl butanoic acid; 6 = pentanoic acid; 7 = 4-methyl pentanoic acid; 8 = hexanoic acid; 9 = heptanoic acid; 10 = octanoic acid; 11 = nonanoic acid; 12 = decanoic acid. Reproduced from Closta, W., Klemm, H., Pospisil, P. *et al.*, The HS-100, an innovative concept for automatic headspace sampling, *Chromatography Newsletter*, **11**, 13–17, 1983.

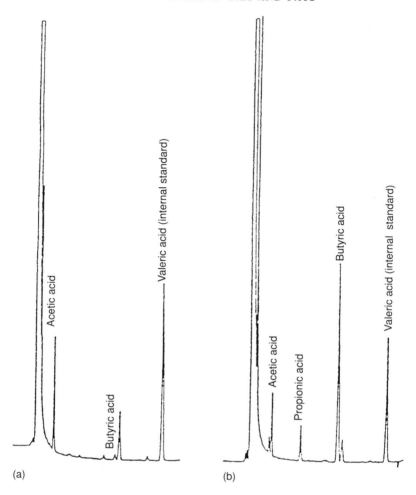

Figure 3.5 Headspace gas chromatography analysis of volatile fatty acids in Dutch cheese after formation of ethyl esters. (a) High-quality cheese. Chromatographic conditions: 10 m × 0.53 mm fused silica column coated with HP-1^{TM}, 60°C (2 min), 10°C min^{-1} to 140°C. (b) Poor-quality cheese. Sample thermostatted at 80°C. Reproduced from Osl, F., Bestimmung der niederen freien Fettsäuren im Hart- und Schnittkäse mit der Head-Space Gaschromatographie, *Deutsche Molkerei Zeitung*, **45**, 1516–18, 1988.

and identified 72 low molecular weight compounds from autoxidized linoleate. Pathways leading to pentane and hexanal, two important indicator substances relating to the oxidative stability of unsaturated lipids, are given in Fig. 3.6. Although hexanal is a good indicator, it is not necessarily the most potent aroma substance compared with the odour threshold values of other carbonyl compounds formed during autoxidation of unsaturated oils (Table 3.1).

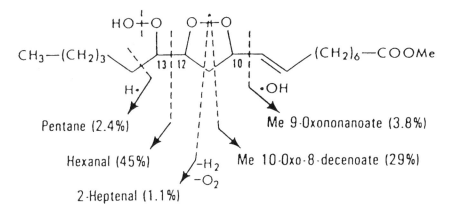

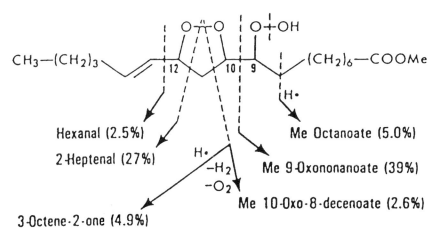

Figure 3.6 Oxidative breakdown of hydroperoxy cyclic peroxides of linoleate leading to pentane and hexanal. Reproduced from Min, D. B., Lee, S. H. and Lee, E. C., Singlet oxygen oxidation of vegetable oils, in *Flavor Chemistry of Lipid Foods* (eds D. B. Min and J. H. Smouse), pp. 57–97; published by The American Oil Chemists' Society, 1989.

The first stage of autoxidation of a pure oil is easily traceable by applying conventional wet chemistry, for example determination of the peroxide value. Peroxides are known to be heat labile compounds, which undergo breakdown at elevated temperatures to form simple hydrocarbons. Thus ethane and pentane are the predominant breakdown products of linolenate and linoleate peroxides, respectively (Evans *et al.*, 1967). Scholz and Ptak (1966) and Evans *et al.* (1969) proposed a gas chromatographic method to measure rancidity in edible oils whereby the undiluted oil is injected directly into the hot injector (250°C). At these temperatures lipid peroxides are

Table 3.1 Odour threshold concentrations of various carbonyl compounds found in autoxidized soya bean oil.

Compound	Odour threshold range (ng l^{-1} air)
1,5 Z-Octadien-3-one	0.01–0.04
2 Z-Nonenal	0.08–0.23
1-Octen-3-one	0.3–0.6
3 Z-Hexenal	0.4–1.1
2 E, 6 Z-Nonadienal	0.6–1.6
2 Z-Octenal	0.9–2.5
3 Z-Nonenal	2.3–5.5
3 E-Nonenal	2.6–5.4
Nonanal	5.2–12.1
Octanal	5.8–13.6
Hexanal	65–98

Source: Ullrich, F. and Grosch, W., Flavour deterioration of soya-bean oil: identification of intense odour compounds formed during flavour reversion, *Fat Science Technology*, **90** 932–6, 1988

instantaneously decomposed and the hydrocarbons formed are separated on a packed Carbowax 20M™ column. Moreover, an excellent correlation of pentane levels with organoleptic panel tests was found. In addition, pentanal and hexanal contents of vegetable oils were also shown to be related to flavour scores (Warner *et al.*, 1978).

Dupuy and co-workers (Dupuy, Fore and Goldblatt, 1973; Dupuy, Rayner and Wordsworth, 1976) increased the sensitivity of the direct gas chromatography method by spreading 500 mg oil onto glass wool contained in an inlet liner. Volatiles are swept by the carrier gas to the chromatographic column packed with Porapak P™, held at 55°C. After a desorption time of 20 min the liner is removed and the temperature programme started. By using this technique volatiles in edible oils can be examined at the 10 ppb level without prior enrichment. The direct gas chromatographic method, although very sensitive, may easily result in contaminated and eventually deteriorated GC columns.

Fairly recently the idea of directly injecting an oil sample has been picked up and applied to the determination of organophosphorous insecticides (Grob, Biedermann and Giuffré, 1994) and to discriminate between conventionally processed edible oils and 'mildly deodorized' oils (Grob *et al.*, 1994). The injector insert employed is shown in Fig. 3.7. Grob *et al.* have termed this sampling technique 'injector-internal HS analysis'.

In order to avoid detrimental effects from involatile sample components on the integrity of the chromatographic system, the thermal decomposition of peroxides can also be performed externally in a HS vial instead of injecting the oil directly (Marsili, 1984; Snyder, Frankel and Selke, 1985). The temperature and atmosphere at which the oil samples are equilibrated prior to HSGC have to be carefully controlled. The incubation temperature

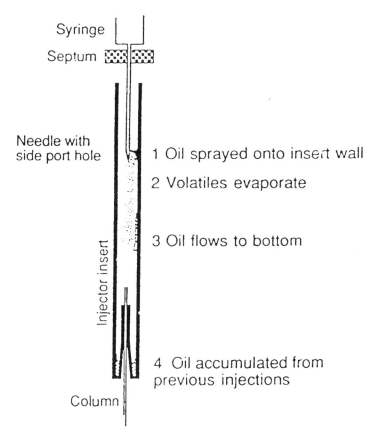

Figure 3.7 Injector insert for injector–internal headspace analysis. Reproduced from Grob, K., Biedermann, M. and Giuffré, A. M., Determination of orgonphosphorus insecticides in edible oils and fats by splitless injection of the oil into a gas chromatograph, *Zeitschrift für Lebensmitteluntersuchung und -forschung*, **198**, 325–8, 1994.

has to be high enough to induce fission of preformed lipid peroxides, but the temperature must not be excessive, otherwise an artificial peroxidation of the sample will be induced. Marsili (1984) used light-exposed soya bean oil as a model to study the interrelationship of these factors and found that the greatest sensitivity (expressed as total peak area) with respect to detecting light abuse was achieved at an incubation temperature of 160°C for 40 min in an air atmosphere [Fig. 3.8(a)]. Flushing the vials with nitrogen before incubation allowed the use of even higher temperatures for incubation, as heat-induced autoxidation is minimized by oxygen depletion [Fig. 3.8(b)]. Precision, expressed as the relative standard deviation of peak areas varied between 1.0 and 8.6. The minimum detectable quantity of one of the major volatiles found, $2(E), 4(E)$-decadienal, was 1 µg per g oil. The potential of the method is illustrated in Fig. 3.9.

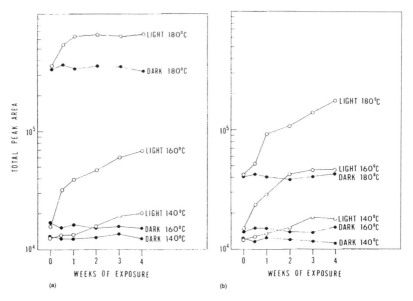

Figure 3.8 Total peak area of either light-abused or dark stored oils, analysed after thermostatting for 40 min at 140°C, 160°C or 180°C in (a) air atmosphere; (b) nitrogen. Reproduced from Marsili, R. T., Measuring light-induced chemical changes in soybean oil by capillary headspace gas chromatography, *Journal of Chromatographic Science*, **22**, 61–7, 1984.

Snyder *et al.* (1988) compared different approaches – direct injection, static HSGC and dynamic HSGC – for volatile analysis of soya bean oil oxidized to different peroxide value levels (Fig. 3.10). The relative percentage volatile composition changed depending on the method used (Table 3.2). Together the 2,4-decadienal isomers represented the major components found by direct injection and also by dynamic HSGC. In these two methods c. 50% of the total peak area was attributable to these substances. In contrast, static HSGC favoured lower molecular weight substances such as acrolein, propanal and pentane. Snyder *et al.* concluded that static HSGC is the method of choice because it is rapid and requires no cleaning between samples.

The HSGC methods described so far rely on the availability of special instrumentation, that is an autosampler capable of equilibrating samples at high temperatures for decomposing lipid peroxides. To allow manual HS sampling at lower temperatures the decomposition step was physically separated from the equilibration step pertinent to HSGC (Ulberth and Roubicek, 1993). Oil samples were heated either in HS vials or in flame-sealed ampoules in a nitrogen atmosphere at 140°C in an oven to split preformed peroxides. Subsequently, the amounts of pentane, hexanal and 2-heptenal present in the sample were determined after equilibration at 60°C. The limit of quantitation was 0.02 ppm, 0.2 ppm and 1 ppm, respectively. Occasionally,

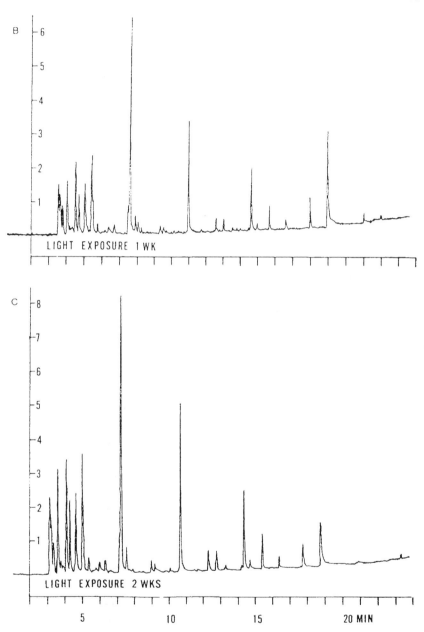

Figure 3.9 Headspace gas chromatography of light-exposed oils. Peak with retention time of 11.2 min is 2(E), 4(E)-decadienal. Chromatographic conditions: 50 m × 0.25 mm fused silica column coated with Carbowax™, 50°C (7 min), then 6°C min^{-1} to 160°C. Sample thermostatted at 160°C in air for 40 min. Reproduced from Marsili, R. T., Measuring light-induced chemical changes in soybean oil by capillary headspace gas chromatography, *Journal of Chromatographic Science*, **22**, 61–7, 1984.

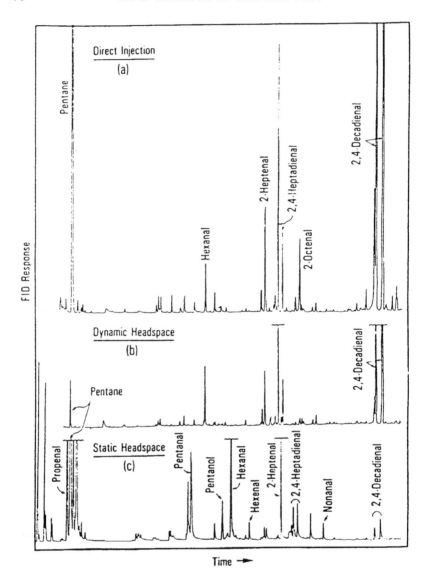

Figure 3.10 Comparison of three gas chromatographic methods for volatile analyses of oxidized vegetable oil (peroxide value 9.5). Chromatographic conditions: 30 m × 0.32 mm fused silica column coated with DB–5^{TM}, 0°C (10 min), then 5°C min^{-1} to 250°C. Reproduced from Snyder, J. M., Frankel, E. N., Selke, E. *et al.*, Comparison of gas chromatographic methods for volatile lipid oxidation compounds in soybean oil, *Journal of the American oil Chemists' Society*, **65**, 1617–20, 1988.

Table 3.2 Headspace analysis of oxidized soya bean oil (peroxide value 9.5) by three methods: direct injection, dynamic headspace gas chromatography (HSGC) and static HSGC. OlOOH = oleic acid hydroperoxide; LoOOH = linoleic acid hydroperoxide; LnOOH = linolenic acid hydroperoxide.

Compound	Area (%)			Origin
	Direct Injection	Dynamic HSGC	Static HSGC	
Acrolein	0.9	0.2	6.6	16-LnOOH
Propanal	0.3	0.0	12.6	16-LnOOH
Pentane	25.5	3.1	23.3	13-LoOOH
Pentene	0.4	0.3	0.1	
Butanal	0.6	1.5	0.7	
1-Pentene-3-ol	0.6	0.5	3.4	
Pentanal	0.3	1.2	6.7	13-LoOOH
2-Pentenal	0.4	1.2	0.7	13-LnOOH
1-Pentanol	0.2	0.4	2.4	13-LoOOH
Hexanal	0.8	4.5	14.9	13-LoOOH
Octane	0.1	0.1	0.4	10-OlOOH
2-Hexenal	0.3	0.7	0.8	13-LoOOH
Octene	0.1	0.2	0.2	
Heptanal	0.1	0.6	0.4	
2-Heptenal	2.0	5.6	5.0	12-LoOOH
1-Octen-3-ol	0.6	0.9	0.2	10-LoOOH
2,4-Heptadienal	4.2	11.1	1.2	12-LoOOH
Pentyl furan	0.0	0.8	0.3	
Octanal	0.0	0.0	0.2	11-OlOOH
2,4-Heptadienal	1.5	4.5	1.5	12-LoOOH
2-Octenal	1.3	0.5	0.5	
Nonanal	0.4	0.7	0.6	9/10-OlOOH
2-Nonenal	0.3	0.2	0.0	
2-Decenal	0.3	0.2	0.0	9-OlOOH
2,4-Decadienal	45.7	53.6	0.8	9-LoOOH
Others	13.1	9.6	16.6	

Source: Snyder, J. M., 1988

outliers were observed when HS vials with aluminium crimp-caps were used. They were found not to be airtight after heating and subsequent cooling down to room temperature. Sealed brown-glass ampoules were found to give superior overall precision of the method. Over a range of peroxide values of 2–12 the precision of the method, expressed as relative standard deviation of peak areas, was 6%–7% for pentane and 4%–6% for hexanal. Further, this method is well suited for assessing the future oxidation stability of an oil. Heating an oil in an air atmosphere to 160°C tests the resistance of the oil to oxidative deterioration (Ulberth and Roubicek, 1992). The quantitative differences of volatiles formed by only thermostatting the oil sample at 60°C, heating it in nitrogen to 140°C and in air to 160°C are illustrated in Fig. 3.11. Pentane concentrations (air atmosphere) are related to the induction period, determined by the Rancimat method (Laeubli and Bruttel, 1986), via a second-order function and are thus able to discriminate between oils of poor and good stability (Fig. 3.12).

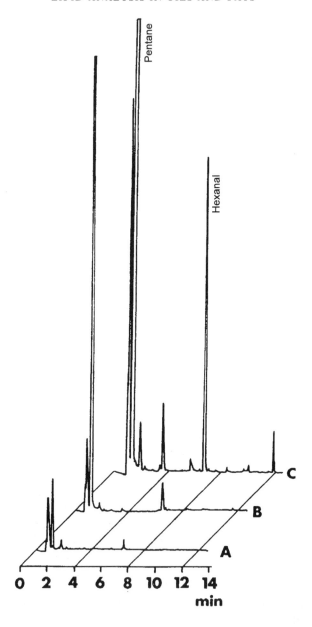

Figure 3.11 Headspace gas chromatography analysis of volatiles of an oxidized vegetable oil (peroxide value 26.2). Trace A: sample thermostatted at 60°C. Trace B: sample heated to 140°C in nitrogen. Trace C: sample heated to 160°C in air. Chromatographic conditions: 25 m × 0.32 mm fused silica column coated with SE-54™, 38°C (3 min), then 6°C min^{-1} to 170°C. Sample thermostatted at 60°C. Reproduced from Ulberth, F. and Roubicek, D., Bestimmung von Pentan als Indikator für die oxidative Ranzigkeit von Ölen, *Fat Science Technology*, **94**, 19–21, 1992.

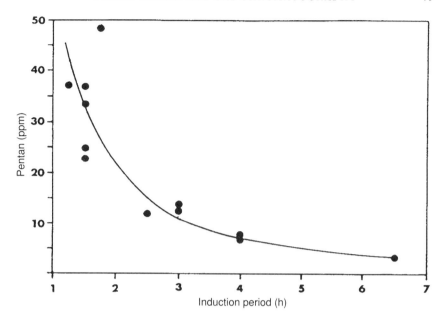

Figure 3.12 Second-order relationship of Rancimat induction periods (test temperature 120°C) and pentane levels (heated to 160°C in air). Reproduced from Ulberth, F. and Roubicek, D., Bestimmung von Pentan als Indikator für die oxidative Ranzigkeit von Ölen, *Fat Science Technology*, **94**, 19–21, 1992.

In fresh oils 0.6–4.5 ppm pentane (Snyder *et al.*, 1985) were found. Oils bought in retail outlets contained 1–18 ppm pentane (Ulberth and Roubicek, 1992). Pongracz (1986) published values which were considerably lower because he did not use a thermal decomposition step in his HSGC method.

The situation becomes more complicated when the oxidative stability of the lipid phase in a complex food is concerned. Lipid extraction, which is tedious and time consuming, is usually the first step followed by some chemical tests (peroxide value, p-anisidine value, thiobarbituric acid value, etc.) to characterize the extent of lipid oxidation (Rossell, 1989). Buttery and Teranishi (1963) were among the first who applied HSGC to lipid stability studies in complex food matrices (dehydrated potatoes and freeze-dried carrots). They added boiling water to the sample in an Erlenmayer flask, covered it with aluminium foil and measured the concentration of hexanal, 2-methylpropanal and 2- and 3-methylbutanal by sampling the HS gas. A similar routine was chosen by Fritsch and Gale (1977) for the determination of hexanal in an oat cereal. The onset of rancid odour was found to occur when the hexanal concentration increased to between 5 ppm and 10 ppm. Static HSGC was also used to assess oxidative stability of cooked chicken meat (Ang and Young, 1989). Significant correlations between three major peaks in the chromatograms and the thiobarbituric acid value were observed.

To follow lipid oxidation and the concomitant changes of volatiles in complex foodstuffs, dynamic HSGC ('purge-and-trap') is mostly preferred to static HSGC, as the former variant offers higher sensitivity (Leahy and Reineccius, 1984). Nevertheless, a combination of steam distillation of the sample and subsequent static HSGC of the distillate has proven to be a simple though powerful option for not so well equipped laboratories. Applied to fatty foods such as whole milk powder (WMP), a sensitive determination of volatile oxidation products is possible (Ulberth and Roubicek, 1995). Equilibration of WMP at high temperatures led to intensive browning of the product. Steam distillation of reconstituted WMP resulted in an enrichment of volatiles in the distillate on the one hand, and decomposition of lipid peroxides on the other. Freshly produced WMP contained < 10 ppb hexanal, whereas in stored samples up to 28 ppm were detected, depending on storage conditions. Short-chain hydrocarbons were not detected with this method, most probably because of losses during steam distillation. As well as aldehydes, some ketones (2-heptanone and 2-octanone) were also reported to occur in WMP stored at 40°C (Fig. 3.13). Hall and Lingnert (1986) investigated the storage stability of WMP at elevated temperatures and noted that the acceleration factor (ratio of zero-order rate constants at 20°C and 50°C) for the formation of 2-heptanone is about 20 times higher than that for hexanal. The ratio of hexanal to 2-heptanone may thus be utilized to differentiate between light-induced and temperature-induced lipid oxidation in WMP.

A HS gas analysis method for the determination of volatile hydrocarbons in sealed containers was described by Löliger (1990). The experimental set-up consists of a gas chromatograph equipped with a gas sampling valve, a vacuum pump and a manometer. The sampling system is evacuated to $c.$ 1 mbar, the tin is punctured by a stainless steel plunger, the gas is left to reach equilibrium and a HS sample is subsequently transferred by means of the valve to a column packed with activated alumina. This system allows estimation of C_1–C_6 hydrocarbons, which represents final and stable secondary oxidation products, and was applied to various food systems ranging from oils and emulsions to frozen meats and dehydrated cereals, potatoes and milk.

3.4.3 Flavour research

Reports about static HSGC and flavour research of fats and oils are scarce. HSGC can be used to good advantage when the primary aim of an analysis is quantification of target components, for example pentane or hexanal in oxidized oils. For an in-depth investigation of lipid-related flavours more sensitive techniques have to be used (Reineccius, 1989). Dynamic HS, simultaneous distillation and solvent extraction and high-vacuum

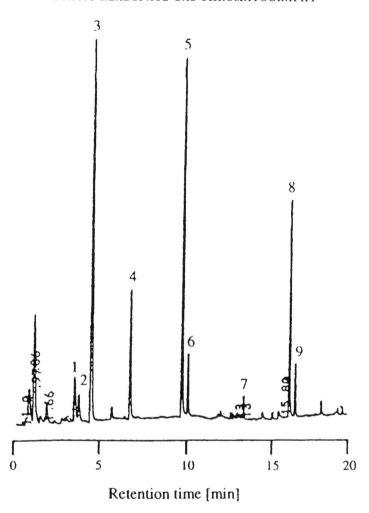

Figure 3.13 Headspace analysis of volatiles of whole milk powder stored in air at 40°C (peroxide value 1.01). 1 = pentanone; 2 = pentanal; 3 = methyl butyrate (internal standard); 4 = hexanal; 5 = heptanone; 6 = heptanal; 7 = octanal; 8 = octanoic; 9 = nonanal. Chromatographic conditions were as in Fig. 3.12. Reproduced from Ulberth, F. and Roubicek, D., Monitoring of oxidative deterioration of milk powder by headspace gas chromatography, *International Dairy Journal*, **5**, 523–31, 1995.

distillation techniques play a dominant role in flavour isolation and enrichment. When speaking about flavour isolation techniques it is worthwhile bearing in mind that the aroma profile isolated depends largely on the technique chosen. Unfortunately, no 'perfect', unbiased and universally applicable flavour method exists.

One 'rare' example of static HSGC in lipid flavour research is given by Perkins (1989) who analysed heated butter fat volatiles (Fig. 3.14).

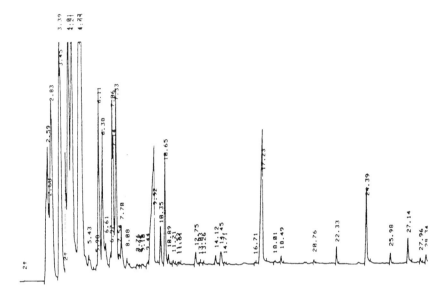

Figure 3.14 Headspace gas chromatography analysis of volatiles of butter heated to 150°C. Chromatographic conditions: 30 m × 0.53 mm fused silica column coated with SPB-5™, programmed from 45°C to 175°C. Reproduced from Perkins, E. G., Gas chromatography and gas chromatography–mass spectrometry of odor/flavor components in lipid foods, in *Flavor Chemistry of Lipid Foods* (eds D. B. Min and T. H. Smouse), pp. 35–56; published by the American Oil Chemists' Society, 1989.

3.4.4 Residues and contaminants

Fat is an excellent solvent for most residues and contaminants of organic origin. Many of them are of comparatively high molecular weight and are thus not traceable by HSGC. Volatile halocarbons (VHC) such as chloroform, carbon tetrachloride and tetrachloroethylene are widely used industrial solvents, intermediates and cleaning agents. In addition, they could be formed during chlorination of water and wastewater. VHC may enter the food chain and accumulate there in the lipid phase of foodstuffs.

Static HSGC has been used extensively for the determination of VHC in oils, lipids and fatty foods (Fig. 3.15). Quantitation of VHC by an electron-capture detector provides the means to achieve superb detection limits. In aqueous foods the limit of detection has been reported to be 1 ppb, whereas in fatty foods it has been reported to be in the 10–50 ppm range (Entz and Hollifield, 1982). With a multiple headspace extraction technique the detection limits for VHC in butter are even lower (1–5 ppb) (Uhler and Miller, 1988). The absolute detection limit for perchloroethylene in olive oil has been found to be 1 pg (Van Rillaer and Beernaert, 1989). A method for the determination of VHC in olive oil by packed column HSGC is described in EC Regulation 2568/91 (CEC, 1991). Precision data of a capillary column

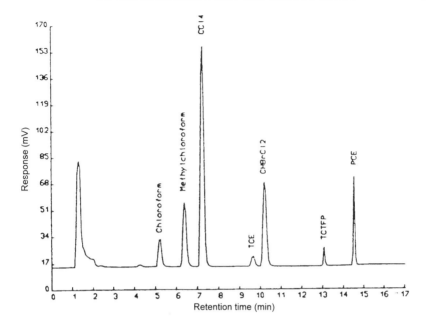

Figure 3.15 Headspace gas chromatographic analysis of a volatile hydrocarbon mixture representing c. 0.3 ng of each component. Chromatographic conditions: 30 m × 0.32 mm fused silica column coated with DB-5^{TM}, 40°C (11 min), then 20°C min^{-1} to 150°C, ^{63}Ni electron-capture detection. Sample thermostatted at 60°C. Reproduced from Uhler, A. D. and Miller, L. J., Multiple headspace extraction gas chromatography for the determination of volatile hydrocarbon compounds in butter, *Journal of Agricultural and Food Chemistry*, **36**, 772–5, 1988.

HSGC method for the analysis of VHC in edible oils were published by the German Health Office (Method L13.04.1; Bundesgesundheitsamt, 1992). At analyte levels of 0.05–0.30 ppm reproducibility was in the range 0.01–0.03 ppm and repeatability in the range 0.02–0.06 ppm.

Another class of toxic volatile pollutants – low molecular weight aromatic hydrocarbons such as benzene and toluene – have also been analysed successfully in fats and oils by HSGC (McCown and Radenheimer, 1989).

Determination of residual hexane in defatted meals is another standard application of HSGC in lipid chemistry (DGF Standard B-II 8a, 1987).

3.4.5 Miscellaneous applications

Apart from pure analytical applications, HSGC has found wide usage among thermodynamicists for the study of mixed phases where it is used to determine physicochemical data on vapour–solid or vapour–liquid systems (Hachenberg and Schmidt, 1977). Van Boekel and Lindsay (1992) studied the partition of some volatiles of cheese in equilibrium with its vapour over milk fat and the aqueous phase. Partition coefficients for hydrogen sulphide, methanethiol and dimethyl sulphide were determined and the impact of the effect of fat content on flavour and HS analysis of low-fat cheese was discussed. In a similar way HSGC was employed to study the influence of a vegetable oil in an emulsion on the volatility of selected aroma compounds (Le Thanh, Pham Thi and Voilley, 1992).

3.5 Conclusions

At first glance HSGC and fats and oils may look incompatible. Despite the fact that the majority of substances related to lipids are involatile, some niches exist where HSGC fits in perfectly. In general, a major weakness of static HSGC – a lack of sensitivity when compared with the dynamic variant – can be overcome by cryofocusing. Such systems are now commercially available. This allows one to present to the gas chromatographic detectors the same vapour mixture as one smells on a sample, except at concentrations which are detectable not only by the human nose but also by the detectors commonly used in GC.

References

Anaerobe Laboratory (1977) *Anaerobe Laboratory Manual*, 4th edn, Anaerobe Laboratory, Virginia Polytechnic Institute and State University, Blacksburg, VA.

Ang, C. Y. W. and Young, L. L. (1989) Rapid headspace gas chromatographic method for assessment of oxidative stability of cooked chicken meat. *Journal of the Association of Official Analytical Chemists*, **72**, 277–81.

Bassette, R. and Ward, G. (1975) Measuring parts per billion of volatile materials in milk. *Journal of Dairy Science*, **58**, 428–9.

Bundesgesundheitsamt (1992) Analysis of foods. Determination of low-boiling halogenated hydrocarbons in edible oils. *Amtliche Sammlung von Untersuchungsverfahren nach Paragraph 35 LMBG, Methode L13.04.1*, Germany.

Buttery, R.G. and Teranishi, R. (1963) Measurement of fat autoxidation and browning aldehydes in food vapors by direct vapor injection gas–liquid chromatography. *Agricultural and Food Chemistry*, **11**, 504–7.

CEC (1991) Commission Regulation (EEC) No. 2568/91 of 11 July 1991 on the characteristics of olive oil and olive-residue oil and the relevant methods of analysis, Commission of the European Communities. Office for Official Publications of the European Communities, Luxembourg.

Chan, H.W.S. (1987) *Autoxidation of Unsaturated Lipids*, Academic Press, London.

Christie, W.W. (1989) *Gas Chromatography and Lipids*, The Oily Press, Ayr.

Closta, W., Klemm, H., Pospisil, P. *et al.* (1983) The HS-100, an innovative concept for automatic headspace sampling. *Chromatography Newsletter*, **11**, 13–17.

Cottyn, B. G. and Boucque, C. V. (1968) Rapid method for the gas chromatographic determination of volatile fatty acids in rumen fluids. *Journal of Agricultural and Food Chemistry*, **16**, 105–7.

Cummings, J. H. and MacFarlane, G. T. (1991) The control and consequences of bacterial fermentation in the human colon. *Journal of Applied Bacteriology*, **70**, 443–59.

DGF Standard B-II 8a (1987) Benzin in Extraktionsschroten. Gaschromatographische Methode. Deutsche Gesellschaft für Fettwissenschaften, Münster.

Dupuy, H. P., Fore, S. P. and Goldblatt, L. A. (1973) Direct gas chromatographic examination of volatiles in salad oils and shortenings. *Journal of the American Oil Chemists' Society*, **50**, 340–2.
Dupuy, H. P., Raynor, E. T. and Wadsworth, J. I. (1976) Correlations of flavour scores with volatiles of vegetable oils. *Journal of the American Oil Chemists' Society*, **53**, 628–31.
Entz, R. C. and Hollifield, H. C. (1982) Headspace gas chromatographic analysis of foods for volatile halocarbons. *Journal of Agricultural and Food Chemistry*, **30**, 84–8.
Ettre, L. S., Jones, E. and Todd, B. S. (1984) Quantitative Analysis with headspace gas chromatography using multiple headspace extraction. *Chromatography Newsletter*, **12**, 1–3.
Evans, C. D., List, G. R., Dolev, A. *et al.* (1967) Pentane from thermal decomposition of lipoxidase-derived products. *Lipids*, **2**, 432–34.
Evans, C.D., List, G. R., Hoffmann, R. L. *et al.* (1969) Edible oils quality as measured by thermal release of pentane. *Journal of the American Oil Chemists' Society*, **46**, 501–4.
Fritsch, C. W. and Gale, J. A. (1977) Hexanal as a measure of rancidity in low fat foods. *Journal of the American Oil Chemists' Society*, **54**, 225–8.
Grob, K., Biedermann, M. and Giuffré, A. M. (1994) Determination of organophosphorous insecticides in edible oils and fats by splitless injection of the oil into a gas chromatograph. *Zeitschrift für Lebensmitteluntersuchung und -forschung*, **198**, 325–8.
Grob, K., Biedermann, M., Bronz, M. *et al.* (1994) Recognition of mild deodorization of edible oils by the loss of volatile components. *Zeitschrift für Lebensmitteluntersuchung und -forschung*, **199**, 191–4.
Grosch, W. (1987) Reactions of hydroperoxides – products of low molecular weight, in *Autoxidation of Unsaturated Lipids* (ed. H. W. S. Chan), Academic Press, London, pp. 95–139.
Hachenberg, H. and Schmidt, A. P. (1977) *Gas Chromatographic Headspace Analysis*, Heiden, London.
Hall, G. and Lingnert, H. (1986) Analysis and prediction of lipid oxidation in foods, in *The Shelf-life of Foods and Beverages* (ed. G. Charalambous), Elsevier Science, Amsterdam, pp. 735–43.
Hamilton, R. J. (1989) The chemistry of rancidity in foods, in *Rancidity in Foods*, 2nd edn (eds J. C. Allen and R. J. Hamilton), Elsevier Applied Science, London, pp. 1–21.
Ioffe, B. V. (1989) Gas-chromatographic head-space analysis: present state and prospects of development. *Fresenius Zeitschrift für Analytische Chemie*, **335**, 77–80.
Kolb, B. and Pospisil, P. (1985) Headspace gas chromatography (HSGC) with capillary columns, in *Sample Introduction in Capillary Gas Chromatography* (ed. P. Sandra), Huethig Verlag, Heidelberg, pp. 191–216.
Kolb, B., Liebhardt, B. and Ettre, L. S. (1986) Cryofocusing in the combination of gas chromatography with equilibrium headspace sampling. *Chromatographia*, **21**, 305–11.
Kolb, B., Pospisil, P. and Auer, M. (1981) Quantitative analysis of residual solvents in food packaging printed films by capillary gas chromatography with multiple headspace extraction. *Journal of Chromatography*, **204**, 371–6.
Kolb, B., Beyaert, G., Pospisil, P. *et al.* (1980) Applications of gas chromatographic head space analysis. Application note 26. Bodenseewerk Perkin Elmer, Überlingen.
Laeubli, M. W. and Bruttel, P. A. (1986) Determination of the oxidative stability of fats and oils: comparison between the active oxygen method (AOCS cd 12–57) and the Rancimat method. *Journal of the American Oil Chemists' Society*, **63**, 792–5.
Larson, L., Mardh, P. A. and Odham, G. (1978) Detection of alcohols and volatile fatty acids by head space gas chromatography in the identification of anaerobic bacteria. *Journal of Clinical Microbiology*, **7**, 23–7.
Le Thanh, M., Pham Thi, S. T. and Voilley, A. (1992) Influence of the presence of vegetable oil in emulsion on the volatility of aroma substances during the course of the extraction. *Sciences des Aliments*, **12**, 587–92.
Leahy, M. M. and Reineccius, G. A. (1984) Comparison of methods for the isolation of volatile compounds from aqueous model systems, in *Analysis of Volatiles* (ed. P. Schreier), de Gruyter, Berlin, pp. 19–47.
Löliger, J. (1990) Headspace gas analysis of volatile hydrocarbons as a tool for the determination of the state of oxidation of foods stored in sealed containers. *Journal of the Science of Food and Agriculture*, **52**, 119–28.

McCown, S. M. and Radenheimer, P. (1989) An equilibrium headspace gas chromatographic method for the determination of volatile residues in vegetables. *LC–GC International*, **2**, 28–31.

Marsili, R. T. (1984) Measuring light-induced chemical changes in soybean oil by capillary headspace gas chromatography. *Journal of Chromatographic Science*, **22**, 61–7.

Min, D. B., Lee, S. H. and Lee, E. C. (1989) Singlet oxygen oxidation of vegetable oils, in *Flavor Chemistry of Lipid Foods* (eds. D. B. Min and T. H. Smouse), The American Oil Chemists' Society, Champaign, IL, pp. 57–97.

Osl, F. (1988) Bestimmung der niederen freien Fettsäuren in Hart- und Schnittkäse mit der Head-Space Gaschromatographie. *Deutsche Molkerei Zeitung*, **45**, 1516–18.

Perkins, E. G. (1989) Gas chromatography and gas chromatography–mass spectrometry of odor/flavor components in lipid foods, in *Flavor Chemistry of Lipid Foods* (eds D. B. Min and T. H. Smouse), The American Oil Chemists' Society, Champaign, IL, pp. 35–56.

Pongracz, G. (1986) Determination of rancidity of edible fats by headspace gas chromatographic detection of pentane. *Fette Seifen Anstrichmittel*, **88**, 383–6.

Reineccius, G. A. (1989) Isolation of food flavors, in *Flavor Chemistry of Lipid Foods* (eds D. B. Min and T. H. Smouse), The American Oil Chemists' Society, Champaign, IL, pp. 26–34.

Rossell, J. B. (1989) Measurement of rancidity, in *Rancidity in Foods*, 2nd edn (eds J. C. Allen and R. J. Hamilton), Elsevier Applied Science, London, pp. 23–52.

Scheppach, W. (1994) Effects of short chain fatty acids on gut morphology and function. *Gut*, **35** (1 Suppl.), S35–8.

Scholz, R. G. and Ptak, L. R. (1966) A gas chromatographic method for measuring rancidity in vegetable oils. *Journal of the American Oil Chemists' Society*, **43**, 596–9.

Schomburg, G. (1985) Temperature programmed sample transfer, in *Sample Introduction in Capillary Gas Chromatography* (ed. P. Sandra), Huethig Verlag, Heidelberg, pp. 55–76.

Snyder, J. M., Frankel, E. N. and Selke, E. (1985) Capillary gas chromatographic analyses of headspace volatiles from vegetable oils. *Journal of the American Oil Chemists' Society*, **62**, 1675–9.

Snyder, J. M., Frankel, E. N., Selke, E. *et al.* (1988) Comparison of gas chromatographic methods for volatile lipid oxidation compounds in soybean oil. *Journal of the American Oil Chemists' Society*, **65**, 1617–20.

Takeoka, G. and Jennings, W. (1984) Developments in the analysis of head-space volatiles: on-column injections into fused silica capillaries and split injections with a low-temperature bonded PEG stationary phase. *Journal of Chromatographic Science*, **22**, 177–84.

Taylor, A. J. (1984) The application of gas chromatographic headspace analysis to medical microbiology, in *Applied Headspace Gas Chromatography* (ed. B. Kolb), John Wiley, Chichester, Sussex, pp. 140–54.

Uhler, A. D. and Miller, L. J. (1988) Multiple headspace extraction gas chromatography for the determination of volatile halocarbon compounds in butter. *Journal of Agricultural and Food Chemistry*, **36**, 772–5.

Ulberth, F. and Roubicek, D. (1992) Bestimmung von Pentan als Indikator für die oxidative Ranzigkeit von Ölen. *Fat Science Technology*, **94**, 19–21.

Ulberth, F. and Roubicek, D. (1993) Evaluation of a static headspace gas chromatographic method for the determination of lipid peroxides. *Food Chemistry*, **46**, 137–41.

Ulberth, F. and Roubicek, D. (1995) Monitoring of oxidative deterioration of milk powder by headspace gas chromatography. *International Dairy Journal*, **5**, 523–31.

Ullrich, F. and Grosch, W. (1988) Flavour deterioration of soya-bean oil: identification of intense odour compounds formed during flavour reversion. *Fat Science Technology*, **90**, 332–6.

Van Boekel, M. A. J. S. and Lindsay, R. C. (1992) Partition of cheese volatiles over vapour, fat and aqueous phases. *Netherlands Milk and Dairy Journal*, **46**, 197–208.

Van Rillaer, W. and Beernaert, H. (1989) Determination of residual tetrachlorethylene in olive oil by headspace-gas chromatography. *Zeitschrift für Lebensmitteluntersuchung und -forschung*, **188**, 221–2.

Warner, K., Evans, C. D., List, G. R. *et al.* (1978) Flavor score correlation with pentanal and hexanal contents of vegetable oils. *Journal of the American Oil Chemists' Society*, **55**, 252–6.

Wylie, P. L. (1986) Headspace analysis with cryogenic focusing: a procedure for increasing the sensitivity of automated capillary headspace analysis. *Chromatographia*, **21**, 251–8.

4 Multinuclear high-resolution nuclear magnetic resonance spectroscopy
B. W. K. DIEHL

4.1 Introduction

The analysis of lipids is dominated by chromatographic methods such as gas chromatography (GC) and high-pressure liquid chromatography (HPLC) and some other physical methods such as thermo analysis. Nuclear magnetic resonance (NMR) spectroscopy covers only a small part of lipid analysis. In most cases it is used for qualitative analysis such as structure elucidation of lipids. Unlike the other methods NMR spectroscopy is coupled with computer technology, and its development is as fast as that of computer software and hardware. In 1991 the Nobel Prize for chemistry was given to Professor Ernst for the development of Fourier Transform NMR spectroscopy (FT-NMR). The capabilities of these methods even for the elucidation of the three-dimensional structures of small molecules up to macromolecules are impressive.

The related technique of NMR tomography has entered the medical domain, whereas high-resolution NMR spectroscopy allows investigations into the lipid metabolism of animals and human beings *in vivo*. Other variations of NMR techniques are solid state NMR and NMR imaging for three-dimensional material control.

But how is the place of NMR spectroscopy defined in lipid analyses? The conventional NMR spectroscopist wants to analyse the structure of purified substances; the classical analytical chemists have to quantify single components in complex, often natural lipid mixtures. It is a misconception both of the NMR spectroscopists and of the analysts that NMR cannot be used as a quantitative method. On the contrary, NMR spectroscopy is very suitable for quantitative analysis, as will be shown later.

I would like to give a survey of qualitative and quantitative NMR spectroscopy in lipid analysis. Within the past 20 years some important papers and reviews have been published (Dennis and Plückthun, 1984; Diehl, 1993; Diehl and Ockels, 1995a; Glonek and Mercant, 1996; Gunstone, 1993a; Meneses and Glonek, 1988; Shoolery, 1976; Wollenberg, 1990). NMR spectroscopy has become an alternative to chromatography in some fields of lipid analysis.

4.2 The instrument

Natural lipids consist of the elements hydrogen, carbon, phosphorus, nitrogen and oxygen. Some isotope nuclei of these elements can be detected by NMR. The low natural abundance of ^{15}N and ^{17}O prevents the routine application of NMR to these elements without use of labelled substances, but ^1H, ^{13}C and ^{31}P NMR spectroscopy are used routinely. Many instruments are equipped with a so-called QNP (quattro nuclei probe) for sequential NMR analysis of ^1H, ^{13}C, ^{31}P and ^{19}F without changing hardware. Modern NMR spectrometers are available up to field strengths of 19.2 T or a resonance frequency of 800 MHz for protons. Routine analysis is made at proton frequencies between 300–500 MHz. Depending on the kind of experiments, the high-field instruments allow analysis of concentrations down to the microgrammes per millilitre level, the usual case being a concentration of 1–100 mg ml^{-1}. The data are recorded using 32 bit ADCs (analog digital converters), which results in high spectral dynamics.

4.3 Principles

NMR enables direct observation of atoms. The integral of an NMR signal is strictly linearly proportional to the number of atoms in the probe volume. The signal is a measure of the molar ratios of molecules, independent of molecular weight. There are no response factors caused by varying rates of extinction dependent on molecular structure, as in ultraviolet (UV) detection, and non-linear calibration curves, as found with light-scattering detectors, are unknown to NMR spectroscopy.

4.4 Spectra

The frequency at which NMR signals appear depends mainly on the magnetic field strength. For example protons have a resonance frequency of 300 MHz at 7.05 T. The chemical environment of an active nucleus leads to a small shift in the resonance frequency – the chemical shift. Functional groups find their expression in the chemical shift. The result is an intensity/frequency diagram – the NMR spectrum.

This collection of more or less separated NMR signals is analogous to intensity/time diagrams in chromatography in which one component is represented by one signal. In ^1H NMR spectroscopy each H atom leads to at least one signal. As most molecules of analytical interest contain more than one H atom, spectra are more complex than chromatograms. Crucial to the information taken from NMR spectra is the spectral dispersion, which is a linear function of the magnetic field strength. The homonuclear spin coupling of protons leads to a low dispersion in ^{1}H NMR spectroscopy. In ^{1}H NMR spectra of complex mixtures it is often not possible to detect single components, but the sum of functional groups in the mixture can be determined. In ^{13}C NMR spectra the dispersion is much higher. The low natural abundance of the NMR-active ^{13}C isotope has a detrimental effect on the sensitivity, but signals are singlets after heteronuclear decoupling. Determination of molecules or structural groups is possible down to 0.1% corresponding to an absolute amount of 10 µg. The high spectral dispersion makes parts of the ^{13}C NMR spectra directly comparable with chromatograms. An example is the carbonyl region in a ^{13}C NMR spectrum of a lipid mixture, where each fatty acid is represented by a specific signal.

^{31}P NMR spectroscopy is the method of choice for phospholipids. Most phospholipids contain only one phosphorus atom, so the ^{31}P NMR spectrum of lecithin reads like an HPLC chromatogram. There are some advantages compared with HPLC: specific detection of the phosphorus nucleus, high dispersion and high dynamics. The role of ^{31}P NMR spectroscopy is discussed in detail later (Section 4.8.3).

4.5 Response

The area of an NMR signal is directly proportional to the molar amount of the detected isotope. The ratio between two different signals of one molecule should be 1:1, the number of represented atoms taken into account. In practice there are differences caused by different relaxation times. This is the time an excited atom needs to fall to the ground state. In the case of heteronuclear decoupling the nuclear Overhouser effect can cause response factors as well. These response problems are influenced by the measuring parameters, with the correct (problem-orientated) choice they disappear or are minimized. Within a family of atoms in a similar chemical surrounding, for example all end-positioned methyl groups in the ^{13}C NMR spectra of fatty-acid-containing material, these effects are negligible. The same applies to corresponding carbonyl groups, but it is not possible to compare areas of carbonyl or methyl signals produced by these groups in a variety of chemical surroundings without appropriate experimental design or experimentally determined response factors. The response factors change from ±10% in ^{1}H NMR spectra up to ±50% in ^{13}C NMR spectra, and this is in fact advantageous when compared with HPLC–UV detection.

4.6 Reproducibility

During an NMR experiment there is no contamination of sample and probe head. The electronic stability of an NMR spectrometer is very good. The spectrum of a stable sample stored in a sealed tube shows reproducible areas over many years with a variation lower than 1%. These facts reduce the need for validation measurements where NMR methods are used.

4.7 Calibration

As for all classical quantitative analysis NMR spectroscopy needs calibration, calibration standards and a validation procedure. Four different techniques are used for calibration: external calibration, the standard addition method, the internal standard method and the tube-in-tube technique. The last is a special NMR calibration method in which a small glass (capillary) tube containing a defined amount of standard is put into the normal, larger, NMR tube filled with the sample to be analysed. In most cases there are slight differences in the chemical shift of corresponding signals of the same molecule in the inner and outer tube. The spectrum shows two signals at different frequencies; evaluation of the signal ratio allows quantification.

4.8 Applications

4.8.1 ^1H NMR spectroscopy of fatty oils

Methyl groups of all fatty acids have resonances between $\delta = 0.8$ ppm and 1.0 ppm (signal H). ω-3 fatty acids show a sharp triplet (signal G) at a lower field. The molar amount of ω-3 fatty acids is determined by comparing the areas of signals H and G. Table 4.1 lists the chemical shifts of different groups in fatty oils. For illustration the ^1H NMR spectra of a plant oil (linola) and a more complex fish oil are shown in Fig. 4.1.

Classical sum parameters such as the iodine number, the percentage of unsaturation and the molar amount of ω-3 fatty acids can be calculated from the corresponding signal areas of ^1H NMR spectra (Diehl and Ockels, 1995a; Manz and Schneider, 1995).

Table 4.1 Chemical shifts of groups in fatty oils

Signal	Chemical shift (ppm)	Proton
A	5.3–5.4	olefinic CH, and CH of glycerol
B	4.1–4.3	CH_2 of glycerol
C	2.8	double allylic CH_2
D	2.3–2.35	α–CH_2–CO–OR
E	2.0–2.1	allylic CH_2
F	1.6–1.7	β–CH_2–CH_2–CO–OR
	1.1–1.5	Methylene
G	1.0	CH_3 in ω-3 fatty acids
H	0.8–0.9	CH_3

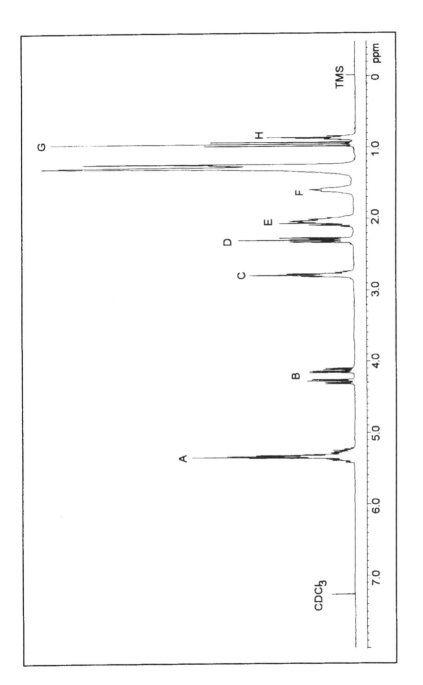

Figure 4.1(a)

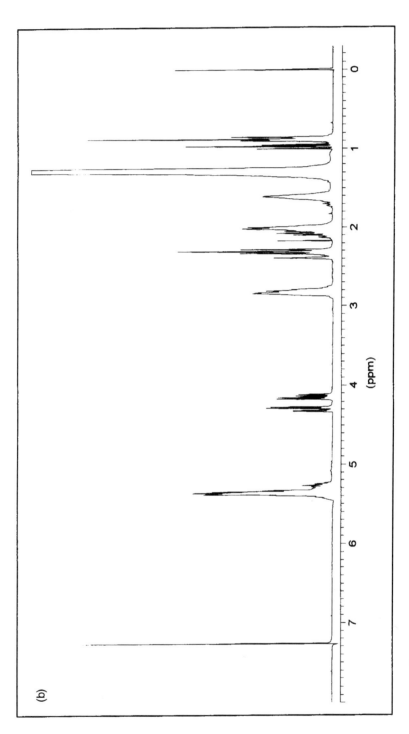

Figure 4.1 ¹H NMR spectrum of (a) plant oil (linseed oil); (b) fish oil (salmon tissue). Signals A–H are described in Table 4.1. Instrument details are given in text, Section 4.10.

Table 4.2 Recordable parameters

Analyte	Structural element	Characteristic chemical shift (ppm)
Total sterol or cholesterol	CH_3	0.75
Monoglyceride and diglyceride	CH_2	3–4
Phosphatidylcholine	$N(CH_3)_3$	3.2
Docosahexaenoic acid (DHA)	$O-CO-CH_2-CH_2-$	2.4
ω-1 fatty acids	$-CH=CH_2$	4.9–5.1; 5.8
oxidation products	$-OOH$	8.5
oxidation products	$-COOH$	9.5
Z,E conjugugated double bonds	$-CH=CH-CH=CH-$	5.5–6.7

There are other parameters of interest which can be calculated in terms of relative or absolute values (Table 4.2; Figs 4.2 and 4.3). The detection limit depends on the field strength of the instrument, the number of accumulated scans and the type of analyte. Without preceding separation it is possible to analyse components down to 0.1%, corresponding to a 10 µg absolute amount (Fig. 4.4).

In many cases a combination of liquid chromatography and NMR spectroscopy allows an exact quantification of small amounts, even trace amounts (ppm). This method is used to quantify neutral lipids in highly enriched phospholipid mixtures (Fig. 4.5).

It was possible to determine 0.2% of intact triglycerides in the presence of 95% phospholipids by the standard addition method (Spectral Service, unpublished). Monoglycerides, sterols or free fatty acids do not disturb the determination and can be quantified by the same method.

The examples shown illustrate only a few of the possible NMR applications. There are other qualitative and quantitative ^{1}H NMR applications, for example the determination of: citric acid, acetic acid, acetone, ethanol, methanol, carbohydrates, ceramides and plasmalogens in lecithin; polyethyleneglycol-based emulsifiers in animal feed (limit 100 ppm); emulsifiers, surfactants and preservatives in cosmetics (Spectral Service, unpublished).

4.8.2 ^{13}C NMR spectroscopy

The second important nucleus in lipids is carbon. The abundance of the NMR active ^{13}C nucleus isotope is only 1.12% of ^{12}C, so the sensitivity of ^{13}C NMR is much lower than that of ^{1}H NMR. The selectivity and the dispersion of ^{13}C NMR spectra are very high. All aliphatic C atoms in the ^{13}C NMR spectrum of tripalmitin are well separated. The spectrum of the carbonyl region (Fig. 4.6) shows two resonances, in the ratio 2:1, assigned to C atoms of carbonyl groups in the α and β position of the glycerol. Each signal provides information on identity and intensity.

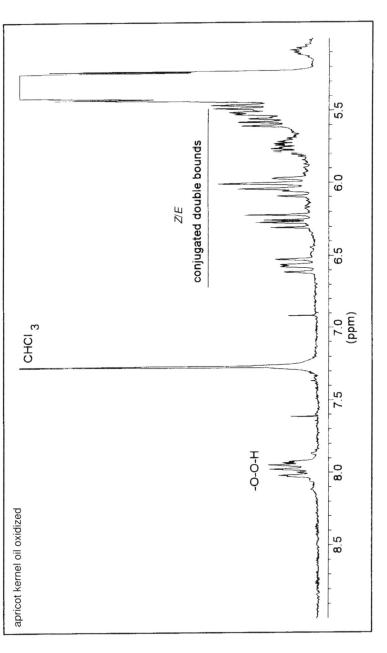

Figure 4.2 ^{1}H NMR spectrum showing conjugated *cis*, *trans* double bonds and hydroperoxides in an oxidized plant oil (oxidized apricot kernel oil). Instrument details are given in text, Section 4.10.

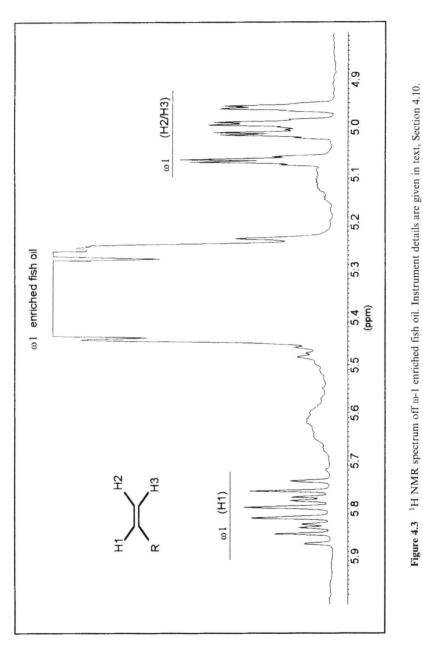

Figure 4.3 ^{1}H NMR spectrum off ω-1 enriched fish oil. Instrument details are given in text, Section 4.10.

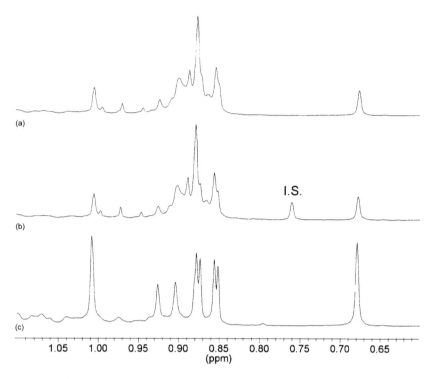

Figure 4.4 ^{1}H NMR spectra: quantification of cholesterol in egg yolk lecithin with use of internal standard (I.S.): (a) pure sample (detail); (b) pure sample plus I.S.; (c) CH$_3$ signals of the cholesterol reference. Instrument details are given in text, Section 4.10. (Source: Spectral Service, unpublished material.)

The chromatogram-like structure of the carbonyl region shows separation of a mixture from different mono-, di- and triglycerides and free fatty acids (Fig. 4.7). This method is used for quality control of olive oils (Sacchi et al., 1990). Integrals of the respective intensities are directly proportional to the molar amount of the component, with one restriction: comparison is only allowed between atoms in the same chemical environment, for example carbonyls only with carbonyls, methyls only with methyls and so on. There is no need for a standard to evaluate the fatty acid distribution, and no calibration and no quality control samples are needed. No chemical modification such as saponification or derivatization is necessary, so degradation of the chemically sensitive polyunsaturated acids is avoided. The material can be recovered unchanged after the measurement. In addition to the general fatty acid distribution a determination of individual distributions is possible.*

---

* For example, see Aitzetmüller et al., 1994; Aursand and Grasdalen, 1992; Aursand, Jørgensen and Grasdalen, 1995; Diehl, 1993; Diehl and Ockels, 1995a,b; Gunstone, 1991, 1993a,b; Herling, 1995; Mallet, Gaydou and Archavlis, 1990; Ng, 1985; Ng and Koh, 1988; Pfeffer, Luddy and Unruh, 1977; Wollenberg, 1990.

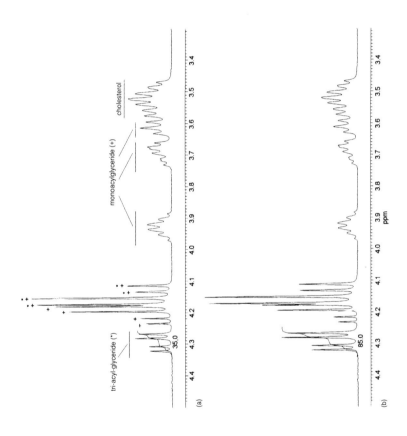

Figure 4.5 ^{1}H NMR spectra: triglyceride determination after liquid chromatography (standard addition method). (a) Mixture of triglyceride, monoglyceride (shown as mono-acyl-glyceride) and cholesterol; (b) mixture after addition of 0.5% triglyceride. Instrument details are given in text, Section 4.10

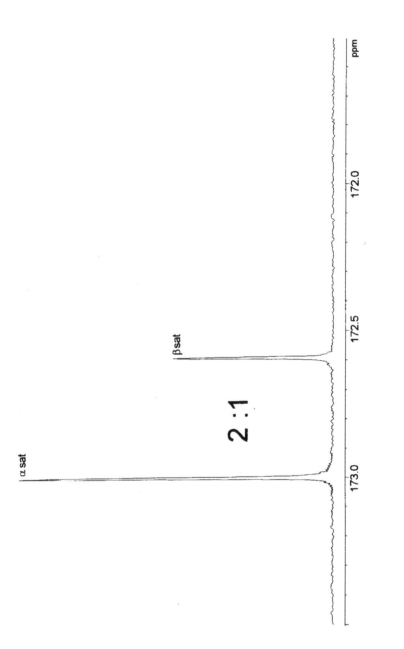

Figure 4.6 ^{13}C NMR spectrum: the carbonyl region of tripalmitin. Instrument details are given in text, Section 4.10.

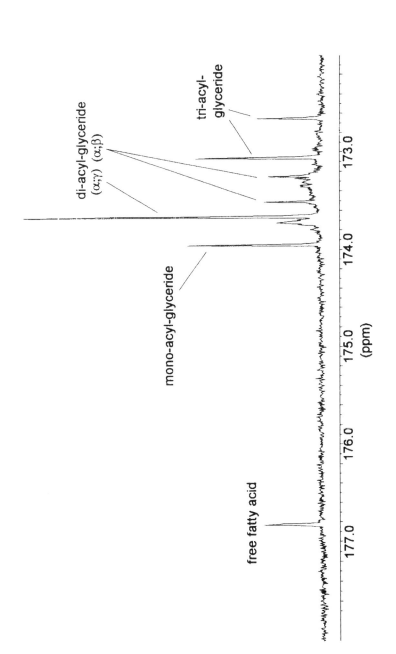

Figure 4.7 ^{13}C NMR spectrum of a mixture of different mono-, di- and triglycerides (labelled in figure as acyl-glycerides) and free fatty acids. Instrument details are given in text, Section 4.10.

Figure 4.8 shows detail of a typical plant oil. There are three signals in two groups (α- and β-carbonyls) in the ratio 2:1. The carbonyl groups 'feel' the double bonds in the chain of the fatty acids, the influence growing with decreasing distance. Saturated and Δ9 fatty acids (18:1, 18:2 and 18:3) are well separated. This spectrum shows impressively the fact that there are no saturated fatty acids in the β position of the glycerol. With higher spectral resolution (described below) even 18:1, 18:2 and 18:3 can be differentiated.

The spectrum shown in Fig. 4.9 is derived from borage oil (signal assignments are given in Table 4.3). There are two additional signals, caused by γ-linolenic acid, which is a ω-6 fatty acid. In a statistical distribution the ratio should be 2:1, but is found to be 1:2, which means that two thirds of all γ-linolenic acid is in the β position of the glycerol, and one third is in the α position (Gunstone, 1990).

Figures 4.10(a) and 4.10(b) show oils from single cells. It is easy to see the difference in chemical shift of the two essential fatty acids, arachidonic acid (ARA) [Fig. 4.10(a)] and 4,7,10,13,16,19-docosahexaenoic acid (DHA) [Fig. 4.10(b)] as well as their asymmetrical distribution.

The spectrum in Fig 4.11(a) shows the carbonyl region of a fish oil [assignments are provided in Table 4.4(a)]. The whole spectrum consists of approximately 300 signals, provided the instrument is well tuned to high resolution. The number of signals has grown obviously, although there are more than 99% triglycerides and nothing else in the sample. I have counted 17 carbonyl signals arising from different fatty acids.

The content of ω-3 acids in salmon and seal oil is in both cases nearly the same. But a look at the carbonyl region shows drastically the differences in distribution of the fatty acid at the α and β positions [Fig. 4.11(b), Table 4.4(b)].

The greater the number of different fatty acids the more complex the spectra become. A good signal resolution is essential. For ultra-high-resolution spectra not only high magnetic fields are necessary but also appropriate parameters for acquisition and processing. Figure 4.11(b)

Table 4.3 Signal assignment of borage oil (the spectrum is illustrated in Fig. 4.9)

Peak (ppm)	Area	Assignment	Mol%
173.0000	49.79	α sat	14.25
172.9947	11.90	αΔ13, 22:1	3.41
172.9877	12.50	αΔ11, 20:1	3.58
172.9724	33.07	αΔ9, 18:1	9.47
172.9633	80.68	αΔ9, 18:2	23.09
172.9584	8.96		2.56
172.8104	32.84	αΔ6, 18:3	9.40
172.5804	14.94	βΔ9, 18:1	4.28
172.5711	50.38	βΔ9, 18:2	14.42
172.4238	54.30	βΔ6, 18:3	15.54

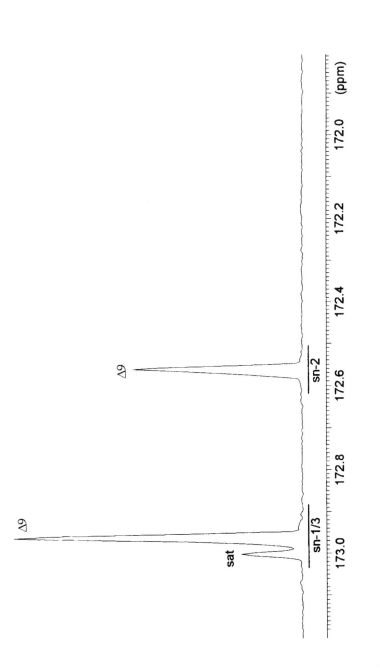

Figure 4.8 ^{13}C NMR spectrum: the carbonyl region of a typical plant oil (linseed oil). Instrument details are given in text, Section 4.10.

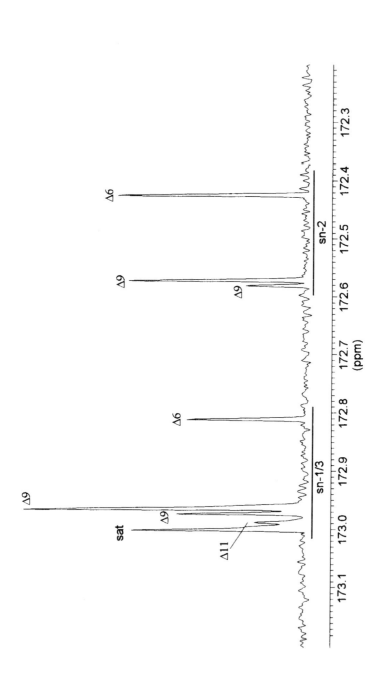

Figure 4.9 ^{13}C NMR spectrum: the carbonyl region of borage oil. Signal assignments are given in Table 4.3. Instrument details are given in text, Section 4.10.

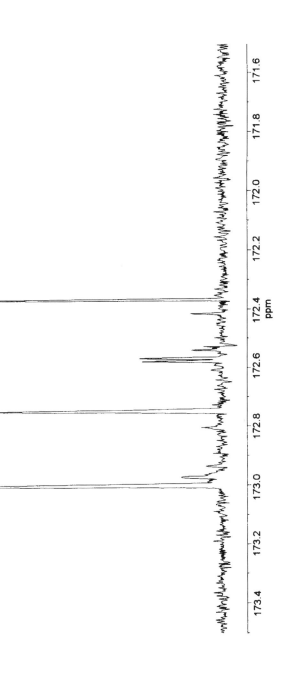

Figure 4.10(a)

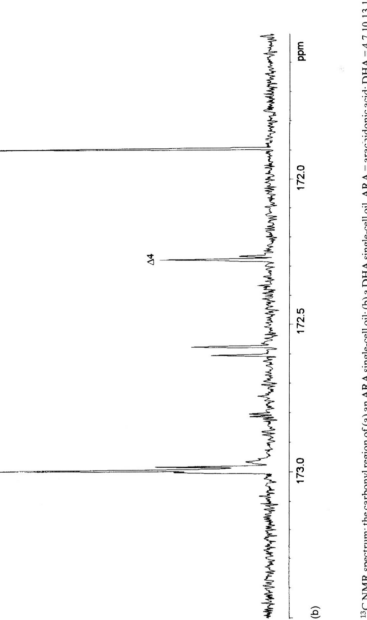

Figure 4.10 ^{13}C NMR spectrum: the carbonyl region of (a) an ARA single-cell oil; (b) a DHA single-cell oil. ARA = arachidonic acid; DHA = 4,7,10,13,16,19-docosahexaenoic acid. Instrument details are given in text, Section 4.10.

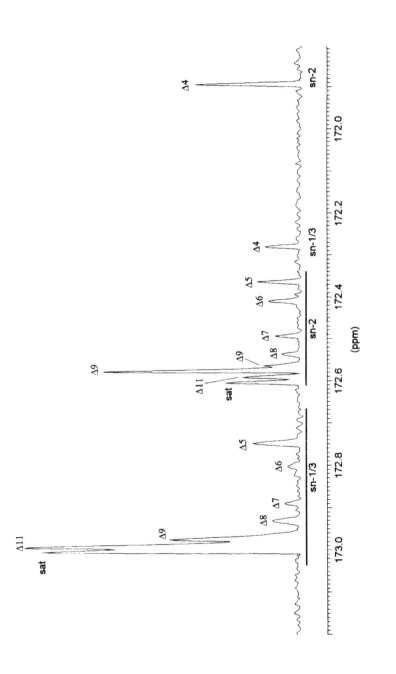

Figure 4.11(a)

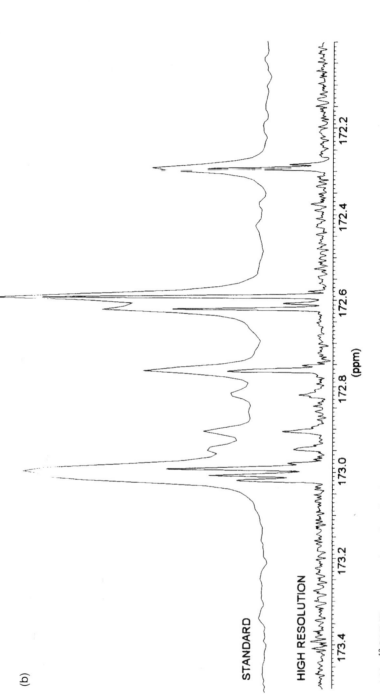

Figure 4.11 ^{13}C NMR spectrum: the carbonyl region of salmon oil (a) and seal oil (b). Peak assignments are given in Table 4.4. Instrument details are given in text, Section 4.10.

Table 4.4 Signal assignment of: (a) salmon oil [spectrum illustrated in Fig. 4.11(a)]; (b) seal oil [spectrum illustrated in Fig. 4.11(b)]. DHA = 4,7,10,13,16,19-docosahexaenoic acid; EPA = 5,8,11,14,17-eicosopentaenoic acid

Peak (ppm)	Area	Assignment	Mol%
(a)			
173.0006	84.36	sat	17.80
172.9888	95.09	△11	20.07
172.9736	76.85	α, (18:1)	16.22
172.9302	13.32	△8	2.81
172.8901	6.15	△7	1.30
172.8026	6.41	△6	1.35
172.7479	18.93	α, EPA	4.00
172.6038	19.69	sat	4.16
172.5910	12.75	△11	2.69
172.5750	53.38	β, 18:1	11.27
172.5650	13.80	β, Rest	2.91
172.5368	6.44	△8	1.36
172.4932	6.78	△7	1.43
172.4097	10.08	△6	2.13
172.3639	11.35	β, EPA	2.40
172.2803	9.86	α, DHA	2.08
171.8932	28.59	β, DHA	6.03
(b)			
172.9995	23.70	α sat	7.74
172.9873	30.98	α△11	10.11
172.9712	49.44	α△9, 18:1	16.14
172.9594	2.84	α△9, Rest	0.93
172.9271	6.42	α△8	2.10
172.8854	13.92	α△7	4.54
172.7999	4.22	α△6	1.38
172.7433	34.21	α△5, (EPA)	11.17
172.7320	0.75	α△5, (EPA)	0.24
172.5997	32.74	β sat	10.69
172.5871	8.82	β△11	2.88
172.5705	64.33	β△9, 18:1	21.00
172.5582	1.15	β△9, Rest	0.38
172.2740	29.40	α△4, DHA	9.60
172.2639	3.36	α△4, DHA	1.10

demonstrates these differences on the carbonyl region of ^{13}C NMR spectra of the same seal oil sample, measured with different parameters. Differences in parameters used for standard compared with high-resolution are given in Table 4.5.

In principle the resolution increases with the field strength, but in some cases the spectra will become more complex with higher fields. At 75 MHz or 100 MHz the effect of a double bond shifts only the 'own' carbonyl group. At higher fields the effect will split the carbonyl atom in the neighboured position too. At ideal conditions each carbonyl signal represents one specific fatty acid combination within a glyceride.

Euonymus oil has a very special composition (Fig 4.12). The acetyl group in the Sn-3 position leads to a splitting of the chemical shift in the Sn-1

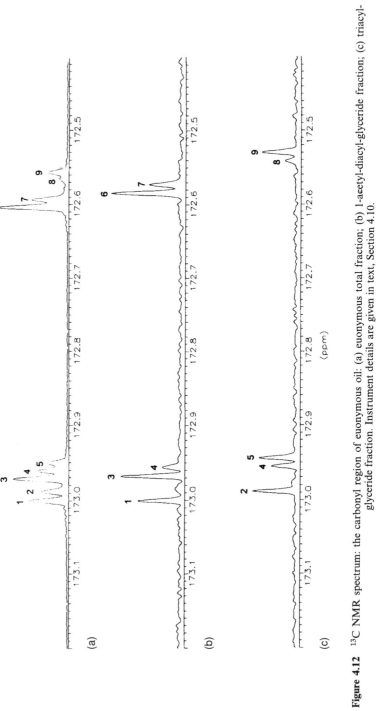

Figure 4.12 ^{13}C NMR spectrum: the carbonyl region of euonymous oil: (a) euonymous total fraction; (b) 1-acetyl-diacyl-glyceride fraction; (c) triacyl-glyceride fraction. Instrument details are given in text, Section 4.10.

Table 4.5 Parameters used to produce the standard and high-resolution ^{13}C NMR spectra illustrated in Fig. 4.11(b)

Parameter	Standard	High-resolution
Sweep width (ppm)	250	5
Acquisition time (s)	1.8	5.4
Relaxation delay (s)	1.2	0.3
Time domain	64 k	4 k
Spectral size	64 k (no zero filling)	32 k (zero filling)
Multiplication	exponential	gaussian[a]

[a] For better line shape.

and Sn-2 positions. This enables one to detect triacylglyceride and 1-acetyldiacylglyceride without chromatographic separation of the oil (Aitzetmüller and Spectral Service, in progress)

The detailed analysis of the spectra of mixtures is not trivial. Beside other parts of the spectrum a very complex region is that attributable to the double bonds. The partial ^{13}C NMR spectrum shown in Fig. 4.13 is from blackcurrant seed oil. All four unsaturated C_{18} fatty acids (18:1, 18:2, α-18:3 and γ-18:3) are differentiated as well as their distribution in the Sn-1/3 and Sn-2 positions.

Carbonyl C atoms 'feel' the distance (Δ) to a double bond, so it is not surprising that the methyl groups also 'feel' the distance (ω) to a double bond, as demonstrated in the methyl group region of ^{13}C NMR spectra of a fish oil. The differentiation of ω-3, ω-6, ω-7, ω-9 and saturated (together with $\omega > 10$) fatty acids as a sum parameter is possible (Fig. 4.14).

The power of the ^{13}C NMR method can be demonstrated by comparison of depot fat from pig and human sources, the latter from myself. The total distribution of ω-3–ω-9 and saturated fatty acids can be evaluated by analysis of the methyl groups (Fig. 4.15). Only slight differences are detectable. The individual distribution in Sn-1/3 and Sn-2 is confirmed by the carbonyl part of the spectra (Fig. 4.16). In this case it can be seen clearly that a human is not a pig.

The method described can be transferred easily to glycolipids (Diehl *et al.*, 1995) and phospholipids (Diehl *et al.*, 1996a).

In Fig. 4.17 the complete ^{13}C NMR spectrum of a digalactosyldiacylglyceride is shown. Figure 4.18 is a detail of this spectrum (carbonyl region) showing the fatty acid distribution in the two different glycerol positions. ^{13}C NMR data have been used to monitor the specifity of enzyme reactions (Diehl *et al.*, 1995).

The region of the sugar α-C atoms between $\delta = 115$ ppm and 90 ppm in carbohydrates shows a high sensitivity to the chemical surroundings. In a complex mixture of glycolipids a differentiation and quantification of different types is possible (Fig. 4.19). The spectrum shown in Fig. 4.19 was taken from a plant glycolipid fraction. The main components are

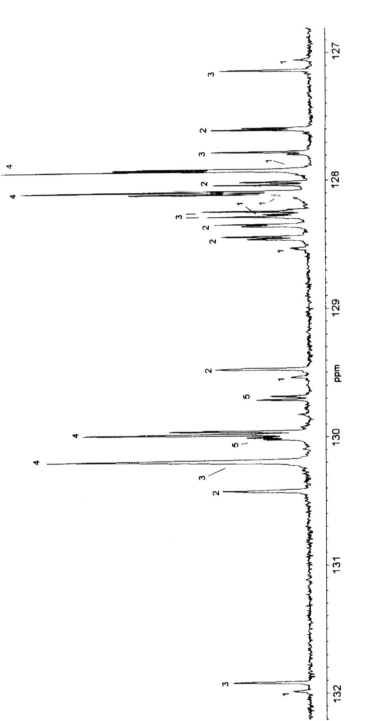

Figure 4.13 ^{13}C NMR spectrum: the double bond region of blackcurrant seed oil. $1 = \Delta 6, 18:4$; $2 = \Delta 6, 18:3$; $3 = \Delta 9, 18:3$; $4 = \Delta 9, 18:2$; $5 = \Delta 9, 18:1$. Instrument details are given in text, Section 4.10.

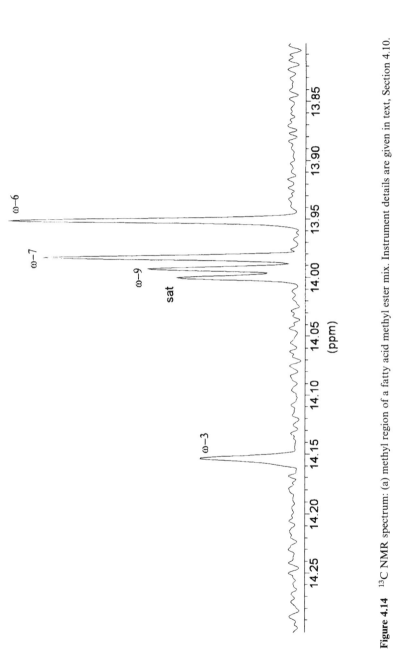

Figure 4.14 ^{13}C NMR spectrum: (a) methyl region of a fatty acid methyl ester mix. Instrument details are given in text, Section 4.10.

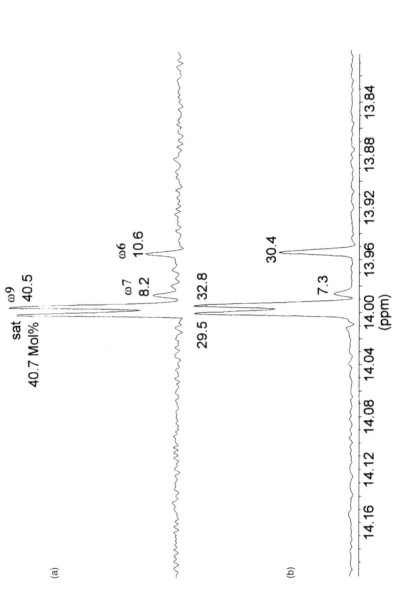

Figure 4.15 ^{13}C NMR spectrum: methyl region of (a) human depot fat; (b) porcine depot fat. Instrument details are given in text, Section 4.10.

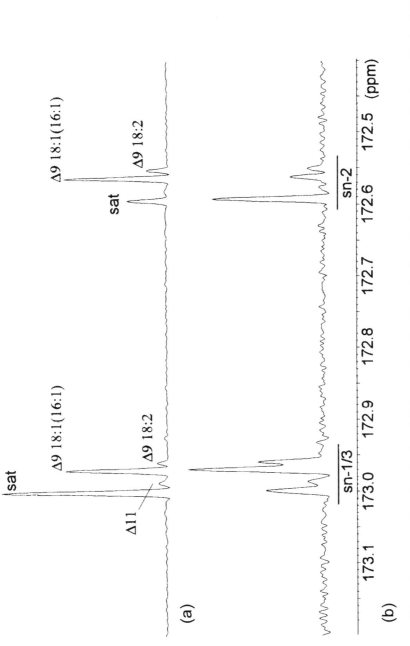

Figure 4.16 ^{13}C NMR spectrum: the carbonyl region of (a) human depot fat; (b) porcine depot fat. Instrument details are given in text, Section 4.10.

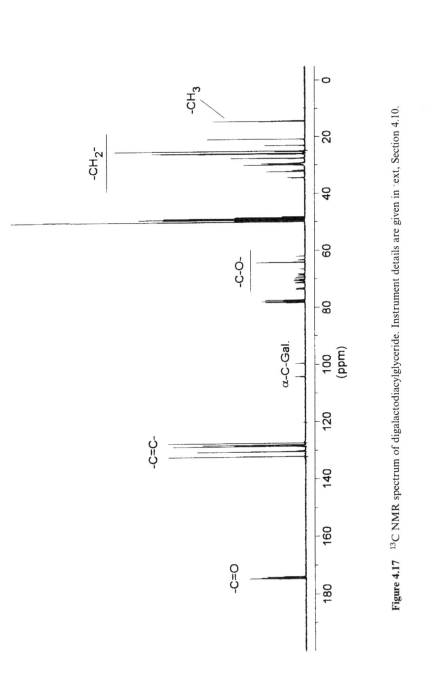

Figure 4.17 ^{13}C NMR spectrum of digalactodiacylglyceride. Instrument details are given in text, Section 4.10.

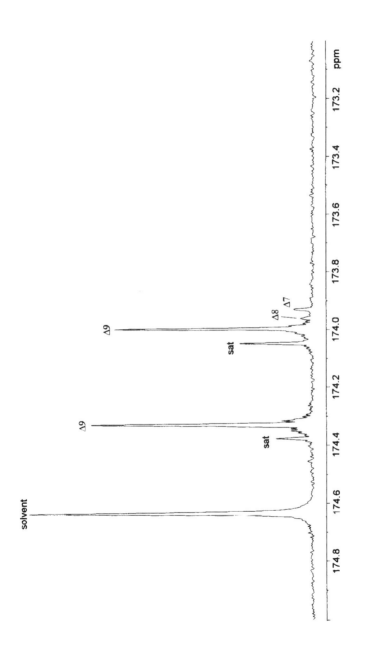

Figure 4.18 The carbonyl region of the digalactodiacylglyceride ^{13}C NMR spectrum. Instrument details are given in text, Section 4.10.

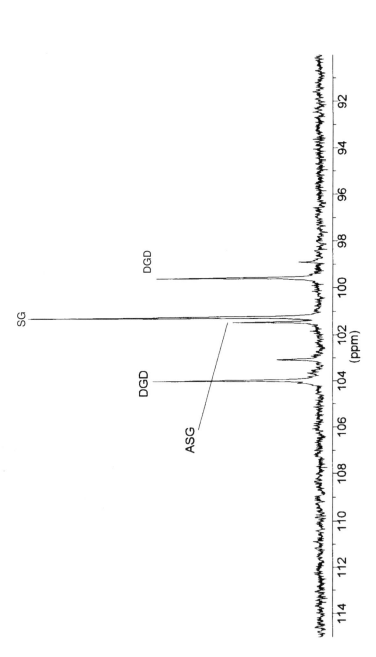

Figure 4.19 ^{13}C NMR spectrum of a glycolipid fraction. ASG = acylsitosterylglycoside; DGD = digalactosyldiacylglyceride; SG = sitosterylglycoside. Instrument details are given in text, Section 4.10.

acylsitosterylglycoside (ASG), sitosterylglycoside (SG) and digalactosyldiacylglyceride (DGD).

The three carbonyl spectra of egg lipids illustrated in Figs 4.20(a)–(c) demonstrate the pathway of ω-3 fatty acids in triglycerides and phospholipids. The eggs are enriched with ω-3 fatty acids by a special hen feed. Nearly all ω-3 fatty acids are found in the phospholipids, the highest amounts in phosphatidylethanolamine (PE) (Diehl *et al.*, 1996a).

4.8.3 ^{31}P NMR spectroscopy of phospholipids

Phospholipids are widespread in nature, in plants as well as in animals. There are medical applications of phospholipids such as infusions for parenteral nutrition or lung surfactants for prematurely born babies. In the food industry they are used as stabilizers for emulsions, for instant preparations or as auxiliary material in the baking process. The cosmetic industry uses phospholipids as emulsifiers and as liposomes to transport active ingredients into skin cells. Well-known phospholipids are phosphatidylcholine (PC) and phosphatidylethanolamine (PE). There are other phospholipids, and including their degradation products about 20 different phospholipids are known. The analysis of these compounds is difficult. The molecules have hydrophobic and hydrophilic parts. The resulting surface activity causes problems in the chromatography of phospholipids; additionally, there are problems with UV detection. Some HPLC methods have been developed, but chromatographic resolution and the dynamics of detection are not satisfactory. For each source of phospholipids special standards are needed because of the different distributions of fatty acids. These standards are expensive and in some cases they are not available. Another problem is represented by the analysis of phospholipids in complex matrices. In many cases separation is not possible or is very difficult because of the surface activity – a positive property in all applications but not in the analysis of these compounds. It is desirable to have a method which is selective in the detection of phospholipids in order to avoid the need to separate them from the matrix. ^{31}P NMR spectroscopy meets this requirement and is described elsewhere.* Within the ILPS (International Lecithin and Phospholipid Society) the ^{31}P NMR has become the reference method (De Kock, 1993, 1995; Diehl and Stein, 1994). It has been tested worldwide by round robin tests in comparison with different HPLC and TLC methods.

^{31}P NMR spectroscopy differentiates between the various phospholipids and is able to quantify very small amounts of phospholipids in the presence of main components. Only phosphorus-containing substances are detected

---

* For example, see Danz, 1993; Diehl, 1993; Diehl and Ockels, 1995a,c; Diehl, Ockels and Woydt, 1994; Genn, 1992; Glonek and Mercant, 1996; Meneses and Glonek, 1988; Sotirhos, Herslöf and Kenne, 1986; Woydt, 1994.

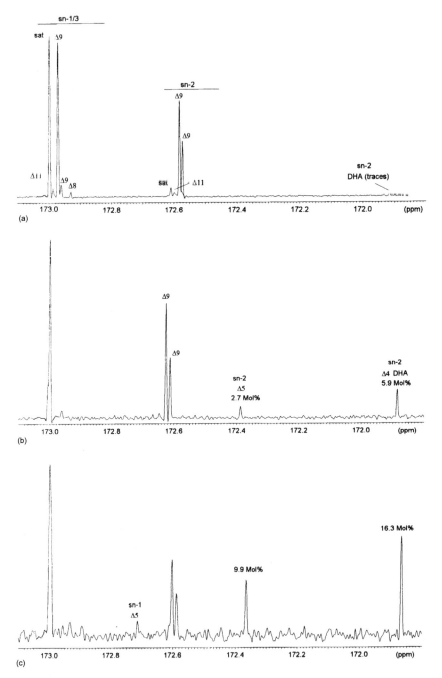

Figure 4.20 ^{13}C NMR spectra of egg lipids: (a) DHA egg oil; (b) phosphatidylcholine fraction of DHA egg oil; (c) phosphatidylethanolamine fraction of DHA egg oil. DHA = 4,7,10,13,16,19- docosahexaenoic acid. Instrument details are given in text, Section 4.10.

by this method, the analysis not being disturbed by other non-phosphorous components.

The resonance frequency of phosphorus depends slightly on the chemical environment within the molecule. Phospholipids with different chemical structures are therefore recorded at distinct frequencies. Frequency can be measured very precisely; even small differences in the chemical structure of phospholipids can be easily detected. Each phospholipid is represented by a single signal. Separation of the various phospholipids is not necessary, the phospholipids being characterized by their different resonance frequencies. Figure 4.21 shows a ^{31}P NMR spectrum of soy lecithin. The intensity of the signals represented by the area is a measure of the quantity. Each phosphorus atom contributes to the area; compounds with two phosphorus atoms produce double the peak area of a compound with one phosphorus atom. Quantitative NMR spectroscopy measures molar amounts. To get percentage by weight the molecular weight of a component has to be known. One measurement produces a qualitative result as well as the relative distribution of phospholipids in mol% related to the sum of all phospholipids in the sample (100% method). Using the internal standard triphenylphosphate (TPP) it is possible to determine the absolute amount of each phospholipid. The method does not need specific standards to cover the different phospholipid sources and the internal standard does not need to be a phospholipid at all. There are no problems with phospholipids from egg yolk, milk, rapeseed, soybean, blood or any other source. Phospholipids from animal sources sometimes contain plasmalogens and other etherlipids. By conventional methods it is not possible to detect them, but with ^{31}P NMR spectroscopy it is no problem. A known degradation product of PC is 2-lysophosphatidylcholine (2-LPC). It can be formed by enzymatic reaction or by chemical hydrolysis. In chemical hydrolysis 1-LPC is formed additionally to 2-LPC in a ratio of 1:9. This cannot be detected by conventional methods but is easily seen with ^{31}P NMR spectroscopy. Even the double lyso-derivative glycerophosphatidylcholine (GPC) is detectable.

The selective quantification of different degradation products makes ^{31}P NMR spectroscopy the ideal method for testing the stability of liposome-containing formulations in pharmaceutical or cosmetic products (Fig. 4.22).

4.8.4 Ultra-high-resolution ^{31}P NMR spectroscopy

The chemical shifts of the phosphorous atoms are very sensitive to changes in the chemical surroundings. To avoid hydrolysis the pH value has to be held between 5 and 9, which is the pK_s region of phospholipids. Even small changes in the pH causes shifts of 0.5–2 ppm. The chemical shift of PC is the most independent of pH because of the inner salt structure; phosphatidic

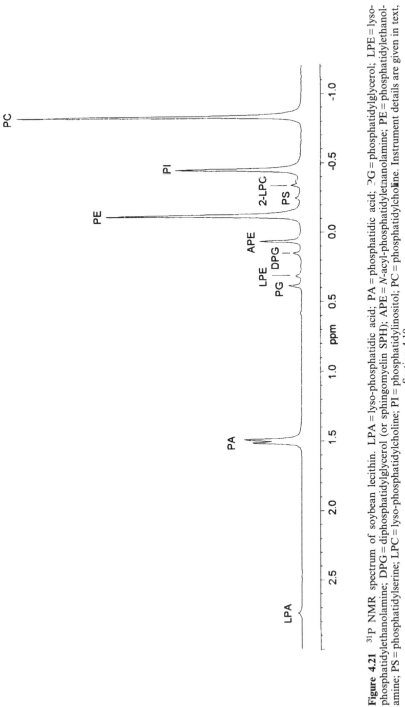

Figure 4.21 ³¹P NMR spectrum of soybean lecithin. LPA = lyso-phosphatidic acid; PA = phosphatidic acid; PG = phosphatidylglycerol; LPE = lyso-phosphatidylethanolamine; DPG = diphosphatidylglycerol (or sphingomyelin SPH); APE = N-acyl-phosphatidyletnanolamine; PE = phosphatidylethanolamine; PS = phosphatidylserine; LPC = lyso-phosphatidylcholine; PI = phosphatidylinositol; PC = phosphatidylcholine. Instrument details are given in text, Section 4.10.

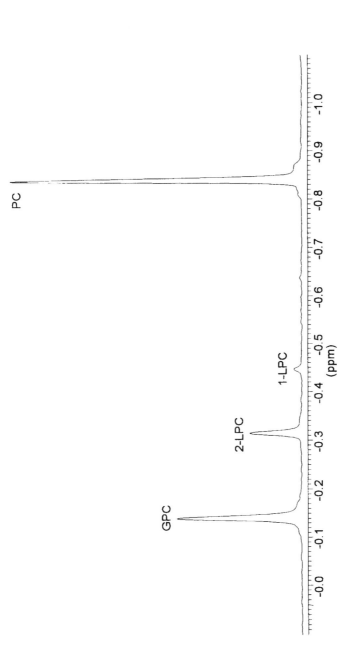

Figure 4.22 ^{31}P NMR spectrum of a liposome formulation containing phosphatidylcholine (PC) and degradation products 1-lyso-phosphatidylcholine (1-LPC), 2-lyso-phosphatidylcholine (2-LPC) and glycerophosphatidylcholine (GPC). Instrument details are given in text, Section 4.10.

acid (PA) (monoester) is the phospholipid most sensitive to pH.* The P—O structure in phospholipids is in a complex interaction with O—H groups forming intramolecular hydrogen bonds. This interaction leads to downfield shifts for the 1-lyso phospholipids, and greater downfield shifts for the 2-lyso phospholipids. In ultra-high-resolution ^{31}P NMR spectra an additional splitting can be observed (Fig. 4.23). This splitting represents the different fatty acid distribution (Hossein Nouri-Sorkhabi et al., 1996; Pearce and Komoroski, 1993). The splitting depends on the type of phospholipid. PA shows the largest effect, PC the smallest. Three major types of phospholipids are detectable: both fatty acids saturated (ss); both fatty acids unsaturated (uu); mixed (su or us).

In diester phospholipids the ss type appears upfield, in the monoester (PA) it appears downfield. At very low and very high pH the splitting goes to zero. The chemical shift itself is a function of the pH, which is not defined in organic solutions, and produces a type of titration curve. Within a diagram of the chemical shift, δ (ppm), versus the splitting (difference between su and uu) of PA from soybean (ppb) there is a maximum at $\delta = 1.6$ ppm relative to internal triphenylphosphate (TPP) at -17.8 ppm (Fig 4.24). The fatty acid splitting shows a maximum at the phospholipid pK_s value (Spectral Service, unpublished).

In the pK_s region the ^{31}P NMR spectrum of soybean PA shows additional splitting due to the combination of different saturated and unsaturated fatty acids (Fig. 4.25). In an ideal case at higher magnetic field strength each individual fatty acid combination in a phospholipid will give a single ^{31}P NMR signal.

4.8.5 Plasmalogens, O-alkyl phospholipids and O-alkenyl-phospholipids

Many phospholipid fractions from animal sources contain ether structures within the glycerol backbone of PE or PC (Fig. 4.26). These 1-O-alkyl-2-acyl or 1-O-alkenyl-2-acyl derivatives (Plasmalogen) can be separated in the ^{31}P NMR spectrum. In both PE and PC the corresponding ether lipids have their resonances downfield to the diacyl-phospholipid. Figure 4.27 shows the high-resolution PC region of bovine lung surfactant. The small signal at $\delta = 0.78$ ppm represents the 1-O-alkyl-2-acyl-PC. The small signals around the PC are caused by the ^{13}C satellites. In the spectrum of PC from bovine heart 1-O-alkenyl-2-acyl-PC is found as a third signal very close to the diacyl-PC. The difference in the chemical shift is only 15 ppb. The differences in chemical shifts of PE derivatives are larger than those observed for PC derivatives (Fig. 4.28).

---

* For example, see Danz, 1993; Dennis and Plückthun, 1984; Diehl, 1993; Diehl and Ockels, 1995a,c; Diehl, Ockels and Woydt, 1994; Genn, 1992; Glonek and Mercant, 1996; Mallet, Gaydou and Archavlis, 1990; Meneses and Glonek, 1988; Ng and Koh, 1988; Woydt, 1994.

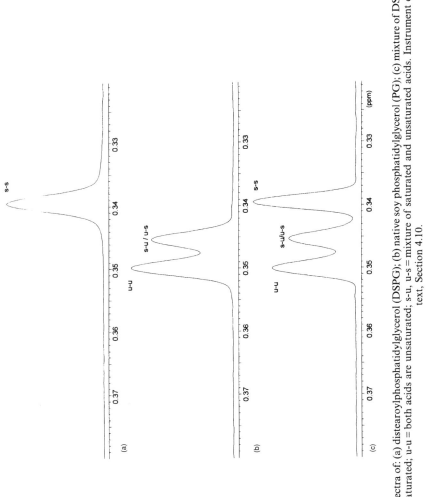

Figure 4.23 ^{31}P NMR spectra of: (a) distearoylphosphatidylglycerol (DSPG); (b) native soy phosphatidylglycerol (PG); (c) mixture of DSPG and native soy PG. s-s = both acids are saturated; u-u = both acids are unsaturated; s-u, u-s = mixture of saturated and unsaturated acids. Instrument details are given in text, Section 4.10.

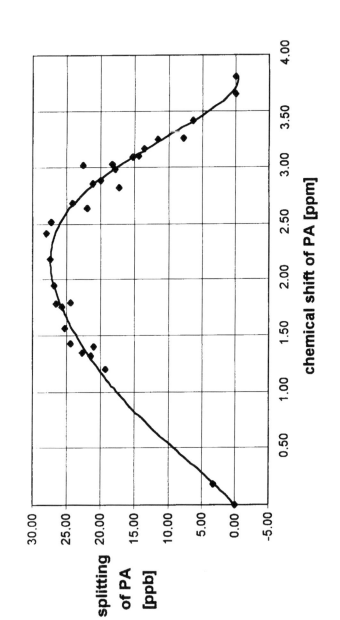

Figure 4.24 The pH dependence of phosphatidic acid splitting in ^{31}P NMR spectroscopy.

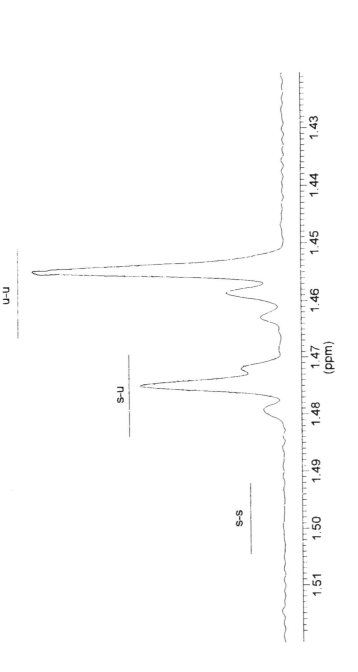

Figure 4.25 Ultra-high-resolution ^{31}P NMR spectrum of soybean phosphatidic acid. Abbreviations are as given in Fig. 4.23. Instrument details are given in text, Section 4.10.

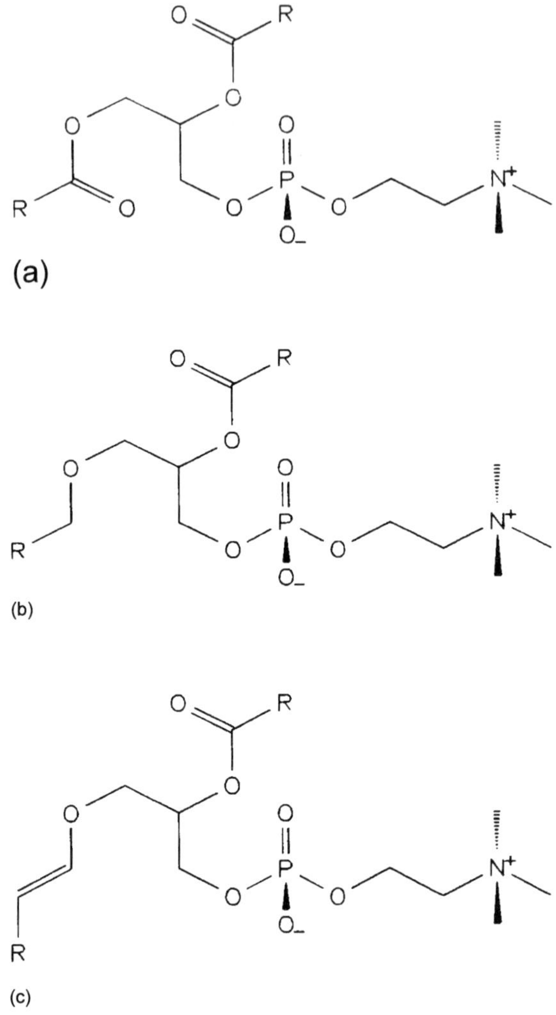

Figure 4.26 Phosphatidylglycerol and ether lipid derivatives: (a) phosphatidylcholine (PC); (b) 1-o-alkyl-PC; (c) 1-o-alkenyl-PC.

4.9 Quantitative determination of phospholipids: validation of the ^{31}P NMR method

A study has been made to show the validity of the quantitative determination of egg phospholipids in a pharmaceutical liposome preparation by means of ^{31}P NMR spectroscopy (Table 4.6; Diehl et al., 1996b). Selectivity and recovery depend on the matrix and have to be tested in each individual

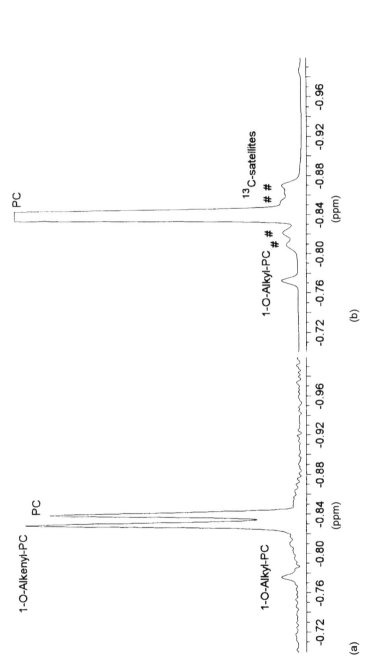

Figure 4.27 The phosphatidylcholine (PC) region of the ^{31}P NMR spectrum of (a) bovine heart phospholipids; (b) bovine lung surfactant. PC = phosphatidylcholine. # = signals caused by ^{13}C satellites. Instrument details are given in text, Section 4.10.

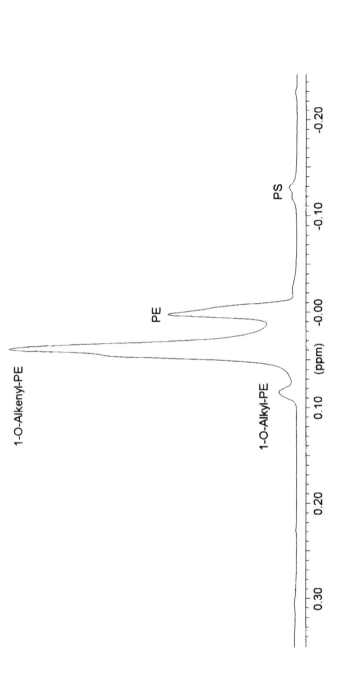

Figure 4.28 The phosphatidylethanolamine (PE) region of bovine lipids. PS = phosphatidylserine. Instrument details are given in text, Section 4.10.

Table 4.6 Subjects of validation

Test	Subject of validation	Description
A	Reproducibility	Seven independent sample preparations
B	Instrument precision	Six measurements of the same sample
C	Repeatability	Three sample preparations on two different days by two different persons
	Robustness:	
D	Variation of pH value	Increase pH from 6.5 to 9.5 in steps of 0.5
E	Variation of the ternary solvent mixture	
F	Amount of sample	Ratio of sample to standard
G	Sample stability	Time between sample preparation and measurement
H	Variation of the number of scans	Dependence on acquisition parameters
I	Data evaluation	Automatic or manual evaluation; different software

case with the material in use. Test substances were pharmaceutical liposome lyophilisate preparations of an egg PC (Fig. 4.29).

4.9.1 Test A: reproducibility

To determine the typical standard deviation for the process with respect to the main component and by-products seven independent samples were prepared under standard conditions.

The percentage standard deviation for the whole process was determined by manual integration with use of the software WIN-NMR to be 0.67% in the case of PC; for sphingomyelin (SPH), which is near the limits of quantitation, the percentage standard deviation was found to be 2.6%.

4.9.2 Test B: instrument precision

To determine the instrument precision the same sample was measured six times in succession. The percentage standard deviation was determined by manual integration with use of the software WIN-NMR to be 0.79% for PC; for SPH, which is near the limits of quantitation, the percentage standard deviation was found to be 3.1%.

4.9.3 Test C: repeatability

To determine the repeatability, on two different days three independent samples were measured; one person prepared the first set of three samples on the one day, and a different person prepared the second set of samples on the other day. Both data groups (C1 and C2, respectively) were compared and no significant deviation was found.

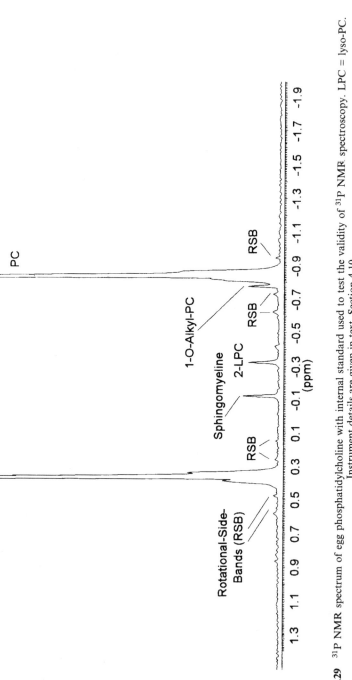

Figure 4.29 ^{31}P NMR spectrum of egg phosphatidylcholine with internal standard used to test the validity of ^{31}P NMR spectroscopy. LPC = lyso-PC. Instrument details are given in text, Section 4.10.

4.9.4 Tests D–I: robustness

Test D: Variation of pH value. Seven samples were prepared according to the standard operational procedure, but the pH value was varied by addition of Na_2CO_3 or HCl.

Within the given limits, variation of the pH value had no influence on the results. The recommended pH value is 8.5, values above 9.5 and below 6.5 lead to hydrolysis of the phospholipids. Additionally, the line shape of the NMR signals broadened under the more acidic conditions and the chemical shift showed large variations.

Test E: Variation of the ternary solvent mixture. According to the standard operation procedure for sample preparation equal amounts (by volume) of $CDCl_3$, methanol and Cs-EDTA (caesium ethylenediaminetetraacetic acid) solution were used. Cs-EDTA was held constant at 1 ml, $CDCl_3$ was changed from 1 ml to 0.6 ml and MeOH from 1 ml to 1.4 ml.

There was no significant influence on the results of changing the composition of the solvent mixture within certain limits. Further changes can lead to line broadening and/or to reduced extraction yield.

Test F: Variation of the amount of sample. The amount of sample was varied from 60 mg to 140 mg, and the amount of internal standard between 20 mg and 40 mg, taking care that even extreme ratios were covered, from 6:4 to 7:2.

The mean value and the standard deviation indicated that there was no significant influence on the results of varying the amount of sample.

Test G: Sample stability. Samples are set into an automated sample changer after preparation and measured one by one. The sample changer holds up to 120 samples; with 20 min measuring time (256 scans) it can take up to 40 h before the last sample has been measured. Therefore a check was made to see whether the time between sample preparation and measurement has an influence on the result. The last measurement was made more than 120 h after sample preparation.

No trend was observed, the results indicating scattering similar to that of all other measured series. Samples were found to be stable for at least 120 h after preparation. Further tests showed that signals caused by lyso-PC near the limit of detection appeared after 3 weeks.

The values of PC and SPH content as determined by ^{31}P NMR spectroscopy for tests A–G are summarized in Table 4.7.

Test H: Variation of the number of accumulated scans. According to the standard operation procedure the number of scans should be greater than 16. In most cases 256 scans are used in routine work. The number of scans has an influence on the signal-to-noise ratio (S/N) and thus on the limits of

Table 4.7 Summary of validation experiments: phosphatidyl (PC) and sphingomyelin (SPH) content in pure egg PC determined over n samples by means of ^{31}P NMR spectroscopy. Manual integration with use of software WIN-NMR was used to determine PC and SPH content. Instrument details are given in text, section 4.10

Test	n	PC (wt%)	SPH (wt%)
A	7	33.248 ± 0.224	0.162 ± 0.004
B	6	33.350 ± 0.265	0.164 ± 0.005
C1	3	33.406 ± 0.243	0.162 ± 0.003
C2	3	33.093 ± 0.205	0.161 ± 0.007
D	7	33.255 ± 0.230	0.158 ± 0.008
E	5	33.233 ± 0.207	0.161 ± 0.007
F	7	33.427 ± 0.196	0.167 ± 0.005
G	5	33.183 ± 0.080	0.158 ± 0.007

detection and quantitation. In general S/N is a square root function of the number of scans; for example to double S/N the number of scans must be increased fourfold. To show the effect the same sample (pure egg PC) was measured eight times, each time doubling the number of scans, except in the last case owing to lack of time. The last measurement was stopped after 6500 pulses (5.5 h). The results are shown in Table 4.8.

In manual integration processing under WIN-NMR variation of number of scans was found to influence S/N only and subsequently the limit of quantitation.

Test I: Integration using different software. All measurements were evaluated (integrated) with the dedicated software DIS-NMR as well as with WIN-NMR in Windows 3.11, which is produced by the spectrometer

Table 4.8 The effect of varying the number of scans on the signal-to-noise ratio (S/N) and on the quantitation of phosphatidylcholine (PC) and sphingomyelin (SPH) content in pure egg PC as determined by ^{31}P NMR spectroscopy. The same sample of pure egg PC was used for each set of measurements. Manual integration (WIN-NMR) was used to determine PC and SPH content. Instrument details are given in text, Section 4.10

Number of scans	PC (wt%)	SPH (wt%)	S/N (SPH signal)
64	94.674	0.450	6.0
128	94.831	0.446	8.5
256	94.852	0.453	10.0
512	94.904	0.465	13.0
1024	94.691	0.441	16.0
2048	94.764	0.469	21.0
4096	94.721	0.444	32.0
6484	94.550	0.468	40.0
mean value	94.748	0.454	
standard deviation	0.107	0.010	

Table 4.9 Comparison of integration procedures in the determination of phosphalidylcholine (PC) and sphingomyelin (SPH) content in pure egg PC. Instrument details are given in text, Section 4.10

Procedure	PC (wt%)	SPH (wt%)
DIS-NMR	33.18 ± 0.20	0.154 ± 0.024
WIN-NMR	33.27 ± 0.22	0.162 ± 0.007

manufacturer. DIS-NMR was used in automatic mode, whereas spectra were integrated manually with WIN-NMR. It was found that DIS-NMR produced significantly lower values than did WIN-NMR. With small random tests the difference is covered up by the scattering, but with larger random tests (>30) the statistical analysis shows a significant deviation (Table 4.9).

The automatic integration led to erroneous results, particularly in cases of more than 512 scans. The integration limits are influenced by S/N; with a high S/N too much of the slope of the NMR signal (Lorentz curve) is added to the integration, leading to increased values. With correct choice of integration limits the results are statistically valid. In routine work 256 pulses are used, so manual as well as automatic integration leads to correct results.

4.9.5 Validation summary

The quantitation of phospholipids by means of ^{31}P NMR spectroscopy is a valid method. Method precision (reproducibility) and instrument precision show for the main component a standard deviation <1%; components near the limit of quantitation have a standard deviation below 5%. In terms of the liposome lyophilisate preparation used the extraction recovery of the phospholipids was 100%.

The analysis method is robust. It is almost insensitive to variation in sample preparation and measurement parameters. Prepared samples can be stored up to 5 days at room temperature (a typical storage time in practical experimental applications would be 24 h) before measurement without negative influence on the results. Manual integration is the preferred method and control of integration limits is recommended if automatic integration is to be used.

4.10 Instrument details

All NMR spectra were measured at Spectral Service GmbH, Cologne, Germany, on an NMR Spectrometer AC-P 300 (Bruker, Karlsruhe, Germany) equipped with automated sample changer and QNP-head for nuclei ^{1}H, ^{13}C, ^{19}F and ^{31}P. Magnetic flux density is 7.05 T; proton frequency is 300.135 MHz; data processing was by Bruker WIN-NMR 5.0 software for PC.

Acknowledgements

This work was financially supported by Spectral Service, Cologne (Germany). I would like to thank my associate Werner Ockels and my coworkers Helmut Herling and Ricarda Unger and Stephanie Winkler, all Spectral Service, Claus C. Becker from Center of Food Research, Steen Balchen from Danish Ministery of Fishery, both Lyngby (Denmark) for making available some fish and seal oils, and Prof. K. Aitzetmüller, Münster (Germany) for the euonymus oil fractions.

References

Aitzetmüller, K., Diehl, B. W. K., Ockels, W. and Herling, H. (1994) Investigation of seed oils containing Δ-5 and Δ-6 fatty acids using ^{13}C-NMR-spectroscopy. Joint Congress of 2nd EUROLIPID and 50th DGF Conference, Münster, Germany, September; copy available from author.
Aitzetmüller, K. and Spectral Service, in production; copy available from author.
Aursand, M. and Grasdalen, H. (1992) Interpretation of the ^{13}C-NMR spectra of omega-3 fatty acids and lipid extracted from the white muscle of Atlantic salmon (*Salmo salar*). *Chemistry and Physics of Lipids*, **62**, 239–51.
Aursand, M., Jørgensen, L. and Grasdalen, H. (1995) Positional distribution of ω-3 fatty acids in marine lipid triacylglycerols by high-resolution ^{13}C-nuclear Magnetic resonance spectroscopy. *J. Am. Oil Chem. Soc.*, **72** (3), 293–7.
Danz, B. (1993) *Storage Stability of Liposomes*. Diploma Thesis, Fachhochschule Magdeburg, Germany.
De Kock, J. (1993) The European Analytical Subgroup of I.L.P.S. – a joint effort to clarify lecithin and phospholipid analysis. *Fat Sci. Technol.*, **95** (9), 352–5.
De Kock, J. (1995) The European Analytical Subgroup of I.L.P.S., in *Characterization, Metabolism and Novel Biological Applications* (eds G. Cevc and F. Paltauf), AOCS Press, Champaign, pp. 22–28.
Dennis, E. A. and Plückthun, A. (1984) Phosphorus-31 NMR of phospholipids in micelles, in *Phosphorus-31 NMR, Principles and Applications*, Academic Press, New York, pp. 423–46.
Diehl, B. W. K. (1993) ^{1}H-, ^{13}C- and ^{31}P-nuclear magnetic resonance spectroscopy of lipids. AOCS Meeting 1993, Anaheim, Inform 4, 513; copy available from author.
Diehl, B. W. K. and Ockels, W. (1995a) Quantitative analysis of emulsifiers by multinuclear high resolution NMR. 21st World Congress and Exhibition of the International Society for Fat Research, The Hague, Netherlands, October 1995; copy available from author.
Diehl, B. W. K. and Ockels, W. (1995b) Fatty acid distribution by ^{13}C-NMR-spectroscopy. *Fat Scientific Technology*, **97** (3), 115–18.
Diehl, B. W. K. and Ockels, W. (1995c) Quantitative analysis of phospholipids, in *Proceeding of the 6th International Colloquium Phospholipids. Characterization, Metabolism and Novel Biological Applications* (eds G. Cevc and F. Paltauf), AOCS Press, Champaign, pp. 29–32.
Diehl, B. W. K. and Stein, J. (1994) Calibration standards for lecithin analysis using HPLC with light scattering detector. 84th AOCS Annual Meeting and Exposition, Atlanta, GA, May 1994; copy available from author.
Diehl, B. W. K., Ockels, W. and Woydt, D. (1994) Kinetics in lecithin degradation. 84th AOCS Annual Meeting and Exposition, Atlanta, GA, May; copy available from author.

Diehl, B. W. K., Herling, H., Heinz, E. and Riedl, I. (1995) ^{13}C-NMR analysis of the positional distribution of fatty acids in plant glycolipids. *Chemistry and Physics of Lipids*, **77**, 147–53.
Diehl, B. W. K., Ockels, W., Herling, H. and Unger, R. (1996a) Das DHA Ei, der Weg der ω-3 Fettsäuren, Multinukleare NMR-Untersuchungen, 51. *Jahrestagung der Deutschen Gesellschaft für Fettwissenschaft*, Bremen, October; copy available from author.
Diehl, B. W. K., Ockels, W., Herling, H. *et al.* (1996b) Quantitative determination of egglecithin in a liposome-preparation, validation of the ^{31}P-NMR-method. 7th International Congress on Phospholipids, Brussels, September; copy available from author.
Genn, B. (1992) *Identification of Phospholipids in Bovine Lung Surfactant*. Diploma Thesis. Fachhochschule Aachen, Abteilung Jülich, Germany.
Glonek, T. and Mercant, T. E. (1996) ^{31}P-nuclear magnetic resonance profiling of phospholipids, in *Advances in Lipid Methodology – Three* (ed. W. W. Christie), The Oily Press, Dundee, pp. 37–75.
Gunstone, F. D. (1990) ^{13}C-NMR spectra of oils containing γ-linolenic acid. *Chemistry and Physics of Lipids*, **56**, 201–7.
Gunstone, F. D. (1991) High resolution NMR studies of fish oils. *Chemistry and Physics of Lipids*, **59**, 83–9.
Gunstone, F. D. (1993a) High resolution ^{13}C-NMR-spectroscopy of lipids, in *Advances in Lipid Methodology – Two* (ed. W. W. Christie), The Oily Press, Dundee, pp. 1–68.
Gunstone, F. D. (1993b) The ^{13}C-NMR spectra of six oils containing petroselinic acid and of aquilegia oil and meadowfoam oil which contain Δ5 acids. *Chemistry and Physics of Lipids*, **58**, 159–67.
Herling, H. (1995) *Fatty Acid Distribution in Lipids*. Diploma Thesis, FH Aachen, Abt. Jülich, Germany; copy available from author.
Hossein Nouri-Sorkhabi, M., Wright, L. C., Sullivan, D. R. and Kuchel, P. W. (1996) Quantitative ^{31}P nuclear magnetic resonance analysis of erythrocyte membranes using detergent. *Lipids*, **31**, 765–70.
Mallet, J. F., Gaydou, E. M. and Archavlis, A. (1990) Determination of petroselinic acid in umbelliflorae seed oils by combined GC and ^{13}C-NMR spectroscopy analysis. *J. Am. Oil Chem. Soc.*, **67** (10), 607–10.
Manz, I. and Schneider, K. (1995) Bestimmung der Jodzahl in epoxidiertem Sojabohnenöl mit automatisierter ^{1}H-NMR Spektroskopie. *GIT Fachzeitschrift für Laboratorien*, **3**, 197–9.
Meneses, P. and Glonek, T. (1988) High resolution ^{31}P-NMR-spectroscopy of extracted phospholipids. *Journal of Lipid Research*, **29**, 679–89.
Ng, S. (1985) Analysis of positional distribution of fatty acids in palm oil by ^{13}C-NMR spectroscopy. *Lipids*, **20**, 778–82.
Ng, S. and Koh, H. F. (1988) Detection of *cis*-vaccenic acid in palm oil by ^{13}C-NMR spectroscopy. *Lipids*, **23**, 140–3.
Pearce, J. M. and Komoroski, R. A. (1993) *Magnetic Resonance in Medicine*, **29**, 724–31.
Pfeffer, P. E., Luddy, F. E. and Unruh, J. (1977) ^{13}C-NMR: a rapid, nondestructive method for determining the *cis*, *trans* composition of catalytically treated unsaturated lipid mixtures. *JOACS*, **54** (9), 380–6.
Sacchi, P., Addeo, F., Giudicianni, I. and Paolillo, L. (1990) Applicazione della spettroscopia ^{13}C-NMR alla determinazione di mono-digliceridi ed acidi grassi liberi nell' olio di oliva di pressione e nei rettificati. *La Rivista Italiana delle sostanze grasse*, **LXVII**, 245–52.
Shoolery, J. N. (1976) Some quantitative applications of ^{13}C-NMR spectroscopy. *Progress in Nuclear Resonance Spectroscopy*, **11**(2), 79–93.
Sotirhos, N., Herslöf, B. and Kenne, L. (1986) Quantitative analysis of phospholipids by ^{31}P-NMR. *Journal of Lipid Research*, **27**, 386–92.
Wollenberg, K. F. (1990) Quantitative high resolution ^{13}C nuclear magnetic resonance of the olefinic and carbonyl carbons of edible vegetable oils. *JAOCS*, **67**, (8), 487–94.
Woydt, D. (1994) *Kinetics in Lecithin Degradation*. Diploma Thesis. Universität Osnabrück, Germany.

5 Cyclic fatty acids: qualitative and quantitative analysis
G. DOBSON

5.1 Introduction

Cyclic fatty acids can be classified into those that are naturally occurring and those that are formed in vegetable oils during heating. The former include cyclopropane, cyclopropene and cyclopentenyl acids. C17 (structure Ia, Fig. 5.1) and C19 (structure Ib) cyclopropane acids are common in many bacteria, for example Lactobacilli and enterobacteria, and mycolic (2-alkyl-3-hydroxy) acids (structure II), with up to about 90 carbons and one or two cyclopropane rings, occur in mycobacteria (Christie, 1970; Minnikin, 1978). Recently, several acids with up to 26 carbons, one or two cyclopropane rings and a double bond in the $\Delta 5$ position were identified in an invertebrate from a deep-water lake (Rezanka and Dembitsky, 1994). C18 (structure Ic) and C19 (structure Id) cyclopropane acids occur in varying amounts in the seed oils of some species of a few plant families including *Malvaceae* and *Sterculiaceae* (Christie, 1970; Sebedio and Grandgirard, 1989). The cyclopropene counterparts (structures IIIa and IIIb) are more widespread in these families, and cyclopentenyl acids (structures IV a–c) are present in the seed oils of the family *Flacourtiaceae*, notably the genus *Hydnocarpus* (Sebedio and Grandgirard, 1989). Fatty acids with six-membered (structure Va; Hippchen, Roell and Poralla, 1981) and, unusually, seven-membered (structure Vb; Poralla and Koenig, 1983) rings have been characterized from the thermoacidophilic bacterium, *Bacillus acidocaldarius*. Complex mixtures of five-membered and six-membered ring acids (Fig. 5.2) are formed in vegetable oils when heated, such as during frying (Le Quere and Sebedio, 1996; Sebedio and Grandgirard, 1989).

The analysis of cyclopropane acids was reviewed by Minnikin (1978) and by Christie (1970), who also covered cyclopropene acids. Subsequently, Sebedio and Grandgirard (1989) discussed naturally occuring cyclic acids and cyclic acids formed in heated vegetable oils. Cyclic acids were the subject of a recent review (Le Quere and Sebedio, 1996). Useful information on gas chromatography (GC), mass spectrometry (MS) and high-pressure liquid chromatography (HPLC) of cyclic fatty acids can be found in books by Christie (1987, 1989). In the present chapter the aim is to concentrate on the more recent literature and those publications considered to be particularly

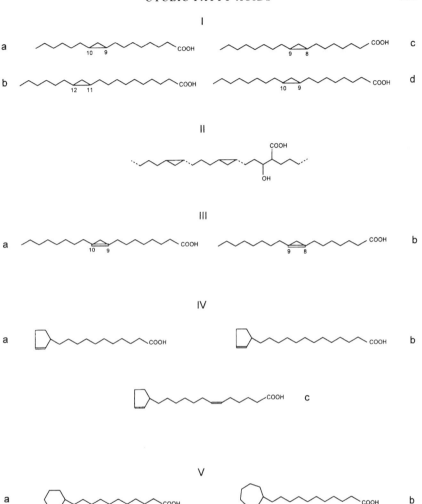

Figure 5.1 Structures of naturally occurring cyclic fatty acids. I, Cyclopropane acids: a, 9,10-methylenehexadecanoic acid; b, 11,12-methyleneoctadecanoic (lactobacillic) acid; c, 8,9-methyleneheptadecanoic (dihydromalvalic) acid; d, 9,10-methyleneoctadecanoic (dihydrosterculic) acid. II, Mycolic (2-alkyl-3-hydroxy) acid. III, Cyclopropene acids: a, 9,10-methyleneoctadec-9-enoic (sterculic) acid; b, 8,9-methyleneheptadec-8-enoic (malvalic) acid. IV, Cyclopentenyl acids: a, 11-cyclopent-2-enyl-undecanoic (hydnocarpic) acid; b, 13-cyclopent-2-enyl-tridecanoic (chaulmoogric) acid; c, 13-cyclopent-2-enyl-tridec-6-enoic (gorlic) acid. V: a, 11-cyclohexyl-undecanoic acid; b, 11-cycloheptylundecanoic acid.

relevant to a contemporary analytical approach. Those procedures which are deemed to be important will be dealt with in some detail. The considerable amount of space devoted to cyclic acids formed in heated oils, in contrast to naturally occurring cyclic fatty acids, reflects recent advances in this area.

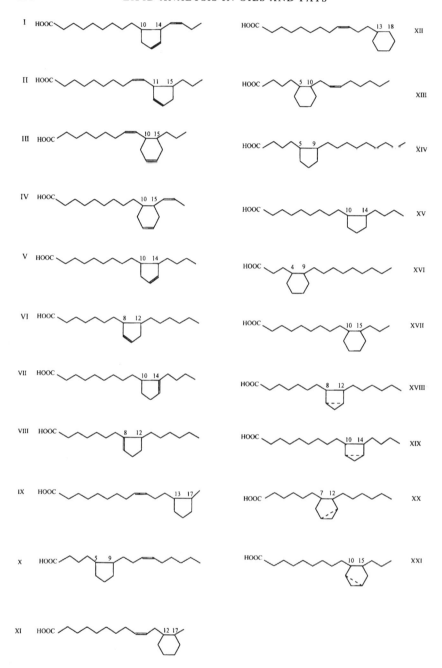

Figure 5.2 Monomeric cyclic fatty acids derived from heated vegetable oils. I–IV, Cyclic dienes from α-linolenate; V–XIII, cyclic monoenes from linoleate; XIV–XVII, saturated monocyclic acids from oleate; XVIII–XXI, saturated bicyclic acids from linoleate. The dashed lines indicate that the positions of the bonds within the rings have not been determined.

5.2 Naturally occurring cyclic fatty acids

5.2.1 Cyclopentenyl fatty acids

In this category, fatty acids with a terminal cyclopentenyl ring are best known (Sebedio and Grandgirard, 1989). There are three main types; hydnocarpic (11-cyclopent-2-enyl-undecanoic) acid (structure IVa, Fig. 5.1), chaulmoogric (13-cyclopent-2-enyl-tridecanoic) acid (structure IVb) and gorlic (13-cyclopent-2-enyl-tridec-6-enoic) acid (structure IVc) and its positional isomers (13-cyclopent-2-enyl-tridec-4-enoate and 13-cyclopent-2-enyl-tridec-9-enoate). They usually occur in varying proportions as major components (up to 90% of total fatty acids) in seed oils along with smaller amounts of saturated, oleic and linoleic acids.

As methyl esters, GC separation from other fatty acids can be achieved on polar capillary columns (Fig. 5.3; Christie, Brechany and Shukla, 1989) and would probably be adequate on non-polar columns, as this is possible for dimethyloxazoline (DMOX) derivatives (Zhang et al., 1989). If necessary, prior isolation of cyclic monoenoic and dienoic fractions, separated from straight-chain saturates, monoenes and dienes, may be obtained by means of silver-ion HPLC (Christie, Brechany and Shukla, 1989). In this way, minor components were concentrated for subsequent GC–MS analysis as the picolinyl (3-hydroxymethylpyridinyl) esters, and the possibility of inadequate resolution from straight-chain esters on non-polar columns, necessary for eluting these relatively involatile derivatives, was avoided. Presumably the use of modern high-temperature polar phases for GC–MS would eliminate possible resolution problems with picolinyl esters.

At this point it is worth spending some time considering the application of GC–MS analyses of picolinyl esters and DMOX derivatives for elucidating fatty acid structure (Dobson and Christie, 1996; Harvey, 1992). As will become clearer in later sections these derivatives are extremely useful as an aid to structural determination not only of cyclopentenyl acids but also of other cyclic fatty acids. One of the most widely used applications of this approach is in the location of double bond position. In the mass spectra of the methyl esters of unsaturated fatty acids diagnostic ions for the position of double bonds are absent because the double bond is ionized, resulting in double bond migration. Conversion of the carboxylic acid group to a picolinyl ester or DMOX derivative, which contain nitrogen functions that are highly favourable charge-sites, minimizes double bond ionization and migration. Simple radical-induced cleavage occurs uniformly along the chain, resulting in a series of ions from the cleavage of each C–C bond, corresponding to fragments containing the derivatized acid group. A spacing of 14 atomic mass units (amu) between consecutive ions is observed for a saturated straight chain. Diagnostic ions occur whenever the aliphatic chain deviates from the normal multimethylene structure. For example, when a double bond is encountered one may observe a spacing of 12 amu (betweenz

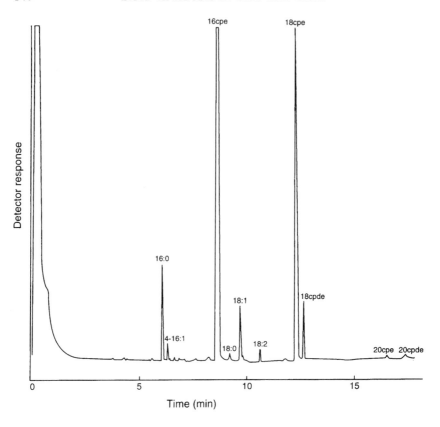

Figure 5.3 Gas chromatogram of fatty acid methyl esters of *Hydnocarpus anthelmintica* on a Silar 5CP 9 (Chrompack UK Ltd, London) (25 m × 0.25 mm internal diameter) capillary column. Temperature was held at 155°C for 3 min, programmed to 195°C at 4°C min^{-1} and held at this point for a further 17 min. Abbreviations: 16cpe, 11-cyclopent-2-enyl-undecanoate; 18cpe, 13-cyclopent-2-enyl-tridecanaote; 18cpde, 13-cyclopent-2-enyl-tridec-4-enoate; 20cpe, 15-cyclopent-2-enyl-pentadecanoate; 20cpde, 15-cyclopent-2-enyl-pentadec-9-enoate. Redrawn from Christie, W. W., Brechany, E. Y. and Shukla, V. K. S., Analysis of seed oils containing cyclopentenyl fatty acids by combined chromatographic procedures, *Lipids*, **24**(2), 116–20, 1989.

the ions corresponding to a fragment of $n-1$ and n carbons, if the double bond is between carbon atoms n and $n+1$) and/or 26 amu (between the ions for a fragment of $n-1$ and $n+1$ carbons); in practice, the 12 amu spacing is often observed for DMOX derivatives but the 26 amu spacing is usually more diagnostic for picolinyl esters. The relative abundance of other characteristic ions, such as allylic cleavage ions for monoenes, are also helpful. In a similar way, other functional groups cause a disruption in the normal pattern and give rise to characteristic fragmentation patterns which often allow the type and position of the function to be determined.

A terminal cyclopentenyl group is evident in the mass spectra of the methyl esters. For example, for methyl hydnocarpate the base peak at *m/z* (mass per charge ratio) 67 was due to the cyclopentenyl ring, and ions at m/z 82 and 185 resulted from cleavage β to the ring (Christie, Rebello and Holman, 1969). When present, double bonds in the chain cannot be located by GC–MS for the methyl esters. Unambiguous identifications of gorlic acid and positional isomers have been readily achieved by GC–MS of picolinyl esters (Christie, Brechany and Shukla, 1989) and DMOX derivatives (Zhang et al., 1989).

In the mass spectra of picolinyl esters (Christie, Brechany and Shukla, 1989) there was a base peak at m/z 67 (cyclopentenyl group) and a strong ion at m/z 302 for the other part of the molecule. In gorlic acid, for example, there was a gap of 26 amu between m/z 192 and 218 to locate the position of the double bond at C-6 (Fig. 5.4). The position of the double bond in the ring could not be determined from the mass spectrum, but the effect of position of double bond in the ring on the pattern of ions between m/z 302 and the molecular ion has not been studied. Deviations from the expected spectra of picolinyl esters occur as the double bond becomes closer to the extremities of the molecule, but diagnostic ions are still often present. In the mass spectrum of the picolinyl ester of the C-4 isomer there were no ions indicative of fragmentation at the double bond. However, a strong ion at m/z 151 was characteristic of straight-chain picolinyl esters with an isolated double bond at C-4. The structure of the isomer was confirmed from the mass spectrum of the bis-dimethyl disulphide adduct.

In a study of DMOX derivatives of cyclopentenyl acids (Zhang et al., 1989), the possible m/z 67 ion was not recorded and, in contrast to the spectra of picolinyl esters, the gap between the molecular M^+ and $[M-67]^+$ ions was not always obvious (e.g. from m/z 331 to 264 for the 4-isomer; Fig. 5.5). However, it appears that the combination of gaps of 12 amu between $[M-67]^+$ and $[M-55]^+$ (m/z 276) and $[M-43]^+$ (m/z 288) ions are characteristic for DMOX derivatives of cyclopentenyl acids. For the 4-isomer, a gap of 26 amu between m/z 126 and 152 together with an unusual odd-number ion at m/z 139 located the double bond position.

Separation of cyclopentenyl-containing triacylglycerols is possible by reversed-phase HPLC using acetonitrile/tetrahydrofuran (68:32 vol./vol.) as the mobile phase and ultraviolet (UV) detection at 220 nm (Shukla and Spener, 1985). The seed oils of *Caloncoba echinata* contained mainly chaulmoogric acid (C) and less gorlic acid (G), and the major triacylglycerols were in decreasing order of amount CCG, CCC, CGG and PCC (where P = palmitate), whereas in *Hydnocarpus anthelminthica*, hydnocarpic acid (H) was predominant, there was less chaulmoogric acid and, correspondingly, the triacylglycerols were in decreasing order of amount HHH, HHC and HCC.

5.2.2 Cyclopropane fatty acids

The two common cyclopropane acids present in bacteria are 9,10-methylenehexadecanoic acid (structure Ia, Fig. 5.1) and 11,12-methyleneoctadecanoic

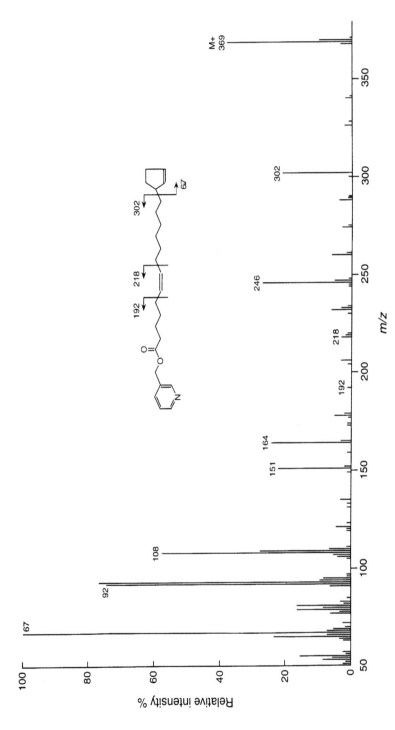

Figure 5.4 Mass spectrum (70 eV) of a picolinyl ester derivative of 13-cyclopent-2-enyl-tridec-6-enoic (gorlic) acid. M^+ = molecular ion. Redrawn from Christie, W. W., Brechany, E. Y. and Shukla, V. K. S., Analysis of seed oils containing cyclopentenyl fatty acids by combined chromatographic procedures, *Lipids*, **24**(2), 116–20, 1989.

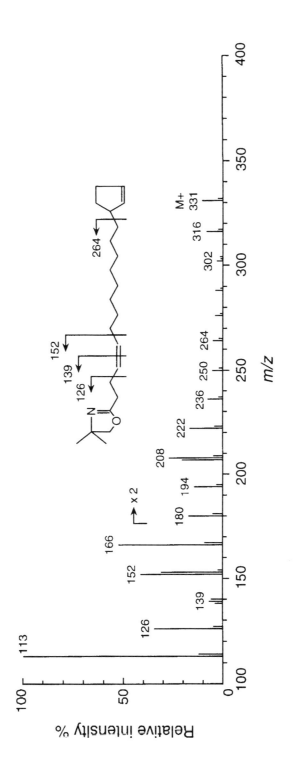

Figure 5.5 Mass spectrum (70 eV) of a dimethyloxazoline derivative of 13-cyclopent-2-enyl-tridec-4-enoate. M⁺ = molecular ion. Redrawn from Zhang, J. Y. Wang, H. Y., Yu, Q. T. et al., The structures of cyclopentenyl fatty acids in the seed oils of *Flacourtiaceae* species by GC–MS of their 4,4-dimethyloxazoline derivatives, *J. Am. Oil Chem. Soc.*, **66**(2), 242–6, 1989.

(lactobacillic) acid (structure Ib), whereas those in plants are 8,9-methyleneheptadecanoic (dihydromalvalic) acid (structure Ic) and 9,10-methyleneoctadecanoic (dihydrosterculic) acid (structure Id) (Christie, 1970). Care has to be taken during isolation and methylation of cyclopropane acids because they are degraded by acidic conditions and, consequently, alkaline hydrolysis/esterification methods should be used (Christie, 1989). However, bacterial material may also contain amide-linked hydroxy fatty acids, which are more efficiently released by acid-catalysed procedures. Recently, a one-tube method involving sequential alkaline–acid hydrolyses followed by mild acid methanolysis gave high yields of both acid types (Mayberry and Lane, 1993).

Early methods for the identification of cyclopropane acids involved either vigorous hydrogenation (forming a straight-chain and two methyl-branched esters) or reaction with boron trifluoride-methanol (forming six isomeric methoxy esters) which could be characterized by GC–MS (Christie, 1970; Minnikin, 1978). Recently, MS was used to show that the seed oil of litchi (*Litchi chinensis* Sonn.) contained dihydrosterculic acid and smaller amounts of lower homologues, notably 7,8-methylenehexadecanoic acid (Gaydou, Ralaimanarivo and Bianchini, 1993). Under specific conditions selective hydrogenation of unsaturated fatty acids could be achieved without degradation of cyclopropane acids (Bussell and Miller, 1979).

Qualitative and quantitative GC analyses of fatty acid methyl ester mixtures, containing cyclopropane esters, are usually straightforward. A common analysis is to examine phospholipid fatty acid methyl esters from environmental samples such as soil as markers of microbial populations (Frostegard, Tunlid and Baath, 1993; Zelles and Bai, 1994). The major fatty acids were saturated straight-chain and branched-chain, cyclopropyl, monoenoic and dienoic, with little or no polyenoic acids. Long (50 m) columns with non-polar (5% phenylmethyl silicone) stationary phases were preferred and cyclopropane acids were resolved from their monounsaturated counterparts, although not always to baseline (Frostegard, Tunlid and Baath 1993). It is worth noting that bacterial cyclopropyl acids are of odd number of carbons, and odd-numbered unsaturated acids, with which they could be misidentified, can be minor components in natural samples and, in some bacteria, can be major acids [e.g. 17:1(9) in *Desulfobulbus*; Taylor and Parkes, 1983]. Separation of these two types of acids, as methyl esters, has been achieved on solid-phase extraction columns converted to the silver-ion form (the cyclopropane acids were in the saturated fraction) and this was found to be a useful step prior to GC–MS analysis (Zelles and Bai, 1994), particularly since their electron impact mass spectra were indistinguishable (Christie and Holman, 1966).

Discrimination between cyclopropane and monounsaturated acid methyl esters was possible by chemical ionization–MS with use of mixtures of methane or isobutane with vinyl methyl ether as reactant gas; only unsaturated esters readily formed vinyl methyl ether adducts (Christopher and Duffield, 1980). However, the position of the cyclopropane ring could not be determined from either the electron impact (Minnikin, 1978) or chemical ionization (Christopher and Duffield, 1980) mass spectra. Fast atom bombardment (FAB) has been used to produce $[M-H]^+$, $[M+Li-H]^+$ ions from free fatty acids and $[M + H]^+$ ions from picolinyl esters which gave collisionally activated dissociation (CAD) spectra by tandem mass spectrometry (Adams, Deterding and Gross, 1987; Tomer, Crow and Gross, 1983; Tomer, Jensen and Gross, 1986). It would appear that the cyclopropane acids studied had characteristic spectra which allowed the position of the ring to be determined (Adams, Deterding and Gross, 1987; Tomer, Jensen and Gross, 1986). However, distinction from monounsaturated acids (Tomer, Crow and Gross, 1983) was based only on the relative intensities of some peaks. Also, because no preseparation technique was employed, mixtures containing fatty acids with the same molecular weight (e.g. monounsaturated and cyclopropane acids) would presumably produce mixed spectra which, coupled with the need for special equipment, rules out the technique for most applications.

GC–MS of nitrogen-containing derivatives, in standard electron-impact mode, is a more practical approach. Pyrrolidides were used first (Minnikin, 1978) but superior spectra were obtained with picolinyl esters (Fig. 5.6) and this should be the choice for absolute identification (Harvey, 1984). A highly characteristic odd-numbered ion (e.g. m/z 247 for 9,10-methyleneoctadecanoate), corresponding to an ester fragment containing the cyclopropane carbon nearest to the ester function, and a prominent ion (m/z 288) 41 amu higher, were present, enabling the position of the cyclopropane ring to be located. The picolinyl esters of acids with one or two cyclopropane rings and a $\Delta 5$ double bond have characteristic odd-numbered ions for each ring together with a gap of 26 amu to locate the position of the double bond (Rezanka and Dembitsky, 1994). The mass spectra of the DMOX derivatives are less distinctive, resembling those of unsaturated derivatives, the only differences being in the intensities of some ions (Zhang, Yu and Huang, 1987).

From the oils of litchi and longan (*Euphoria longana*) seeds, methyl esters of dihydrosterculic and 7,8-methylenehexadecanoic acids were individually separated from normal saturated and unsaturated esters by semi-preparative reversed-phase HPLC (Grondin *et al.*, 1993). On subsequent examination by ^1H and ^{13}C nuclear magnetic resonance (NMR), all signals were assigned, but considering the other analytical options available the possibility of using this approach for identification purposes is not practical.

Alcohols can be derivatized to nicotinates, equivalent to picolinyl esters of fatty acids. A preliminary study involving GC–MS of the nicotinoyl derivative of 1,3-di(9,10 methyleneoctadecanoyl)glycerol showed a prominent odd-numbered ion due to fragmentation through the ring of one of the

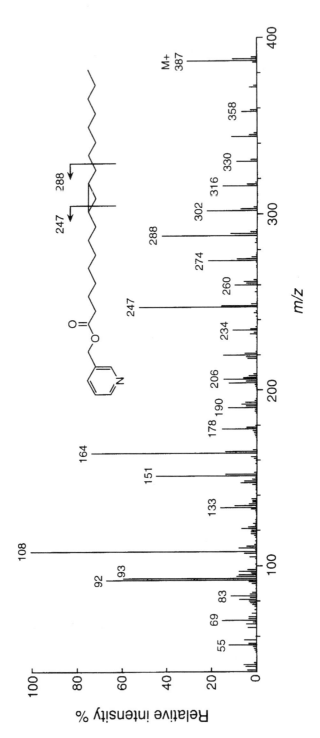

Figure 5.6 Mass spectrum (25 eV) of picolinyl ester derivative of 9,10-cyclopropaneoctadecanoic (dihydrosterculic) acid. Redrawn from Harvey, D. J., Picolinyl derivatives for the characterization of cyclopropane fatty acids by mass spectrometry, *Biomed. Mass Spectrum.*, **11** (4), 187–92, 1984.

cyclopropane chains, and one 41 amu higher, analogous to those for cyclopropane acid picolinyl esters (Zollner and Schmid, 1996). This approach seems to have potential for structural determination of complex lipids containing acyl chains with cyclopropane rings and other groups.

5.2.3 Cyclopropene fatty acids

Cyclopropene acids may range from about 1% to over 70% of the total fatty acids in seed oils of the families *Bombaceae, Malvaceae, Sapicidaceae, Sterculiaceae* and *Tiliaceae* (Sebedio and Grandgirard, 1989). There are two main types, sterculic (9,10-methyleneoctadec-9-enoic) acid (structure IIIa, Fig. 5.1) and malvalic (8,9-methyleneheptadec-8-enoic) acid (structure IIIb), although 2-hydroxy-sterculic and sterculynic (8,9-methyleneoctadec-8-en-17-ynoic) acids have also been detected. There has been substantial interest in these acids because of the physiological effects they produce in animals; one notable effect is that, as minor components of cottonseed (a component of animal meal), they are toxic if not totally removed during refining (Nixon *et al.*, 1974). Analytical difficulties arise because of their instability, and consequently numerous methods have been developed for their quantification.

Traditional procedures for quantification of total cyclopropene acid content have been reviewed by Christie (1970) and include titration with hydrobromic acid and GC of the methyl mercaptan derivatives, products from reaction with silver nitrate/methanol and from methanethiol addition. However, a colorimetric test involving reaction with sulphur/carbon disulphide (the Halphen test) was found to be one of the most reliable methods for determining the small levels of cyclopropene acids present in oils containing cottonseed oil (Coleman, 1970). Another approach is GC of the relatively stable cyclopropane acids after hydrogenation with hydrazine (Conway, Ratnayake and Ackman, 1985)

As with cyclopropane acids, the ring is degraded under acidic conditions and similar extraction and derivatization methods to those recommended for cyclopropyl acids should be used (Section 5.5.2). A more serious problem is potential thermal degradation of the methyl esters on GC analysis, hence the use of methods involving GC after conversion to more stable products. As reviewed by Christie (1989) and Sebedio and Grandgirard (1989), direct GC analysis of methyl esters has been possible with inert supports and silicone liquid phases, but small levels of decomposition products may be missed and some may co-chromatograph with other fatty acids. However, it appears that the best GC method to date, developed for cottonseed oil, involves conventional non-polar [DB-5™ (J & W Scientific Inc., Rancho Cordova, CA); 30 m × 0.25 mm inner diameter, 0.1 µm film thickness] fused-silica capillary column, split injection and a specific temperature programme (150°C for 4 min, then programmed to 230°C at 3°C min^{-1}) to give precise quantification of methyl malvalate and sterculate down to about 0.03% of total fatty acids (Park and Rhee, 1988). Under these conditions, the thermal instability of cyclopropene acids was found to be insignificant and resolution was satisfactory; incomplete resolution of cyclopropene from straight-chain saturated esters was observed for a moderately polar column [Supelcowax™ 10 (Supelco Inc., Bellefonte, PA); 30 m × 0.25 mm, 0.25 µm]. The overlap of α-linolenate and oleate was not of concern for cottonseed oil, because it contained only negligible amounts of α-linolenate. If required, quantification of these acids could be achieved by a separate analysis on a more polar column.

A popular method is to analyse the total fatty acid methyl esters by GC after reaction with silver nitrate/anhydrous methanol for 2 h at 30°C (Bianchini, Ralaimanarivo and Gaydou, 1981; Eisele *et al.*, 1974; Gaydou, Bianchini and Ralaimanarivo, 1983; Ralaimanarivo, Gaydou and Bianchini, 1982). Most fatty esters remain unchanged, but cyclopropene esters are converted to later-eluting methoxy ether derivatives and small amounts of ketone derivatives. Two partially resolved peaks, those from sterculic acid eluting later, are observed for each type of derivative from each cyclopropene acid and can be quantified and used to determine the proportion of sterculic and malvalic acids in the untreated oil. Verification of the identities of the acids can be determined from GC–MS of the methyl esters of the products (Ahmad *et al.*, 1979; Eisele *et al.*, 1974) but prominent allylic ions in the mass spectra of the DMOX derivatives of the methoxy ethers are more readily interpretable to reveal the positions of the cyclopropene rings (Spitzer, 1995).

In the ^{1}H NMR spectrum of the total fatty acid methyl esters from seed oils, a specific signal at 0.76 ppm for the cyclopropene methylene protons has been used, relative to the protons of either the terminal methyl or methyl ester, to give quantification of total cyclopropene acids down to about the 1% level (Boudreaux, Bailey and Tripp, 1972; Pawlowski, Nixon and Sinnhuber, 1972). However, because of incomplete resolution of signals, quantitation becomes increasingly problematic as the proportion of cyclopropene acids decreases.

In the infra-red (IR) spectrum a band at 1010 cm^{-1} characteristic of the cyclopropene system has been used to determine high levels of cyclopropene acids (Bailey, Boudreaux and Skau, 1965). There is another specific band at 1870 cm^{-1} which is weak in the IR spectrum but intense in the Raman spectrum (Kint *et al.*, 1981). Using peak height ratios of this band relative to that at 1745 cm^{-1} (for triacylglycerol carbonyls), quantification of cyclopropene acids down to 0.03% was possible. The method may have potential for detecting vegetable oil contamination with, for example, cottonseed oil, but information is needed on its accuracy and precision.

HPLC methods for quantification do not suffer from the same problems as GC methods, and cyclopropene acids can be analysed directly as methyl or other esters (Gaydou, Bianchini and Ralaimanarivo, 1983; Loveland *et al.*, 1983; Wood, 1986a, 1986b). As detailed by Sebedio and Grandgirard (1989), the problem with some HPLC methods (Gaydou, Bianchini and Ralaimanarivo 1983; Loveland *et al.*, 1983) appears to be the incomplete resolution of sterculate and malvalate from other methyl esters (oleate or palmitate), resulting in potentially significant errors when cyclopropene acids are present in relatively low amounts as in, for example, cottonseed oil. One method (Wood, 1986a, 1986b) used C_{18} and C_8 reversed-phase columns in series to obtain almost baseline separation of malvalic, sterculic and dihydrosterculic acids as phenacyl esters (Fig. 5.7), and quantification down to 0.3% cyclopropene acids was obtained. This method would appear to have potential for accurately quantifying cyclopropene acids over a range of concentrations, but it remains to be tested to see whether very low levels (< 0.1%), which can be quantified by GC (Park and Rhee, 1988), can be determined. The partial resolution of sterculate from *trans*-18:1(9) is also a potential problem for accurate quantification in some oils.

An early approach to confirming identities of cyclopropene acids was to form stable methanethiol adducts (of the double bond) which gave informative mass spectra locating the position of the ring (Hooper and Law, 1968); possibilities of incomplete derivatization preclude the method for quantitative analysis. More recently, malvalic and sterculic acids have been identified by GC–MS as their picolinyl esters (Spitzer *et al.*, 1994) and DMOX derivatives (Spitzer, 1991), although neither give particularly distinct spectra. The diagnostic ions for picolinyl esters were of low abundance, although a gap of 66 amu between two ions resulting from allylic fragmentation either side of the ring located the position of the ring. The mass spectra of the DMOX derivatives had a gap of 10 amu (between m/z 196 and 206 corresponding to C8 and C9 fragments for sterculate; Fig. 5.8A) similar to acetylenic acids but were distinguished from these acids by the relative intensities of other ions. The DMOX derivative of 2-hydroxy-sterculic acid gave analogous ions (m/z 212 and 222) but the presence of the 2-hydroxyl group caused the low-mass ions normally at m/z 113 and 126 to shift to m/z 129 and 142, respectively (Fig. 5.8B). Considering that the methyl esters of cyclopropene acids may be degraded by GC (Christie, 1989; Sebedio and Grandgirard, 1989) it would be worth testing if low levels of the DMOX derivatives or picolinyl esters could be analysed by GC–MS. It is interesting that, although DMOX derivatization involved heating at 170°C, the percentage of cyclopropene acids (*c.* 16% determined by GC) from seeds of *Pachira aquatica* was only slightly less as DMOX derivatives than as methyl esters (Spitzer, 1991).

Analysis of cyclopropene acid-containing triacylglycerols is not possible by GC because they are polymerized at high temperatures. One approach, applied to the seed oils of munguba (*Bombax munguba* Mart; Schuch, Ahmad and Mukherjee, 1986) and *Sterculia foetida* (Pasher and Ahmad,

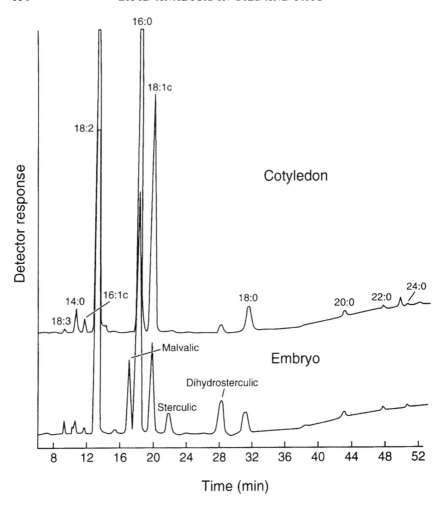

Figure 5.7 Reversed-phase high-pressure liquid chromatogram of phenacyl ester derivatives of the fatty acids from cottonseed cotyledons and embryos. C18 (50 mm × 4.5 mm, 3 μm particle size) and C8 (150 mm × 4.5 mm, 5 μm) were used in tandem. The mobile phase was composed of acetonitrile and water, and the flow rate was 1.5 ml min^{-1}. An initial mixture of acetonitrile/water (80:20 vol./vol.) was maintained for 25 min, followed by a gradient to acetonitrile/water (85:15) for a further 10 min and then to 100% acetonitrile for a final 10 min. A variable-wavelength detector was employed. Redrawn from Wood, R., Comparison of the cyclopropene fatty acid content of cottonseed varieties, glanded and glandless seeds, and various seed structures, *Biochem. Arch.*, **2**, 73–80, 1986.

1992), was to convert the cyclopropene rings exclusively to α,β-unsaturated ketones with silver nitrate in acetonitrile-acetone, followed by thin-layer chromatography (TLC) separation of the ketones according to the number of keto groups (0–3, corresponding to the number of cyclopropene-containing acyl moieties). For each fraction, the fatty acid composition

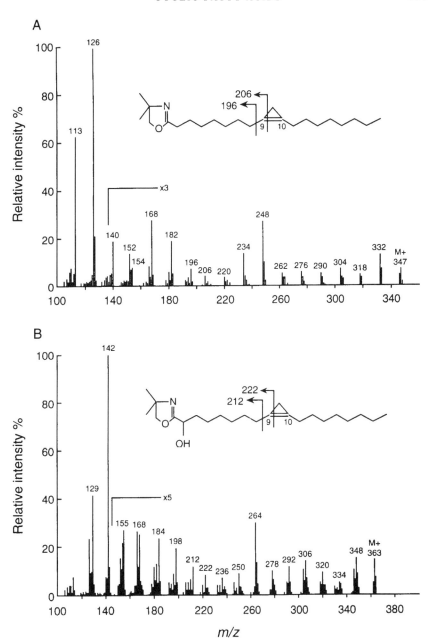

Figure 5.8 Mass spectra (70 eV) of dimethyloxazoline derivatives of: A, 9,10-methyleneoctadec-9-enoic (sterculic) acid; B, 2-hydroxy-sterculic acid. Redrawn from Spitzer, V., GC–MS characterization (chemical ionization and electron impact modes) of the methyl esters and oxazoline derivatives of cyclopropenoid fatty acids, *J. Am. Oil Chem. Soc.*, **68**(12), 963–9, 1991.

was determined by GC, and pancreatic lipase treatment (to produce 2-acylglycerols) was used to ascertain the composition at the sn-2 and sn-1/3 positions. Additionally, for munguba, the molecular species of the major fractions with either 0 or 1 keto group were analysed directly by GC. In munguba seeds, both sterculoyl and malvaloyl moieties were mainly at the sn-2 position, but in *Sterculia foetida* sterculoyl was only marginally predominant at sn-2 and malvaloyl was mainly at sn-1/3. A similar conclusion for *Sterculia foetida* was reached by ^{13}C NMR (using carbonyl and olefinic regions), which was used to determine directly the distribution of sterculate and malvalate at the sn-2 and sn-1/3 positions (Howarth and Vlahov, 1996).

5.3 Cyclic fatty acids formed in heated vegetable oils

During heating of vegetable oils, triacylglycerols may undergo a variety of chemical changes, including hydrolysis (to form free fatty acids, monoacylglycerols and diacylglycerols), oxidation (to form hydroperoxides, conjugated dienoic acids, epoxides, hydroxides, ketones and various volatiles) and formation of new carbon–carbon bonds in the absence of oxygen (White, 1991). New bonds between two different fatty acids can result in dimeric acids, either within or between the triacylglycerol molecules, and further cross-linking produces polymers; cyclic fatty acids are the products of new bond formation within one fatty acid molecule. The nature of cyclic fatty acid monomers (CFAMs) has been the subject of study for over 40 years (Le Quere and Sebedio, 1996; Sebedio and Grandgirard, 1989). Interest in these compounds was enhanced by the finding that when fed to pregnant rats there was an increased mortality in the offspring; the biological properties have already been critically reviewed (Sebedio and Grandgirard, 1989). CFAMs are formed in low amounts (0.01%–0.7%) during deep frying (Sebedio and Grandgirard, 1989) and inevitably form part of the human diet. They have been detected in a human milk sample (Chardigny *et al.*, 1995), but the potential toxicity to humans is unknown.

CFAMs occur as complex mixtures in heated vegetable oils. They are eighteen carbons long with either a five-membered or six-membered ring. The position of the ring, the geometry of the side chains about the ring and the number, position (including within the ring or in the side chains) and configuration of double bonds are all variables which give a large number of potential structures (Fig. 5.2). The degree of unsaturation of the acid reflects the origin. Saturated (Dobson, Christie and Sebedio, 1996a), monoenoic (Christie *et al.*, 1993; Sebedio *et al.*, 1987) and dienoic (Dobson *et al.*, 1995; Mossoba *et al.*, 1994, 1995b, 1996b; Sebedio *et al.*, 1987) acids are derived from oleic, linoleic and α-linolenic acids, respectively, one double bond being lost upon cyclization. CFAMs are formed in greater abundance from α-linolenic acids than from linolenic acids (Sebedio, Prevost and

Grandgirad, 1987) and although those from oleic acid might be expected to be formed at the lowest levels there are no supportive data available. It is clear that the type of vegetable oil affects the types and quantities of CFAMs. For example, oleic acid is the major acid in olive oil and high-oleic sunflower oil, and linoleic acid is also significant in these oils and is the predominant acid in normal sunflower oil and soybean oils. α-Linolenic acid is a significant component of soybean and rapeseed oil and is the major acid in linseed oil, although the latter is not used for human consumption.

By using linseed oil and high-oleic and normal sunflower oils as models, only recently have researchers elucidated the complete CFAM structures from the three unsaturated precursors, as described in the following section. It is apparent that a complete analysis of CFAMs in a heated oil or in a fried food is a far from routine task and involves detailed study. The methods available for quantification of CFAMs are often multistep and do not always give satisfactory results, although continuous improvements have been made over many years. This aspect is considered in Section 5.3.2.

5.3.1 *Structural analysis of monomeric cyclic fatty acids*

Earlier studies on the structure of CFAMs formed in heated oils have been covered in considerable detail by Sebedio and Grandgirard (1989) and more recently by Le Quere and Sebedio (1996). Sunflower and linseed oils were used as models. First, a monomeric cyclic fatty acid fraction was isolated by preparing methyl esters from the heated oil, collecting a non-polar fraction by column chromatography, and after two urea fractionations the CFAMs were recovered in the non-adduct fraction (Sebedio, Prevost and Grandgirard, 1987). Further purification of CFAMs from heated sunflower oil by means of reversed-phase HPLC was necessary. The CFAM fractions were complex mixtures and could be simplified by hydrogenation, which eliminated positional and geometrical double bond isomers. The skeletal structures were then elucidated by GC–MS of the methyl esters and some were confirmed by comparison with synthetic standards (Sebedio *et al.*, 1989). The geometry of the side chains was determined on the basis of GC retention times compared with synthetic standards (Rojo and Perkins, 1989a; Vatele, Sebedio and Le Quere, 1988).

A novel approach to studying hydrogenated CFAMs was to use FAB GC–MS/MS (Le Quere *et al.*, 1991). Gas-phase carboxylate ions were generated from electron-capture ionization of pentafluorobenzyl CFAM esters. CAD–MIKE (mass-analysed ion kinetic energy) spectra of the carboxylate ion were characterized by charge-remote fragmentation α to the ring. In this way, compared with the previous study (Sebedio *et al.*, 1989), a number of additional skeletal CFAM structures from heated linseed and sunflower oils were proposed. CAD–MIKE spectra of the carboxylate anions generated from synthetic cyclopentyl and cyclopentenyl acids were also examined.

Fragmentation α to the ring and allylic cleavage to double bonds in the straight chain were observed but, although there seems to be potential for determining the exact structures of intact CFAMs, including positions of double bonds, the methodology has not been applied to native complex CFAM mixtures from heated oils.

The degree of unsaturation of the native acids was determined by GC–MS of the methyl esters and, by comparison with the mass spectra of synthetic standards, some structures were proposed (Sebedio *et al.*, 1987). However, interpretation of the mass spectra of the methyl esters of intact acids was difficult. An on-line hydrogenation method in conjunction with GC–MS was developed to correlate intact CFAMs and their hydrogenated analogues (Le Quere *et al.*, 1989). A hydrogenation capillary reactor was connected to the end of the GC column and was fed into the ion source of the mass spectrometer; CFAMs were examined with and without hydrogenation. Correspondingly, ten CFAMs from heated linseed oil were examined and interpretation of the mass spectra was aided by knowing the skeletal structures from the more easily interpretable spectra of the hydrogenated counterparts. Some tentative dienoic structures, with either cyclopentenyl or cyclohexenyl rings and a double bond of unknown location in the side chain, were suggested.

Some of the earliest information obtained on a number of CFAMs from heated sunflower and linseed oils was on the nature of the geometry of the double bonds, determined by GC–FTIR (Fourier transform infra-red) spectroscopy of the methyl esters (Sebedio *et al.*, 1987). More recently, this information has been expanded and correlated to other structural features of CFAMs from sunflower oil (Christie *et al.*, 1993) and especially linseed oil (Dobson *et al.*, 1995; Mossoba *et al.*, 1995a, 1995b, 1996a). Mossoba and colleagues have examined methyl esters (Mossoba *et al.*, 1995a, 1995b) and DMOX derivatives (Mossoba *et al.*, 1996a; Fig. 5.9) by GC–MI (matrix isolation)–FTIR spectroscopy to obtain greater sensitivity. The essential features of the spectra were =C—H stretch and, more distinctively, deformation bands for Z double bonds in cyclopentenyl (3058–3063 cm^{-1} and 711–720 cm^{-1}; Figs 5.9A, B) and cyclohexenyl (3030–3032 cm^{-1} and 662–664 cm^{-1}; Fig. 5.9C) rings; E double bonds cannot occur in rings. A strong deformation band (970–979 cm^{-1}; Figs 5.9B, C) was highly characteristic for E double bonds in the chain. A band around 723 cm^{-1} was suggested to be indicative of a cyclohexadiene moiety (Mossoba *et al*, 1994, 1995a, b, 1996a).

Until recently, a complete analysis of CFAMs from heated oils was a daunting task: first, because of the complexities of the mixtures and, second, because of the difficulties in obtaining full structural information, especially with respect to double bond positions. However, recently this has been made possible by combining fractionation of a CFAM mixture (obtained by urea adduction) by silver-ion HPLC with GC–MS analysis of the picolinyl esters or DMOX derivatives of the fractions (Fig. 5.10; Christie *et al.*, 1993;

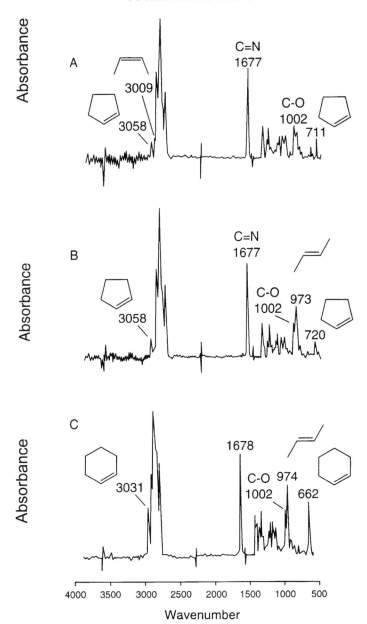

Figure 5.9 Matrix-isolation Fourier transform infra-red spectra of dimethyloxazoline derivatives of: A, 9-(2'-but-*cis*-1-enyl-cyclopentenyl)nonanoic acid; B, 9-(2'-but-*trans*-1-enyl-cyclopentenyl)nonanoic acid; C, 9-(2'-prop-*trans*-1-enyl-cyclohex-*cis*-4-enyl)nonanoate from heated linseed oil. Redrawn from Mossoba, M. M., Yuracwecz, M. P., Roach, J. A. G. *et al.*, Analysis of cyclic fatty acid monomer 2-alkenyl-4,4-dimethyloxazoline derivatives by gas chromatography–Fourier transform infra-red spectroscopy, *J. Agric. Food Chem.*, **44**, 3193–6, 1996.

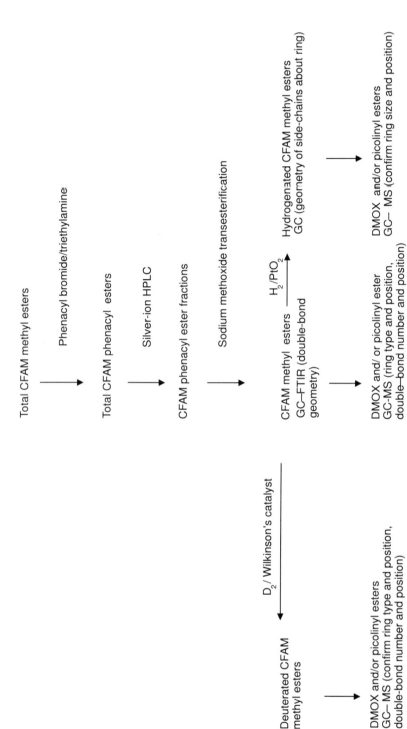

Figure 5.10 Scheme for characterization of cyclic fatty acid monomers (CFAMs) from heated vegetable oils. HPLC = high-pressure liquid chromatography; GC = gas chromatography; FTIR = Fourier transform infra-red (spectroscopy); MS = mass spectrometry; DMOX = dismethyloxazoline.

Dobson, Christie and Sebedio, 1996a, 1997; Dobson *et al.*, 1995). GC–FTIR was used to determine the geometry of the double bonds, and the configuration of the side chains about the ring was determined by comparison with published values on the order of the hydrogenated acid methyl esters by GC (Sebedio *et al.*, 1989; Table 5.1). These techniques were necessary to complete the structures because there were no significant differences in the mass spectra of geometrical isomers.

Silver-ion HPLC separated the acids, as phenacyl esters (because of improved separation compared with methyl esters), according to the degree of unsaturation and the geometry of the double bonds (E isomers eluted before Z isomers) (Christie *et al.*, 1993; Dobson *et al.*, 1995) and, for cyclic dienes, also on size of the ring (six-membered before five-membered) and configuration of the side chains about the ring (Dobson *et al.*, 1995). In this way, several fractions containing only a limited number of usually well-resolved GC components, some of which were unresolved from other peaks in the total mixture, were obtained. It should be noted that reversed-phase HPLC gave poor separation of CFAMs and indeed was used to separate a total CFAM fraction (Sebedio, Prevost and Grandgirard, 1987; Sebedio *et al.*, 1994).

GC–MS analysis of the picolinyl esters or DMOX derivatives of CFAM fractions gave information on the positions of the double bonds in the straight chains (but not always in the ring) and also on the type and position of the ring. In general, the mass spectra of the two types of derivatives were analogous and could be easily interpreted using a few basic principles without the need for standards. The skeletal structures were confirmed by GC–MS of the derivatives of the hydrogenated fractions and, moreover, GC–MS of the derivatized deuterated fractions also verified positions of double bonds in the chains; Wilkinson's catalyst, which does not give scrambling of double bonds, and deuterium, were used to deuterate double bonds. Only by analysing simple acid mixtures produced by silver-ion HPLC was it possible to correlate mass spectral information about individual acids which were intact and had been hydrogenated and deuterated.

As an illustration of the approach, CFAMs from linseed oil heated to 275°C, shown to be a complex mixture by GC (Fig. 5.11), were separated into nine fractions, and the cyclic dienes occurred in fractions 3–8 (Fig. 5.12), each fraction containing 1 to 5 acids (Dobson *et al.*, 1995). By correlating the information from subsequent GC–MS and FTIR analyses, and the order of the hydrogenated methyl esters by GC, 16 cyclic dienoic acids (a–p; Fig. 5.11; Table 5.1) were characterized. There were about equal amounts of eight cyclopentenyl and eight cyclohexenyl acids. The cyclopentenyl acids were represented by two basic structures, one with a ring from C-10 to C-14 of the parent α-linolenic acid with a double bond at C-15 (structure I, Fig. 5.2) and the other with a ring from C-11 to C-15 and a double bond at C-9

Table 5.1 Nature and properties of monomeric cyclic fatty acids in heated vegetable oils

Peak[a]	Structure (Fig. 5.2)	Fatty acid[b]	Ring (double bond position)[c]	ECL[d] CP-Wax 52CB	ECL[d] BPX70	Hydrogenated fatty acid[e]	Configuration of side chains	ECL[f] CP-Wax 52CB
α-Linolenate-derived (linseed oil)								
a	I	9-(2'-but-cis-1-enyl-cyclopentenyl)nonanoate	10–14(12Z, 15Z)	18.49	18.83	9-(2'-butyl-cyclopentyl)ronanoate	trans	18.12
b	I	9-(2'-but-trans-1-enyl-cyclopentenyl)nonanoate	10–14(12Z, 15E)	18.46	18.61			
c	II	10-(2'-propyl-cyclopentenyl)dec-cis-9-enoate	11–15(9Z, 12Z)	18.38	18.80	10-(2'-propyl-cyclopentyl)decanoate	trans	18.28
d	II	10-(2'-propyl-cyclopentenyl)dec-trans-9-enoate	11–15(9E, 12Z)	18.46	18.68			
e	I	9-(2'-but-cis-1-enyl-cyclopentenyl)nonanoate	10–14(12Z, 15Z)	18.88	19.34	9-(2'-butyl-cyclopentyl)nonanoate	cis	18.51
f	I	9-(2'-but-trans-1-enyl-cyclopentenyl)nonanoate	10–14(12Z, 15E)	18.58	18.84			
g	II	10-(2'-propyl-cyclopentenyl)dec-cis-9-enoate	11–15(9Z, 12Z)	18.84	19.39	10-(2'-propyl-cyclopentyl)decanoate	cis	18.7
h	II	10-(2'-propyl-cyclopentenyl)dec-trans-9-enoate	11–15(9E, 12Z)	18.58	18.95			
i	III	9-(2'-propyl-cyclohex-cis-4'-enyl)non-cis-8-enoate[g]	10–15(8Z, 12Z)	19.01	19.64	9-(2'-propyl-cyclohexyl)nonanoate	trans	18.54
j	III	9-(2'-propyl-cyclohex-cis-4'-enyl)non-trans-8-enoate	10–15(8E, 12Z)	19.12	19.65			
k	IV	9-(2'-prop-cis-1-enyl-cyclohex-cis-4'-enyl)nonanoate[h]	10–15(12Z, 16Z)	19.6	20.21			
l	IV	9-(2'-prop-trans-1-enyl-cyclohex-cis-4'-enyl)nonanoate	10–15(12Z, 16E)	19.43	19.87			

m	III	9-(2'-propyl-cyclohex-cis-4'-enyl)non-cis-8-enoate[g]	10-15(8Z, 12Z)	19.26	19.94	cis	9-(2'-propyl-cyclohexyl)nonanoate	18.76
n	III	9-(2'-propyl-cyclohex-cis-4'-enyl)non-trans-8-enoate	10-15(8E, 12Z)	19.01	19.55			
o	IV	9-(2'-prop-cis-1-enyl-cyclohex-cis-4'-enyl)nonanoate[h]	10-15(12Z, 16Z)	19.84	20.51			
p	IV	9-(2'-prop-trans-1-enyl-cyclohex-cis-4'-enyl)nonanoate	10-15(12Z, 16E)	19.52	19.98			

Lionleate-derived (normal sunflower oil); excluding bicyclic fatty acid monomers (Dobson, Christie and Sebedio, 1997)

	V	9-(2'-butyl-cyclopent-cis-3'-enyl)nonanoate[i]	10-14(12Z)[i]	ND	ND	trans	9-(2'-butyl-cyclopentyl)nonanoate	18.12
	VII	9-(2'-butyl-cyclopent-cis-2'-enyl)nonanoate	10-14(13Z)[i]	ND	ND	cis	9-(2'-butyl-cyclopentyl)nonanoate	18.51
	VI	7-(2'-hexyl-cyclopent-cis-4'-enyl)heptanoate[i]	8-12(9Z)[i]	ND	ND	trans	7-(2'-hexyl-cyclopentyl)heptanoate	17.96
	VIII	7-(2'-hexyl-cyclopent-cis-5'-enyl)heptanoate[i]	8-12(8Z)[i]	ND	ND	cis	7-(2'-hexyl-cyclopentyl)heptanoate	18.34
	IX	12-(2'-methyl-cyclopentyl)dodec-cis-9-enoate[j]	13-17(9Z)	ND	ND	ND	12-(2'-methyl-cyclopentyl)dodecanoate	ND
	X	4-(2'-non-cis-3-enyl-cyclopentyl)butanoate[j]	5-9(12Z)	ND	ND	ND	4-(2'-nonyl-cyclopentyl)butanoate	ND
	XI	11-(2'-methyl-cyclohexyl)undec-cis-9-enoate[j]	12-17(9Z)	ND	ND	ND	11-(2'-methyl-cyclohexyl)undecanoate	ND
	X'	11-(2'-methyl-cyclohexyl)undec-trans-9-enoate	12-17(9E)	ND	ND	ND		
	XII	12-cyclohexyldodec-cis-9-enoate	13-18(9Z)	ND	ND	trans	4-(2'-octyl-cyclohexyl)butanoate	18.07
	XIII	4-(2'-oct-cis-2-enyl-cyclohexyl)butanoate[j]	5-10(12Z)	ND	ND	cis	4-(2'-octyl-cyclohexyl)butanoate	18.28
	XIII	4-(2'-oct-trans-2-enyl-cyclohexyl)butanoate[j] Not detected	5-10(12E)	ND	ND	ND	Not detected 10-(2'-ethyl-cyclohexyl)decanoate	ND

Oleate-derived (high-oleic sunflower oil)

q	XIV	4-(2'-nonyl-cyclopentyl)butanoate	5-9	17.9[k]	18.19	ND	4-(2'-nonyl-cyclopentyl)butanoate[l]	ND
r	XIV	4-(2'-nonyl-cyclopentyl)butanoate	5-9	18.27[k]	18.67			

Table 5.1 Continued

Peak[a]	Structure (Fig. 5.2)	Fatty acid[b]	Ring (double bond position)[c]	ECL[d] CP-Wax 52CB	ECL[d] BPX70	Hydrogenated fatty acid[e]	Configuration of side chains[f]	ECL[f] CP-Wax 52CB
s	XV	9-(2'-butyl-cyclopentyl)nonanoate	10–14	18.11[k]	18.43	9-(2'-butyl-cyclopentyl)nonanoate[l]	trans	18.12
t	XV	9-(2'-butyl-cyclopentyl)nonanoate	10–14	18.51[k]	18.93	9-(2'-butyl-cyclopentyl)nonanoate[l]	cis	18.51
u	XVI	3-(2'-nonyl-cyclohexyl)propanoate	4–9	18.17[k]	18.59	Not detected[l]	ND	ND
v	XVI	3-(2'-nonyl-cyclohexyl)propanoate	4–9	18.43[k]	18.93			
w	XVII	9-(2'-propyl-cyclohexyl)nonanoate	10–15	18.54[k]	19.02	9-(2'-propyl-cyclohexyl)nonanoate[l]	trans	18.54
x	XVII	9-(2'-propyl-cyclohexyl)nonanoate	10–15	18.77	19.31	9-(2'-propyl-cyclohexyl)nonanoate	cis	

[a] Refers to peaks in GC profiles of total CFAM from heated linseed oil (Fig. 5.11) and of saturated CFAM from normal and high-oleic sunflower oils (Fig. 5.16).
[b] Determined by Dobson et al. (1995) and Mossoba et al. (1994, 1995b, 1996a) for CFAM from α-linolenate, by Christie et al. (1993) for CFAM from linoleate and by Dobson, Christie and Sebedio (1996a) for CFAM from oleate.
[c] Numbered according to positions in parent straight-chain unsaturated acid.
[d] Equivalent chain lengths according to Dobson, Christie and Sebedio (1996a, 1996c).
[e] Determined by Sebedio et al. (1989) and Le Quere et al. (1991) by means of total hydrogenated CFAM mixtures from heated linseed and normal sunflower oils.
[f] Determined by Sebedio et al. (1989) by means of total hydrogenated CFAM mixtures from heated linseed and normal sunflower oils; configurations for individual intact acids determined by Dobson, Christie and Sebedio (1996a) and Dobson et al. (1995).
[g] Geometry of double bond not confirmed by GC–FTIR.
[h] According to Dobson et al. (1995), but proposed by Mossoba et al. (1994, 1995a, 1995b, 1996a, 1996b) to have cyclohexadiene structures.
[i] For structures V and VI, the double bond in the ring may be alternatively at the 4'- and 3'- positions (C-11 and C-10 of parent linoleate), respectively. For structures VII and VIII, double bond may be at alternative substituted positions. Upon hydrogenation, structures VII and VIII would also contribute towards the trans isomers of the side chains about the ring.
[j] More than one acid of each type were detected but the configurations of the side chains about the rings were not determined.
[k] ECL value at 180 °C.
[l] From in total hydrogenated CFAM mixtures from heated normal sunflower oil (Sebedio et al., 1989; Le Quere et al., 1991).
CFAM = cyclic fatty acid monomers; ECL = equivalent chain length; FTIR = Fourier transform infra-red (spectroscopy); GC = gas chromatography; ND = not determined.

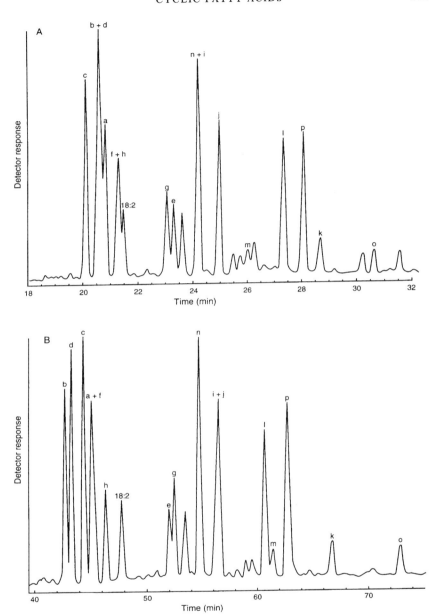

Figure 5.11 Partial gas chromatograms of total cyclic fatty acid methyl esters derived from linolenic acid using: A, Cp-Wax 52CB™ (Chrompack UK Ltd, London) (25 m × 0.25 mm inner diameter, 0.20 μm film thickness) capillary column and an initial temperature of 160°C for 5 min followed by a programme at 0.5°C min^{-1} to 180°C; B, BPX70™ (SGE, Milton Keynes, UK) (50 m × 0.22 mm, 0.25 μm) capillary column and an isothermal temperature of 160°C. Peaks are identified in Table 5.1. Redrawn from Dobson, G., Christie, W. W. and Sebedio, J.-L., Gas chromatographic properties of cyclic dienoic acids formed in heated linseed oil, *J. Chromatogr. A*, **723**, 349–54, 1996.

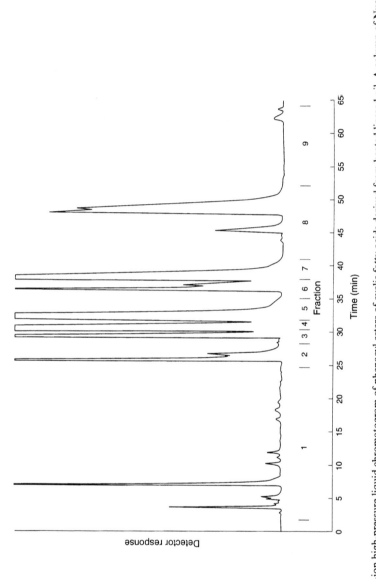

Figure 5.12 Silver-ion high-pressure liquid chromatogram of phenacyl esters of cyclic fatty acids derived from heated linseed oil. A column of Nucleosil™ 5SA (HPLC Technology Ltd, Macclesfield, UK) (250 mm × 4.6 mm inner diameter; 5 μm particle size) was used in the silver-ion form. The mobile phase was composed of dichloromethane/dichloroethane (50:50 vol./vol. solvent A) and dichloromethane/dichloroethane/acetonitrile (49:49:2 vol./vol./vol. solvent B), and the flow rate was 1 ml min^{-1}. There was a linear gradient from 100% solvent A to solvent B (75%:25%) over 50 min, then to 100% solvent B over a further 5 min. An evaporative light-scattering detector was employed. Redrawn from Dobson, G. *et al.*, 1995.

(structure II). The double bond in the ring presumably remained at the C-12 position (and must have a Z configuration), and one double bond was lost upon ring formation. Similarly, there were two basic cyclohexenyl structures, both with rings between C-10 and C-15. In one, the double bond was at C-8 (structure III) and in the other it was at C-16 (structure IV), having shifted from the original positions at C-9 and C-15 in α-linolenic acid. Again the double bond in the ring remained at C-12 (proved by the mass spectra; see below) and one double bond was lost upon cyclization. The carbon atoms involved in cyclization are illustrated in Fig. 5.13A. There were four acids of each basic structure representing all possible combinations of geometrical configurations of the double bond in the straight chain and of the straight chains about the ring. For the cyclopentenyl fatty acids, there were similar amounts of Z and E double bond isomers and a predominance of *trans* isomers of the chains about the ring. For the cyclohexenyl acids, there was a predominance of E double bond isomers and similar amounts of isomers of the straight chains about the ring.

A.

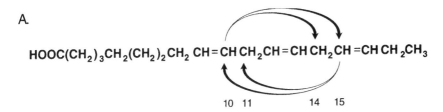

B.

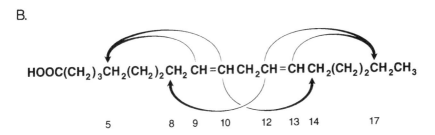

C.

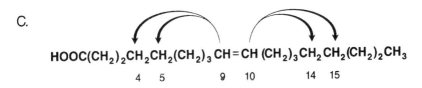

Figure 5.13 Carbon atoms involved in cyclization of: A, α-linolenic acid; B, linoleic acid; C, oleic acid. Redrawn from Dobson, G., Christie, W. W. and Sebedio, J.-L., The nature of cyclic fatty acids formed in heated vegetable oils, *Grasas y Aceites*, **4**, 26–33, 1996.

To illustrate the mass spectra of picolinyl esters, an acid with a ring from C-10 to C-14 and a double bond at C-15 is used as an example (Fig. 5.14A). The molecular ion was at m/z 369, regular gaps of 14 amu from m/z 164 to 248 indicated a saturated straight chain up to C-9, and a gap of 66 amu between m/z 248 and 314 located the cyclopentenyl ring between C-10 and C-14. A gap of 26 amu between m/z 314 and 340 suggested a double bond at C-15. There were no diagnostic ions for locating the position of the double bond in the ring, but it was assumed that it remained at the C-12 position. The mass spectrum of the hydrogenated ester gave a molecular ion 4 amu higher, at m/z 373; a gap of 68 amu between m/z 248 and 316 verified the position of the ring; and gaps of 14 amu between m/z 316 and the $[M-15]^+$ ion at m/z 358 indicated the saturated straight-chain nature of the remainder of the chain. In the spectrum of the deuterated ester, the molecular ion was a further 4 amu higher, at m/z 377; a gap of 70 amu between m/z 248 and 318 verified two deuterium atoms in the ring; and gaps of 15 amu from m/z 318 to 333 and m/z 333 to 348 indicated deuterium atoms at C-15 and C-16. Likewise, the mass spectra of the derivatives of the cyclopentenyl acids with rings between C-11 and C-15 (gap of 66 amu from m/z 260 to 326 for picolinyl esters; Fig. 5.14B) and a double bond at C-9 (gap of 26 amu from m/z 234 to 260) could be interpreted easily. The mass spectra of the DMOX derivatives had molecular ions at m/z 331, 335 and 339 for the native, hydrogenated and deuterated acids, and the diagnostic ions were analogous to those of the picolinyl esters.

The mass spectra of cyclohexenyl acid derivatives with a double bond at C-8 were also readily interpretable. There was a gap of 80 amu (from m/z 208 to 288 for the DMOX derivatives; Fig. 5.15A) for the ring from C-10 to C-15 and a gap of 26 amu (from m/z 182 to 208) located the double bond in the straight chain. The spectra of the hydrogenated derivatives had gaps of 82 amu (from m/z 210 to 292) for the ring, and gaps of 84 amu (from m/z 212 to 296; Fig. 5.15B) and 30 amu (from m/z 182 to 212) for the deuterated derivatives, confirming the positions of the ring and double bond, respectively. One additional feature in all cyclohexenyl spectra was the presence of an ion (m/z 315 for picolinyl esters, and m/z 277 for DMOX derivatives; Figs 5.14C and 5.15A) due to a retro Diels–Alder fragmentation which could arise only if the double bond in the ring was located at C-12. An analogous ion had previously been observed for the methyl esters (Sebedio et al., 1987).

Interpretation of the mass spectra of derivatives of the native cyclohexenyl acids with a double bond at C-16 was equivocal. For the picolinyl ester (Fig. 5.14C) a gap of 78 amu between m/z 248 and 326 suggested a cyclohexadiene structure, yet the presence of a retro Diels–Alder fragment (m/z 315) indicated only one double bond in the ring. The mass spectrum of the deuterated ester had gaps of 84 amu from m/z 248 to 332 and of 30 amu from m/z 332 to 362, showing clearly that there was only one double bond in the ring, with

CYCLIC FATTY ACIDS 165

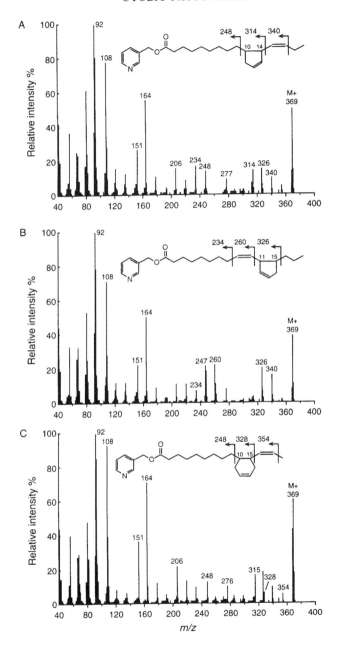

Figure 5.14 Mass spectra (70 eV) of picolinyl ester derivatives of cyclic dienoic acids derived from α-linolenic acid in heated linseed oil. A, 9 (2'-but-*cis*-1-enyl-cyclopentenyl)nonanoate; B, 10-(2'-propyl-cyclopentenyl)dec-*cis*-9-enoate; C, 9-(2'-prop-*trans*-1-enyl-cyclohex-*cis*-4-enyl)-nonanoate. Redrawn from Dobson, G., Christie, W. W., Brechany, E. Y. *et al.*, Silver ion chromatography and gas chromatography–mass spectrometry in the structural analysis of cyclic dienoic acids formed in frying oils, *Chem. Phys. Lipids*, **75**, 171–82, 1995.

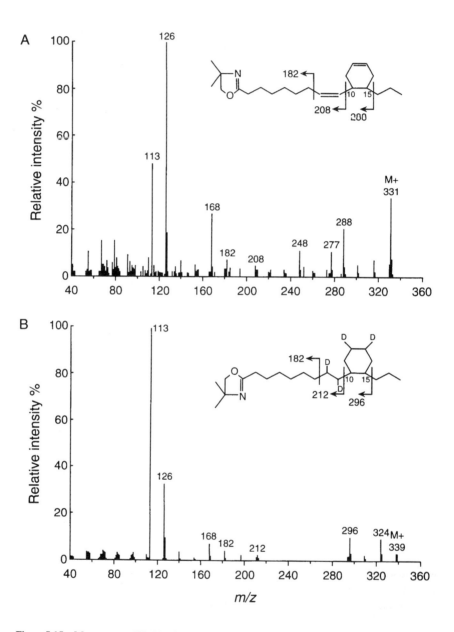

Figure 5.15 Mass spectra (70 eV) of dimethyloxazoline derivatives of 9-(2'-propyl-cyclohex-*cis*-4-enyl)non-*trans*-8-enoate derived from α-linolenic acid in heated linseed oil: A, before, B, after deuteration. Redrawn from Dobson, G., Christie, W. W., Brechany, E. Y. *et al.*, Silver ion chromatography and gas chromatography–mass spectrometry in the structural analysis of cyclic dienoic acids formed in frying oils, *Chem. Phys. Lipids*, **75**, 171–82, 1995.

the other at C-16. This example clearly illustrates the necessity for examining the deuterated as well as the native acid derivatives.

Another group (Mossoba et al., 1994, 1995b) studying a similar CFAM sample but analysing the DMOX derivatives of the total mixture by GC–MS, characterized thirteen acids; three minor components were not detected, illustrating the benefits of fractionation prior to analysis. Of the thirteen acids, there was agreement with the other study (Dobson et al., 1995) on the majority of structures (including geometry of the double bonds by GC–FTIR, but the configuration of the chains about the rings was not determined), and there was a discrepancy in the identities of two acids (k and o, Fig. 5.11; Table 5.1). It would appear that the interpretation of the mixed mass spectra of deuterated DMOX derivatives, claimed to be supportive of cyclohexadiene structures, was erroneous (Mossoba et al., 1996b). On the other hand, cyclohexenyl structures with *cis* double bonds at the C-16 positions (structure IV, Fig. 5.2), proposed by the other group (Dobson et al., 1995), were substantiated by the mass spectra of deuterated DMOX derivatives of fractions containing only a few acids. Two C18 bicyclic acids with a six-membered ring fused to a five-membered ring and containing one double bond with either a methyl or an ethyl substituent on the ring were also characterized (Mossoba et al., 1995b, 1996b).

The cyclic monoenoic acids from linoleate sunflower oil heated to 275°C were analysed by the silver-ion HPLC/GC–MS approach outlined above (Christie et al., 1993). Several acids were characterized and the carbon atoms involved in cyclization are illustrated in Fig. 5.13B. Equal amounts of two cyclopentenyl acids, one with the ring from C-10 to C-14 (structure V, Fig. 5.2) and the other from C-8 to C-12 (structure VI), were the major CFAMs. The position of the double bonds could not be confirmed from the mass spectra but were presumably in the original C-12 and C-9 positions, respectively. Only one isomer, with respect to the configuration of the side chains about the ring, of each acid was detected. In a study on the hydrogenated acids of a similar mixture (Sebedio et al., 1989), the *trans* isomers of the skeletal analogues of these acids predominated and therefore it can only be assumed that the native acids have the same configuration. In the intact mixture, two similar acids (structures VII and VIII, Fig. 5.2) to the major ones were detected but, on the basis that upon hydrogenation they each gave two products (presumably both isomers of the side chains about the ring), it was postulated that the double bond must be at one of the two substituted carbons; these are the only examples in which the double bond in the ring has shifted from the original position in the parent straight-chain acid. These acids would account for the smaller amounts of the two *cis* isomers present in the hydrogenated mixture (Sebedio et al., 1989). Interestingly, CFAMs with cyclohexenyl rings were not detected.

Of the remaining acids there were several with a *cis* double bond in the straight chain at C-9 or C-12 and either cyclopentyl [from C-13 to C-17 (structure IX) or C-5 to C-9 (structure X), respectively] or cyclohexyl [from C-12 to C-17 (structure XI), C-13 to C-18 (structure XII) or C-5 to C-10 (structure XIII), respectively] rings. Analogous CFAMs with a *trans* double bond occurred only as cyclohexyl (C-5 to C-10 and C-12 to C-17) acids.

In general, there was a correlation between the compositions of the intact CFAMs (Christie *et al.*, 1993) and hydrogenated CFAM mixtures (Sebedio *et al.*, 1989; Le Quere *et al.*, 1991) from heated sunflower oil (Table 5.1). However, the native monoenoic counterparts of cyclohexyl acids with rings from C-10 to C-15 and C-11 to C-16, detected in the hydrogenated mixture, were not encountered. Whereas the latter acid cannot obviously be explained, the former acid is presumably present as a saturated acid in the intact mixture and is derived from oleate, present at lower abundance than linoleate in sunflower oil (Dobson, Christie and Sebedio, 1996a; see below).

It is interesting that whereas cyclization of linoleate (Fig. 5.13B) can involve both double bonds and can occur towards both the centre and the ends of the molecule (Christie *et al.*, 1993), cyclization of α-linolenate (Fig. 5.13A) is directed only towards the centre of the molecule, and the double bond at C-12 is never involved (Dobson *et al.*, 1995).

It was stated above that the position of the double bond in the rings of cyclopentenyl acids, present in CFAMs from linolenic and α-linolenic acids, could not be determined from their mass spectra. One approach to locating double bonds, including those in rings, has been to carry out oxidative ozonolysis of total CFAM mixtures from heated sunflower and linseed oils in boron trifluoride/methanol (Le Quere *et al.*, 1991). Double bonds were cleaved to give a mixture of short-chain dimethyl and trimethyl esters which were analysed by GC-MS. It was difficult to piece together most of the information to postulate structures for the intact CFAMs. However, in the products from heated linseed oil CFAMs, dimethyl nonanedioate and dimethyl 2-carbomethoxy-3-propylglutarate were detected, indicating that they were derived from a cyclopentenyl acid with a ring from C-11 to C-15, and double bonds in the straight chain at C-9 and in the ring at C-12. This was confirmed by synthesis, and it was not surprising that the double bond remained in the original C-12 position. Similar information for the other cyclopentenyl acids might well be obtained if the products from simpler CFAM fractions (from silver-ion HPLC) were to be examined.

The saturated CFAMs isolated from a high-oleate and a normal sunflower oil used in a small-scale frying operation of potato strips were also analysed by a silver-ion HPLC/GC-MS approach (Dobson, Christie and Sebedio, 1996a, 1997). The first HPLC fraction contained the saturated components, and subsequent analysis of the methyl esters by GC revealed two major groups of CFAMs (Fig. 5.16). The first group, presumably derived from oleate, constituted a greater proportion of the total CFAMs in high-oleate (about two-thirds of total CFAMs) compared with normal (about one-

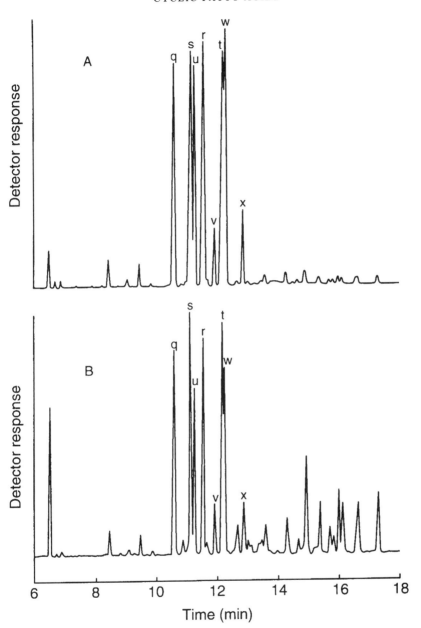

Figure 5.16 Partial gas chromatograms of saturated cyclic fatty acid methyl esters derived from: A, high-oleate sunflower oil; B, normal sunflower oil. A Cp-Wax 52CB™ (Chrompack UK Ltd, London) (25 m × 0.25 mm inner diameter, 0.20 μm film thickness) capillary column was used. There was an initial temperature of 170°C for 5 min followed by a programme at 2°C min^{-1} to 210°C. Peaks are identified in Table 5.1. Redrawn from Dobson, G., Christie, W. W. and Sebedio, J.-L., Monocyclic saturated fatty acids formed from oleic acid in heated sunflower oils, *Chem. Phys. Lipids*, **82**, 101–10, 1996.

quarter) sunflower oil (Dobson, Christie and Sebedio, 1996a). On the basis of the mass spectra of the DMOX derivatives, eight monocyclic acids (q–x, Fig. 5.16; Table 5.1) were detected. There were four basic structures, two cyclopentyl acids with a ring either from C-5 to C-9 (structure XIV, Fig. 5.2) or C-10 to C-14 (structure XV), and two cyclohexyl acids with a ring either from C-4 to C-9 (structure XVI) or C-10 to C-15 (structure XVII). Cyclopentyl acids were about twice as abundant as cyclohexyl acids. The carbon atoms involved in cyclization are summarized in Fig. 5.13C. Each structure was represented by two geometrical isomers of the side chains about the ring although the configuration could be determined for only half of the acids.

The second group of peaks (Fig. 5.16) comprised a greater proportion of the total CFAMs in normal (about 14%) than in high-oleate (about 4%) oil and were probably derived from linoleate (Dobson, Christie and Sebedio, 1997). There were several peaks with a molecular weight (m/z 331 for DMOX) 2 amu less than saturated monocyclic acids; they were unchanged after hydrogenation, verifying the saturated nature. The mass spectra of the DMOX derivatives and picolinyl esters suggested five-membered and six-membered ring structures with bonds across the ring to give bicyclic structures; the position of the bond could not be determined from the mass spectra. Again there were four basic structures, two five-membered ring [C-8 to C-12 (structure XVIII, Fig. 5.2) or C-10 to C-14 (structure XIX)] acids and two six-membered ring [C-7 to C-12 (structure XX) or C-10 to C-15 (structure XXI)] acids. The former acids had analogous structures to the major cyclopentenyl CFAMs derived from linoleate in laboratory-heated sunflower oil (Christie *et al.*, 1993) and a common origin was suggested. The latter acids were more abundant and their formation may be favoured to such an extent that the cyclohexenyl counterparts were not detected (Christie *et al.*, 1993).

Several CFAMs were characterized from a lightly partially hydrogenated soybean oil, both before and after heating, by GC–MS of the methyl esters of hydrogenated non-urea adduct mixtures (Rojo and Perkins, 1987). The mixtures would be expected to be particularly complicated because oleate (45%), linoleate (37%) and α-linolenate (2%) were all present in the starting oil. Although the relative amount of α-linolenate was low it has been shown that, under similar conditions, CFAMs are formed at about ten times the rate in α-linolenate than they are in linoleate (Sebedio, Prevost and Grandgirard, 1987). The profile may have been complicated even more by CFAMs from geometrical and positional double bond isomers formed by partial hydrogenation. Some of the expected products (cyclohexyl acids with rings from C-10 to C-15, and cyclopentyl acids with rings from C-10 to C-14 and C-11 to C-15) were detected, but others, notably cyclopentyl acids with rings from C-8 to C-12 (major product from linoleate) and C-5 to C-9 (from oleate) and cyclohexyl acids with rings from C-4 to C-9 (from oleate), were not found. Conversely, some cyclohexyl acids (with rings from C-8 to C-13, C-9 to C-14 and C-11 to C-16) were identified but the intact counterparts have not been encountered. A re-examination of similar mixtures by GC–MS using DMOX derivatives and picolinyl esters, whose mass spectra are easier to interpret than those of methyl esters, would be worthwhile.

The types of hydrogenated CFAMs from peanut and soybean oils, used in frying of frozen pre-fried french fries, were essentially similar, with only differences in the relative proportions of components (Sebedio et al., 1996a). At first sight this may seem surprising because only soybean oil contains significant amounts of α-linolenic acid, but similar hydrogenated CFAMs may be derived from different intact CFAMs (Table 5.1). However, one striking anomaly was the presence of methyl 10-(2-propylcyclopentyl)-decanoate from peanut oil because it would appear that this acid can be derived only from α-linolenic acid.

Information on variation of CFAM composition with heating conditions is limited. However, in one study the same types of CFAMs seemed to be present when sunflower and linseed oils were heated at 200°C under air or at 275°C under nitrogen; only the relative proportions and concentrations in the oil varied (Sebedio et al., 1987).

It is interesting that of the major hydrogenated CFAMs 10-(2'-propylcyclopentyl)decanoate, 7-(2'-hexylcyclopentyl)heptanoate and 3-(2'-nonylcyclohexyl)propanoate appear to be derived specifically from α-linolenate, linoleate and oleate, respectively (Table 5.1). Considering that use of hydrogenated CFAMs simplifies the analytical task, this information may be useful for determining the contribution to CFAM formation of the three major unsaturated acids from a variety of sources.

5.3.2 Quantification of monomeric cyclic fatty acids

Quantitative analysis of total CFAMs in oils has been reviewed by Sebedio and Grandgirard (1989) and has been updated (Le Quere and Sebedio, 1996). Earlier work can be summarized as follows. As pointed out by Le Quere and Sebedio (1996), direct GC analysis of total fatty acid methyl esters was not possible because of similar equivalent chain length values of CFAMs to C18 straight-chain unsaturated esters and their geometrical isomers, which are present in deodorized and/or heated vegetable oils. Hydrogenation simplified the CFAM mixture and, moreover, converted all C18 straight-chain unsaturated esters to stearate, but even on polar capillary columns CFAMs may be present on the tail of methyl stearate, making quantification difficult (Gente and Guillaumin, 1977; Grandgirard and Julliard, 1983; Sebedio, 1985).

A urea adduction step to concentrate CFAMs was introduced so that stearate was removed and low levels of CFAMs, encountered in many samples, could be analysed (Potteau, 1976). However, this approach had the drawbacks that a substantial proportion of CFAMs was lost and there was incomplete removal of stearate (Gere, Gertz and Morin, 1984). Also, although urea may eliminate some impurities which might interfere by GC, for small samples contamination may still be a problem. These problems were partially addressed in subsequent adaptations. Phenanthrene was added as an internal standard instead of the usual heptadecanoate prior to urea adduction because it would be expected to behave in a similar way to CFAMs with respect to adduction and therefore would take into account any CFAM losses (Gere, Gertz and Morin, 1984). A step for cleaning up the non-urea adduct, by means of reversed-phase C18 solid-phase extraction columns, was developed to remove contaminants from the CFAMs but there were inconsistent recoveries of phenanthrene, and it was suggested that a cyclic fatty acid standard should be considered (Rojo and Perkins, 1989b). This approach has potential particularly when small amounts of CFAMs are present and errors arising from the presence of contaminants might be considerable.

In short, it is now clear that a successful method requires a hydrogenation step, followed by a concentration step and finally a GC analysis of CFAMs. Polar columns such as CP-Sil 84 (Chrompack, Middelburg, The Netherlands) and BPX70 (SGE, Milton Keynes, UK) are the best choice because hydrogenated CFAMs are more clearly resolved from stearate than on less polar columns such as Carbowax (Sebedio, 1985). The troublesome step is the concentration of CFAMs. Ideally, a method should allow recovery of all CFAMs and totally exclude saturated fatty acids, particularly stearate, and other impurities which may interfere with subsequent GC.

A more promising approach was recently developed by Sebedio *et al.* (1994) with use of HPLC as the enrichment step. The procedure, as outlined in Fig. 5.17, can be applied both to vegetable oils and to biological samples and involves preparation of the methyl esters of the oil, or the extracted lipids of a biological sample, followed by addition of an internal standard. The sample is then hydrogenated and subjected to semi-preparative reversed-phase HPLC, and a fraction containing CFAMs and the internal standard between palmitate and stearate is collected (Fig. 5.18). The CFAMs were then quantified by GC on a polar capillary column and analysed by GC–MS to verify the presence of CFAMs and lack of contaminants. Using stearate and lipids from biological samples, spiked with added CFAM fractions from heated linseed and sunflower oils, the method was reproducible and accurate down to 50 ppm CFAMs. Also, unlike urea-adduction, HPLC fractionation was shown not to alter the CFAM profile.

The method appears to offer relative accuracy, sensitivity and speed of analysis over previous methods, but it would be of interest to test its robustness with a variety of samples, including vegetable oils. Although the procedure is applicable to CFAMs from sunflower and linseed oils,

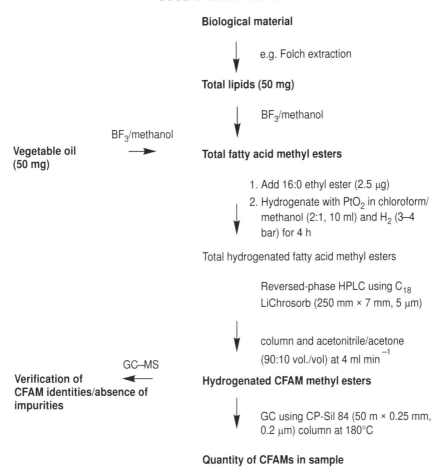

Figure 5.17 Scheme for quantification of cyclic fatty acid monomers (CFAMs) from vegetable oils and biological material. Adapted from Sebedio *et al.* (1994).

that is, essentially from linoleic and α-linolenic acids, in the light of recent findings that saturated CFAMs are formed from oleic acid (Dobson, Christie and Sebedio, 1996a) it is important to show that the HPLC fraction also includes these acids. The choice of internal standard is also worth mentioning. Either methyl heptadecanoate or, when this is present in significant amounts in the sample, ethyl hexadecanoate are recommended. It would seem advisable always to use ethyl hexadecanoate because low levels of odd-chain fatty acids, which may well be present at similar levels to the CFAMs, are often present. It may also be advisable to include a work-up of the sample without internal standard added so that it can be checked that impurities that may co-chromatograph by GC with the internal standard are absent.

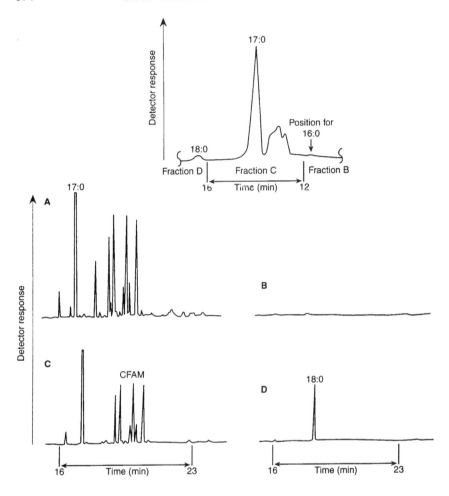

Figure 5.18 Reversed-phase high-pressure liquid chromatographic (HPLC) fractionation of a mixture of methyl heptadecanoate and methyl esters of hydrogenated cyclic fatty acid monomers isolated from a heated sunflower oil. A C18 (250 mm ×7 mm internal diameter, 5 μm particle size) column was used with acetonitrile/acetone (90:10 vol./vol.) mobile phase at a flow rate of 4 ml min^{-1}. A refractive index detector was employed. Gas chromatograms of the mixture (A) and HPLC fractions (B, C and D) were obtained by means of a CP-Sil 84™ (Chrompack), Middleburg. The Netherlands (50 m × 0.25 mm inner diameter, 0.2 μm film thickness) capillary column at an isothermal temperature of 180°C. Redrawn from Sebedio, J.-L., Prevost, J., Ribot, E. and Grandgirard, A., Utilization of high-performance liquid chromatography as an enrichment step for the determination of cyclic fatty acid monomers in heated fats and biological samples, *J. Chromatogr.*, **659**, 101–9, 1994.

In an industrial operation for production of pre-fried french fries using palm oil, the HPLC method was used to show that total CFAMs present increased from 0.02% (formed during deodorization) in the starting oil to a maximum of 0.08% and 0.10% in the oil and french fries, respectively

(Sebedio et al., 1991). Although not all hydrogenated CFAMs were identified, the profile, not surprisingly, appears to resemble the saturated CFAM profile derived from oleate in high-oleate sunflower oil (Dobson et al., 1996a), with some contribution from linoleate-derived CFAMs. In another study involving the deep frying of french fries in peanut and soybean oils, the quantity of CFAMs increased with number of frying operations and increasing temperature (180°C to 220°C) (Sebedio et al., 1996a). There were no quantitative differences between soybean and peanut oil, although the former contained significant amounts of α-linolenic acid which has been found to be more reactive in terms of CFAM production compared with linoleate (Sebedio, Prevost and Grandgirard, 1987).

The types of CFAMs in heated fish oils include C20 and C22 acids of unknown structure, derived from eicosapentaenoate and docosahexaenoate, respectively, together with those from C18 fatty acids (Sebedio and De Rasilly, 1993). An adaptation of the described HPLC method (Fig. 5.17; Sebedio et al., 1994) included additional fractions between 18:0 and 20:0 and between 20:0 and 22:0 (using 19:0 and 21:0 as internal standards) encompassing C20 and C22 CFAMs, respectively.

5.3.3 Incorporation of monomeric cyclic fatty acids into biological material

Recently, there have been a few studies concerned with the incorporation of CFAMs from heated linseed oil into biological tissue (Chardigny et al., 1995; Ribot et al., 1992; Sebedio et al., 1996b). In one study involving assimilation into rat heart cell cultures (Ribot et al., 1992), CFAMs were isolated from neutral and phospholipid fractions by transesterification, hydrogenation and urea adduction followed by GC–MS analysis for identification and determination of percentage composition of hydrogenated CFAMs. There was increased incorporation of cyclopentyl acids relative to cyclohexyl acids into both fractions. Attempts to examine individual phospholipid classes were successful only with the major component, phosphatidylcholine, because of CFAM losses from minor phospholipids during urea adduction. CFAMs, incorporated into rat heart cell cultures from media of varying CFAM concentrations, were quantified by means of the HPLC enrichment method described in Section 5.3.2 (Fig. 5.17; Sebedio et al., 1994). The same method was used to show that CFAMs were present in a human milk sample (Chardigny et al., 1995).

GC separation of all sixteen CFAMs from heated linseed oil can be achieved by using two different columns (Fig. 5.11; Dobson, Christie and Sebedio, 1996c). In a study where only small amounts of material were available, this approach was used to show that in pregnant rats fed a diet containing CFAMs from heated linseed oil, a cyclohexenyl acid with an E double bond at C-8 was preferentially incorporated into the liver phospholipids of the dam and the pups (Sebedio et al., 1996b).

5.4 Summary

Since the last review (Sebedio and Grandgirard, 1989) there have not been substantial advances in the development of new methodologies or in the novel application of existing methodologies in the analysis of natural cyclic fatty acids, and this trend is likely to continue. One area that seems to require attention is in the quantification of cyclopropene fatty acids. The absolute establishment of a simple method is needed, and a rigorous assessment of a direct GC method (Park and Rhee, 1988) would seem appropriate. There have only been a few publications concerned with the analysis of cyclic fatty acid-containing lipids such as diacylglycerols and triacylglycerols, and further advances in this area would not be surprising.

In contrast, there has been considerable progress in the characterization of cyclic fatty acids formed in heated vegetable oils. The structures of CFAMs from α-linolenic, linoleic and oleic acids have now been almost completely elucidated. This has been possible by the use of silver-ion HPLC for simplifying the mixtures, followed by GC–MS characterization of nitrogen-containing derivatives and GC–FTIR. Still to be determined are the configurations of side chains about the rings for some acids (especially for CFAMs from linoleate), confirmation of positions of double bonds in cyclopentenyl rings, detection of other possible minor components suggested from studies of hydrogenated mixtures and further characterization of complex mixtures of bicyclic acids. Apart from some information on heating at different temperatures and presence or absence of air (Sebedio *et al.*, 1987) little is known about the effects of frying conditions on the types and quantities of CFAMs.

It has been speculated that CFAMs are formed by mechanisms involving allylic radicals (Christie *et al.*, 1993; Dobson, Christie and Sebedio, 1996a, 1996b, 1997; Dobson *et al.*, 1995) and it would be of interest to test out these proposals experimentally. Finally, the state of present knowledge is such that it would be appropriate to build on an initial study (Sebedio *et al.*, 1996b) and examine the incorporation of individual CFAMs in biological systems and, moreover, to study the toxicological effects of those CFAMs.

References

Adams, J., Deterding, L. J. and Gross, M. L. (1987) Tandem mass spectrometry for determining structural features of fatty acids. *Spectros. Int. J.*, **5**, 199–228.

Ahmad, M. S., Ahmad, M. U., Osman, S. M. and Ballantine, J. A. (1979) *Eriolaena hookeriana* seed oil: a rich source of malvalic acid. *Chem. Phys. Lipids*, **25**, 29–38.

Bailey, A. V., Boudreaux, G. J. and Skau, E. L. (1965) Determination of cyclopropenoid fatty acids. VI. A direct infrared absorption method. *J. Am. Oil Chem. Soc.*, **42** (7), 637–8.

Bianchini, J.-P., Ralaimanarivo, A. and Gaydou, E. M. (1981) Determination of cyclopropenoic and cyclopropanoic fatty acids in cottenseed and kapok seed oils by gas–liquid chromatography. *Anal. Chem.*, **53**, 2194–201.

Boudreaux, G. J., Bailey, A. V. and Tripp, V. W. (1972) High resolution NMR for purity determination of cyclopropenoid concentrates. *J. Am. Oil Chem. Soc.*, **49**, 278–80.
Bussell, N. E. and Miller, R. A. (1979) Analysis of hydroxyl, unsaturated, and cyclopropane fatty acids by high pressure liquid chromatography. *J. Liquid Chromatogr.*, **2** (5), 697–718.
Chardigny, J.-M., Wolff, R. L., Mager, E. *et al.*, (1995) Trans mono- and polyunsaturated fatty acids in human milk. *Eur. J. Clin. Nutr.*, **49**, 523–31.
Christie, W. W. (1970) Cyclopropane and cyclopropene fatty acids, in *Topics in Lipid Chemistry, Vol. 1* (ed. F. D. Gunstone), Logos Press, London, pp. 1–49.
Christie, W. W. (1987) *HPLC and Lipids: A Practical Guide*, Pergamon Press, Oxford.
Christie, W. W. (1989) *Gas Chromatography and Lipids: A Practical Guide*, The Oily Press, Dundee.
Christie, W. W. and Holman, R. T. (1966) Mass spectrometry of lipids. 1. Cyclopropane fatty acid esters. *Chem. Phys. Lipids*, **1** (3), 176–82.
Christie, W. W., Brechany, E. Y. and Shukla, V. K. S. (1989) Analysis of seed oils containing cyclopentenyl fatty acids by combined chromatographic procedures. *Lipids*, **24** (2), 116–20.
Christie, W. W., Rebello, D. and Holman, R. T. (1969) Mass spectrometry of derivatives of cyclopentenyl fatty acids. *Lipids*, **4** (3), 229–31.
Christie, W. W., Brechany, E. Y., Sebedio, J.-L., and Le Quere, J.-L. (1993) Silver ion chromatography and gas chromatography–mass spectrometry is the structural analysis of cyclic monoenoic acids formed in frying oils. *Chem. Phys. Lipids*, **66**, 143–53.
Christopher, R. K. and Duffield, A. M. (1980) Use of vinyl methyl ether as a chemical ionization reagent gas for gas chromatographic chemical ionization mass spectrometric discrimination between cyclopropanoid and monoenoic fatty acid methyl esters. *Biomed. Mass Spectrom.*, **7** (10), 429–32.
Coleman, E. C. (1970) Evaluation of five methods for the quantitative determination of cyclopropenoid fatty acids. *J. Ass. Offic. Anal. Chem.*, **53** (6), 1209–13.
Conway, J., Ratnayake, W. M. N. and Ackman, R. G. (1985) Hydrazine reduction in the gas liquid chromatographic analysis of the methyl esters of cyclopropenoic fatty acids. *J. Am. Oil Chem. Soc.*, **62** (9), 1340–3.
Dobson, G. and Christie, W. W. (1996) Structural analysis of fatty acids by mass spectrometry of picolinyl esters and dimethyloxazoline derivatives. *Trends Anal. Chem.*, **15**(3), 130–7.
Dobson, G., Christie, W. W. and Sebedio, J.-L. (1996a) Monocyclic saturated fatty acids formed from oleic acid in heated sunflower oils. *Chem. Phys. Lipids*, **82**, 101–10.
Dobson, G., Christie, W. W. and Sebedio, J.-L. (1996b) The nature of cyclic fatty acids formed in heated vegetable oils. *Grasas y Aceites*, **4**, 26–33.
Dobson, G., Christie, W. W. and Sebedio, J.-L. (1996c) Gas chromatographic properties of cyclic dienoic acids formed in heated linseed oil. *J. Chromatogr. A*, **723**, 349–54.
Dobson, G., Christie, W. W. and Sebedio, J.-L. (1997) Saturated bicyclic fatty acids formed in heated sunflower oils. *Chem. Phys. Lipids*, **87**, 137–47.
Dobson, G., Christie, W. W., Brechany, E. Y. *et al.* (1995) Silver ion chromatography and gas chromatography–mass spectrometry in the structural analysis of cyclic dienoic acids formed in frying oils. *Chem. Phys. Lipids*, **75**, 171–82.
Eisele, T. A., Libbey, L. M., Pawlowski, N. E. *et al.* (1974) Mass spectrometry of the silver nitrate derivatives of cyclopropenoid compounds. *Chem. Phys. Lipids*, **12**, 316–26.
Frostegard, A., Tunlid, A. and Baath, E. (1993) Phospholipid fatty acid composition, biomass, and activity of microbial communities from two soil types experimentally exposed to two different heavy metals. *Appl. Environ. Microbiol.*, **59** (11), 3605–17.
Gaydou, E. M., Bianchini, J.-P. and Ralaimanarivo, A. (1983) Determination of cyclopropenoic fatty acids by reversed-phase liquid chromatography and gas chromatography. *Anal. Chem.*, **55**, 2313–17.
Gaydou, E. M., Ralaimanarivo, A. and Bianchini, J.-P. (1993) Cyclopropanoic fatty acids of litchi (*Litchi chinensis*) seed oil. A reinvestigation. *J. Agric. Food Chem.*, **41** (6), 886–90.
Gente, M. and Guillaumin, R. (1977) Determination of cyclic monomers. *Rev. Fr. Corps Gras*, **24** (4), 211–18.
Gere, A., Gertz, C. and Morin, O. (1984) Methodological studies on the quantitation of cyclic monomers formed in fats during heating. *Rev. Fr. Corps Gras*, **31** (9), 341–6.

Grandgirard, A. and Julliard. F. (1983) Determination of cyclic monomers in heated oils – a critical study. *Fr. Corps Gras*, **30** (3), 123–9.

Grondin, I., Smadja, J., Farines, M. and Soulier, J. (1993) The lipids of litchi and longan seeds. Study of cyclopropanoic acids by NMR. *Oleagineux*, **48** (10), 425–8.

Harvey, D. J. (1984) Picolinyl derivatives for the characterization of cyclopropane fatty acids by mass spectrometry. *Biomed. Mass Spectrom.*, **11** (4), 187–92.

Harvey, D. J. (1992) Mass spectrometry of picolinyl and other nitrogen-containing derivatives of lipids, in *Advances in Lipid Methodology – One* (ed. W. W. Christie), The Oily Press, Dundee, pp. 19–80.

Hippchen, B., Roell, A. and Poralla, K. (1981) Occurrence in soil of thermo-acidophilic bacilli possessing ω-cyclohexane fatty acids and hopanoids. *Arch. Microbiol.*, **129** (1), 53–5.

Hooper, N. K. and Law, J. H. (1968) Mass spectrometry of derivatives of cyclopropene fatty acids. *J. Lipid Res.*, **9**, 270–5.

Howarth, O. W. and Vlahov, G. (1996) ^{13}C nuclear magnetic resonance study of cyclopropenoid triacylglycerols. *Chem. Phys. Lipids*, **81**, 81–5.

Kint, S., Lundin, R. E., Waiss, A. C. and Elliger, C. A. (1981) Analysis of cyclopropenoid fatty acids by Raman spectroscopy. *Anal. Biochem.*, **118**, 364–70.

Le Quere, J.-L. and Sebedio. J.-L. (1996) Cyclic monomers of fatty acids, in *Deep Frying: Chemistry, Nutrition and Practical Applications* (eds E. G. Perkins and M. D. Erickson), AOCS Press, Champaign, IL, pp. 49–88.

Le Quere, J.-L., Sebedio, J.-L., Henry, R. *et al.* (1991) Gas chromatography–mass spectrometry and gas chromatography–tandem mass spectrometry of cyclic fatty acid monomers isolated from heated fats. *J. Chromatogr.*, **562**, 659–72.

Le Quere, J.-L., Semon, E., Lanher, B. and Sebedio, J.-L. (1989) On-line hydrogenation in GC–MS analysis of cyclic fatty acid monomers isolated from heated linseed oil. *Lipids*, **24** (4), 347–50.

Loveland, P. M., Pawlowski, N. E., Libbey, L. M., *et al.* (1983) HPLC analysis of cyclopropenoid fatty acids. *J. Am. Oil Chem. Soc.*, **60** (10), 1786–8.

Mayberry, W. R. and Lane, J. R. (1993) Sequential alkaline saponification/acid hydrolysis/esterification: a one-tube method with enhanced recovery of both cyclopropane and hydroxylated fatty acids. *J. Microbiol. Methods*, **18**, 21–32.

Minnikin, D. E. (1978) Location of double bonds and cyclopropane rings in fatty acids by mass spectrometry. *Chem. Phys. Lipids*, **21**, 313–47.

Mossoba, M. M., Yuracwecz, M. P., Lin, H. S., *et al.* (1995a) Application of GC–MS–FTIR spectroscopy to the structural elucidation of cyclic fatty acid monomers. *Am. Lab. (Shelton, Conn.)*, **27**, 16K–O.

Mossoba, M. M., Yuracwecz, M. P., Roach, J. A. G. *et al.* (1994) Rapid determination of double bond configuration and position along the hydrocarbon chain in cyclic fatty acid monomers. *Lipids*, **29** (12), 893–6.

Mossoba, M. M., Yuracwecz, M. P., Roach, J. A. G. *et al.* (1995b) Elucidation of cyclic fatty acid monomer structures. Cyclic and bicyclic ring sizes and double bond position and configuration. *J. Am. Oil Chem., Soc.*, **72** (6), 721–7.

Mossoba, M. M., Yuracwecz, M. P., Roach, J. A. G. *et al.*, (1996a) Analysis of cyclic fatty acid monomer 2-alkenyl-4,4-dimethyloxazoline derivatives by gas chromatography–matrix isolation–Fourier transform infrared spectroscopy. *J. Agric. Food Chem.*, **44**, 3193–6.

Mossoba, M. M., Yuracwecz, M. P., Roach, J. A. G. *et al.* (1996b) Confirmatory mass-spectral data for cyclic fatty acid monomers. *J. Am. Oil Chem. Soc.*, **73** (10), 1317–21.

Nixon, J. E., Eisele, T. A., Wales, J. H. and Sinnhuber, R. O. (1974) Effect of subacute toxic levels of dietary cyclopropenoid fatty acids upon membrane function and fatty acid composition in the rat. *Lipids*, **9** (5), 314–21.

Park, S. W. and Rhee, K. C. (1988) A capillary GC determination of cyclopropenoid fatty acids in cottonseed oils. *J. Food Sci.*, **53** (5), 1497–502.

Pasha, M. K. and Ahmad, F. (1992) Analysis of triacylglycerols containing cyclopropene fatty acids in *Sterculia foetida* (Linn.) seed lipids. *J. Agric. Food Chem.*, **40**, 626–9.

Pawlowski, N. E., Nixon, J. E. and Sinnhuber, R. O. (1972) Assay of cyclopropenoid lipids by nuclear magnetic resonance. *J. Am. Oil Chem. Soc.*, **49**, 387–92.

Poralla, K. and Koenig, W. A. (1983) The occurrence of ω-cycloheptane fatty acids in a thermoacidophilic bacillus. *FEMS Microbiol. Lett.*, **16** (2–3), 303–6.

Potteau, B. (1976) Presence of monomeric acids of cyclic structure type in the milk of the female rat fed thermopolymerized linseed oil. *Ann. Nutr. Aliment.*, **30** (1), 89–93.

Ralaimanarivo, A., Gaydou, E. M. and Bianchini, J.-P. (1982) Fatty acid composition of seed oils from six *Adansonia* species with particular reference to cyclopropane and cyclopropene acids. *Lipids*, **17** (1), 1–10.

Rezanka, T. and Dembitsky, V. M. (1994) Identification of unusual cyclopropane fatty acids from the deep-water lake invertebrate *Acanthogammarus grewingkii*. *Comp. Biochem. Physiol.*, **109B** (2/3), 407–13.

Ribot, E., Grandgirard, A., Sebedio, J.-L. *et al.* (1992) Incorporation of cyclic fatty acid monomers in lipids of rat heart cell cultures. *Lipids*, **27** (1), 79–81.

Rojo, J. A. and Perkins, E. G. (1987) Cyclic fatty acid monomer formation in frying fats. 1. Determination and structural study. *J. Am. Oil Chem. Soc.*, **64** (3), 414–21.

Rojo, J. A. and Perkins, E. G. (1989a) Chemical synthesis and spectroscopic characteristics of C_{18} 1,2-disubstituted cyclopentyl fatty acid methyl esters. *Lipids*, **24** (6), 467–76.

Rojo, J. A. and Perkins, E. G. (1989b) Cyclic fatty acid monomer: isolation and purification with solid phase extraction. *J. Am. Oil Chem. Soc.*, **66** (11), 1593–5.

Schuch, R., Ahmad, F. and Mukherjee, K. D. (1986) Composition of triacylglycerols containing cyclopropene fatty acids in seed lipids of Munguba (*Bombax munguba* Mart.). *J. Am. Oil Chem. Soc.*, **63** (6), 778–83.

Sebedio, J.-L. (1985) Application of methoxy-bromomercuric-adduct fractionation to the study of cyclic fatty acid monomers from a heated linseed oil. *Fette Seifen Anstrichm.*, **87** (7), 267–73.

Sebedio, J.-L. and De Rasilly, A. (1993) Analysis of cyclic fatty acids in fish oil concentrates, in *Proceedings of the 17th Nordic Lipid Symposium*, Imatra, Finland, pp. 212–16.

Sebedio, J.-L. and Grandgirard, A. (1989) Cyclic fatty acids: natural sources, formation during heat treatment, synthesis and biological properties. *Prog. Lipid Res.*, **28**, 303–36.

Sebedio, J.-L., Prevost, J. and Grandgirard, A. (1987) Heat treatment of vegetable oils I. Isolation of the cyclic fatty acid monomers from heated sunflower and linseed oils. *J. Am. Oil Chem. Soc.*, **64** (7), 1026–32.

Sebedio, J.-L., Catte, M., Boudier, M. A. *et al.* (1996a) Formation of fatty acid geometrical isomers and of cyclic fatty acid monomers during the finish frying of frozen prefried potatoes. *Food Res. Internat.*, **29** (2), 109–16.

Sebedio, J.-L., Chardigny, J.-M., Juaneda, P. *et al.* (1996b) Nutritional impact and selective incorporation of cyclic fatty acid monomers in rats during reproduction, in *Proceedings of the 21st World Congress of the International Society for Fat Research, The Hague, 1995*, PJ Barnes & Associates, Bridgwater, pp. 307–10.

Sebedio, J.-L., Kaitaranta, J., Grandgirard, A. and Malkki, Y. (1991) Quality assessment of industrial prefried french fries. *J. Am. Oil Chem. Soc.*, **68** (5), 299–302.

Sebedio, J.-L., Le Quere, J.-L., Semon, E. *et al.* (1987) Heat treatment of vegetable oils. II. GC–MS and GC–FTIR spectra of some isolated cyclic fatty acid monomers. *J. Am. Oil Chem. Soc.*, **64** (9), 1324–33.

Sebedio, J.-L., Le Quere, J.-L., Morin, O. *et al.* (1989) Heat treatment of vegetable oils. III. GC–MS characterization of cyclic fatty acid monomers in heated sunflower and linseed oils after total hydrogenation. *J. Am. Oil Chem. Soc.*, **66**, 704–9.

Sebedio, J.-L., Prevost, J., Ribot, E. and Grandgirard, A. (1994) Utilization of high-performance liquid chromatography as an enrichment step for the determination of cyclic fatty acid monomers in heated fats and biological samples. *J Chromatogr.*, **659**, 101–9.

Shukla, V. K. S. and Spener, F. (1985) High-performance liquid chromatography of triglycerides of *Flacourtiaceae* seed oils containing cyclopentenyl fatty acids (chaulmoogric oils). *J. Chromatogr.*, **348**, 441–6.

Spitzer, V. (1991) GC–MS characterization (chemical ionization and electron impact modes) of the methyl esters and oxazoline derivatives of cyclopropenoid fatty acids. *J. Am. Oil Chem. Soc.*, **68** (12), 963–9.

Spitzer, V. (1995) The mass spectra of the 4,4-dimethyloxazoline derivatives of the methoxymethyl olefins of malvalic and sterculic acids. *J. Am. Oil Chem. Soc.*, **72** (3), 389–90.

Spitzer, V., Marx, F., Maia, J. G. S. and Pfeilsticker, K. (1994) The mass spectra of the picolinyl ester derivatives of malvalic and sterculic acid. *Fat Sci. Technol.*, **10**, 395–6.

Taylor, J. and Parkes, R. J. (1983) The cellular fatty acids of the sulphate-reducing bacteria, *Desulfobacter* sp., *Desulfobulbus* sp. and *Desulfovibrio desulfuricans*. *J. Gen. Microbiol.*, **129**, 3303–9.

Tomer, K. B., Crow, F. W. and Gross, M. L. (1983) Location of double bond position in unsaturated fatty acids by negative ion MS/MS. *J. Am. Chem. Soc.*, **105**, 5487–8.

Tomer, K. B., Jensen, N. J. and Gross, M. L. (1986) Fast atom bombardment and tandem mass spectrometry for determining structural modification of fatty acids. *Anal. Chem.*, **58**, 2429–33.

Vatele, J. M., Sebedio, J.-L. and Le Quere, J.-L. (1988) Cyclic fatty acid monomers: synthesis and characterization of methyl ω-(2-alkylcyclopentyl) alkenoates and alkanoates. *Chem. Phys. Lipids*, **48**, 119–28.

White, P. J. (1991) Methods for measuring changes in deep-fat frying oils. *Food Technol.*, **45** (2), 75–80.

Wood, R. (1986a) High-performance liquid chromatography analysis of cyclopropene fatty acids. *Biochem. Arch.*, **2**, 63–71.

Wood, R. (1986b) Comparison of the cyclopropene fatty acid content of cottonseed varieties, glanded and glandless seeds, and various seed structures. *Biochem. Arch.*, **2**, 73–80.

Zelles, L. and, Bai, Q. Y. (1994) Fatty acid patterns of phospholipids and lipopolysaccharides in environmental samples. *Chemosphere*, **28** (2), 391–411.

Zhang, J. Y., Wang, H. Y., Yu, Q. T. *et al.* (1989) The structures of cyclopentenyl fatty acids in the seed oils of *Flacourtiaceae* species by GC–MS of their 4,4-dimethyloxazoline derivatives. *J. Am. Oil Chem. Soc.*, **66** (2), 242–6.

Zhang, J. Y., Yu, Q. T. and Huang, Z. H. (1987) 2-Substituted 4,4-dimethyloxazolines as useful derivatives for the localization of cyclopropane rings in long-chain fatty acids. *Mass Spectros.*, **35** (1), 23–30.

Zollner, P. and Schmid, R. (1996) Utility of nicotinoyl derivatives in structural studies of mono- and diacylglycerols by gas chromatography/mass spectrometry. *J. Mass Spectrom.*, **31**, 411–17.

6 Mass spectrometry of complex lipids
A. KUKSIS

6.1 Introduction

Recent developments in mass spectrometry (MS) designed to accommodate high molecular weight molecules such as polypeptides and polynucleotides have also revolutionized the MS of complex lipid molecules. Soft ionization MS techniques such as fast atom bombardment (FAB), thermospray (TS) and electrospray (ES) have the ability to ionize lipid molecules without causing extensive fragmentation. This generally leads to mass spectra characterized by $[M + 1]^+$ (positive ion mode) or $[M - 1]^-$ (negative ion mode) ions. Fragmentation of the molecular ion species produced by soft ionization processes is achieved in a second mass spectrometer (MS/MS) by collision-induced dissociation (CID). The FAB/MS and TS/MS methods with or without on-line high-pressure liquid chromatography (HPLC) provided the initial successes in unravelling the structure of complex lipids. Subsequently ES/MS and ES/MS/MS with flow injection (Han and Gross, 1994; Kerwin, Tuininga and Ericsson, 1994) and ES/MS with HPLC (LC/ES/MS) (Kim, Wang and Ma, 1994; Myher et al., 1994) proved superior to the earlier approaches and are currently being applied in the analyses of molecular species from minimal amounts of complex lipids without derivatization. The ES/MS methods have proven especially well suited for analyses of oxygenated cholesteryl esters and oxygenated glycerolipids, including the hydroperoxides and ozonides, which until recently had escaped MS characterization. This chapter reviews recent applications of both positive and negative ion ES/MS in combination with HPLC and ES/MS/MS with flow injection, as well as recent applications of FAB/MS/MS for the rapid and sensitive identification and quantification of molecular species of complex lipids in synthetic and natural mixtures. The use of other ionization techniques in recent studies of complex lipid molecules are also discussed.

6.2 Equipment and principles of soft ionization mass spectrometry

The basic hardware employed in soft ionization MS and MS/MS of lipids has been discussed in detail by Le Quere (1993) and Murphy (1993) and the subject has been updated by Kuksis and Myher (1995) and Myher and

Kuksis (1995). This review is to be confined largely to the quadrupole instruments, which have proven best-suited for computer-controlled operation. The discussion is divided according to the methods used for sample introduction into direct probe/MS, GC/MS and LC/MS operation.

6.2.1 Direct probe mass spectrometry

Direct probe insertion has been the standard method for sample introduction into the FAB/MS system, whereas for TS/MS and ES/MS flow injection has often been used for this purpose. The direct probe method requires an appropriate matrix for sample suspension in order to effect proper ionization by FAB (Murphy and Harrison, 1994). No derivatization or sample pre-fractionation is required with any of these methods. Usually, in FAB/MS molecular and fragment ions can be observed. These provide sufficient information for the identification of the structure of the solute. In principle, the flow injection of samples employed in TS/MS and ES/MS is similar to direct probe injection as solute mixtures are subjected to ionization without prior fractionation (Myher and Kuksis, 1995). However, because TS/MS and ES/MS methods require solvent evaporation certain limits exist on the liquids that can be employed for these analyses. The milder ionization methods yield essentially molecular ions, $[M + 1]^+$ or $[M - 1]^-$, depending on the ionization mode. As these methods do not yield fragment ions it is not possible to determine the structure of the solutes in the absence of chromatographic pre-fractionation. In order to ascertain the origin of the fragment ions in FAB/MS, and the structure of the molecular ions in flow injection TS/MS or ES/MS, the molecular ions may be subjected to collision-induced dissociation (CID) in a second mass spectrometer (MS/MS). The most widely used pieces of equipment for MS/MS are the triple quadrupole instruments in which the first and third quadrupoles are operated in the usual way for mass analysis by combination of radio frequency (r.f.) and direct current (d.c.) voltages. The second quadrupole is operated in the r.f.-only mode. The purpose of the second quadrupole is to serve as a high-pressure CID cell and to focus the fragment ions scattered from the central axis of the ion beam by the collision process. The FAB/MS of lipids has been reviewed by Kim and Salem (1993), Murphy (1993) and Murphy and Harrison (1994).

6.2.2 Gas chromatography/mass spectrometry

Volatile lipids or lipid derivatives are conveniently introduced into the ion source of the mass spectrometer by a gas chromatograph (Kuksis and Myher, 1989). The modern capillary columns can be inserted directly into the ion source thus avoiding the use of an interface. This method of sample introduction is widely employed for fatty acid and diacylglycerol (DG) analysis, but other higher molecular weight solutes may also be introduced in this manner into the ion source (Kuksis and Myher, 1989). The GC/MS system ordinarily employs electron-impact ionization (EI), but chemical ionization (CI) may be achieved by using hydrogen as a carrier gas or by a post-column addition of other appropriate ionization gases. The GC/MS system may be expanded to provide additional fragmentation of any molecular or daughter ions in a second mass spectrometer (GC/MS/MS).

6.2.3 Liquid chromatography/mass spectrometry

Both LC/MS and LC/MS/MS are now widely employed for the analysis of polar lipids without derivatization. Although the choice of solvents is somewhat limited and large volumes of liquid have to be evaporated, the advantages gained by this approach justify the expense and extra effort. Whereas the high flow rates of conventional HPLC can be more or less easily accommodated by FAB and TS ionization, ES ionization can accommodate flow rates of $1-20\,\mu l\,min^{-1}$ only, and post-column flow splitting is necessary in order to combine it with conventional HPLC. The problem of excessive flow rates can be minimized by using either pneumatic or ultrasonic-assisted nebulization (Voyksner, Linder and Keever, 1994). Nebulizer-assisted ES is sometimes called ion-spray, but in this review the term ES will be used for both conventional and nebulizer-assisted ES. The high volume of the effluent requires the use of high temperatures for solvent evaporation, which may affect the stability of the more labile molecules. The use of narrow bore columns or microcolumns avoids large effluent volumes and permits operation at much lower evaporation temperatures.

A pseudo ES/MS/MS effect may be obtained in a single quadrupole ES/MS system equipped with a dual-stage ES ionization source (Voyksner and Pack, 1991). With a low voltage (50-120 V) applied to the capillary exit, the molecular ionic species remain intact and the molecular mass of the analyte is obtained. However, when higher voltages are applied to the capillary exit, extensive and reproducible fragmentation of the molecular or molecular adduct ion can be effected. The fragmentation arises by means of a CID process, which is identical to that which takes place in the collision chamber of an MS/MS system. According to Voyksner and Pack (1991) this procedure can produce CID spectra of singly charged species with greater sensitivity than can be achieved with ES/MS/MS systems because the fragment ions produced in the ES/CID region are more effectively transferred into the quadrupole mass analyser with much less off-axis scatter than the transfer of fragment ions from the CID process within the MS/MS lens system. Thus, for LC/ES/CID/MS, an HPLC system is used instead of a mass analyser to carry out the separation of a single component in the mixture. The ES source is used as the collision chamber, where CID takes place, and, finally, a single quadrupole (quad 1) is used to mass analyse the product ions. In many experiments, HPLC is a better method for separation than the use of a mass filter, particularly for isomeric compounds such as positional isomers, *cis/trans* isomers, *syn/anti* isomers, diastereomers and enantiomers for which MS/MS spectra are not sufficiently different to distinguish their individual identities. Voyksner and Pack (1991) have concluded that the claim that MS/MS has greater specificity than has single MS does not apply when comparing ES/MS/MS to LC/ES/CID/MS. The principles of ES/MS techniques for the analysis of lipids have been recently reviewed (Kuksis and Myher, 1995; Myher and Kuksis, 1995).

6.3 Applications

This review is restricted largely to discussion of LC/MS and LC/MS/MS combinations for glycerolipid, glycerophospholipid and sphingolipid analyses. Recent applications to complex fatty acids and other non-glycerolipids are discussed as required for structural analyses of glycerolipids and sphingolipids.

6.3.1 Fatty acids

Fatty acids are readily separated and identified by GC/MS following addition of methyl (Me), pentafluorobenzyl (PFB), or other alkyl groups to the carboxyl moieties, and trimethylsilyl (TMS) groups to any hydroxyl functions, and methoxime groups to carbonyl functions prior to analysis by EI or CI (Murphy, 1993; Murphy and Zirrolli, 1994; Couderc, 1995).

Unsaturated and cyclic fatty acids. The determination of double bond positions in long alkyl chains of fatty acids and similar compounds has been a long-standing problem in the field of MS, although the preparation of derivatives containing pyridine rings, for example picolinyl and nicotinoyl esters, have been helpful (Harvey, 1991). Detailed reviews of selected topics in this area have been prepared by Harvey (1992). Recently, Zoellner, Lorbeer and Remberg (1994) have shown that in the mass spectra of the nicotinoyl derivatives of saturated monoacylglycerols (MGs) and diacylglycerols (DGs) the structure of the alkyl chains is reflected clearly by fragmentation patterns which are induced by the nicotinoyl moiety. Zoellner and Lorbeer (1995) have reported that the nicotinoyl derivatives of MGs and DGs are suitable for structural studies by GC/MS. In their EI mass spectra, each methylene group and double bond of the alkyl chain was reflected by a fragmentation pattern in the high-mass region, which was caused by radical-induced cleavage of the alkyl chains following random hydrogen abstraction by the pyridine nucleus. An accurate determination of double bond positions in MGs and DGs was made possible by characteristic spacings between abundant diagnostic ions in this fragmentation pattern. Zoellner and Schmid (1996) have since extended this approach to the positional location of methyl branches and epoxy and cyclopropyl rings in MGs and DGs by noting characteristic features in the fragmentation patterns. The nicotinoyl fragmentation pattern of 1-mono-(16-methylheptadecanoyl)glycerol could be observed in the high-mass region between the McLafferty ion (m/z 344) and the molecular ion (m/z 568). All methylene groups of the alkyl chains were reflected by peak clusters separated by 14 amu spacings. The methyl branching at position 16 of the fatty acid alkyl chain produced striking characteristics in this regular series of fragment ions. As ion formation at the branch point requires the rupture of two C—C bonds, the ion at m/z 539 was greatly reduced in relative abundance. The presence of epoxy rings in the MGs and DGs was clearly indicated in the mass spectra of the nicotinoyl derivatives by an abundant $[M - 1]^-$ ion (loss of hydrogen atoms) and an abundant $[M - 18]^-$ ion (loss of water molecules). The nicotinoyl-induced fragment ion series extends in the case of the 1-mono(9,10-epoxyoctadecanoyl)glycerol from the McLafferty ion at m/z 344 up to the $[M - 1]^-$ ion at m/z 581. The position of the epoxy ring could be easily determined in the mass spectrum by the remarkably high abundance of four ions at m/z 483, 469, 427 and 413, which reflected carbon–carbon bond cleavages on both sides of the epoxy ring. Oxidative drying of linoleic and linolenic acid ester films leads to formation of cross-links. Muizebelt and Nielsen (1996) have used direct chemical ionization (DCI) FAB, field desorption (FD), ES and EI/MS to study the oligomer mixtures resulting from cross-linking. Consistent results were obtained with all techniques. Oligomers yielded signals consisting of groups of peaks 16 amu apart, pointing to a series of oxygenated homologues. With ES/MS and EI/MS, oligomer formation up to the tetramer (linoleic) or hexamer (ricinoic) was observed. From the detailed ES/MS and DCI/MS spectra, a difference in the cross-linking mechanism between conjugated and non-conjugated fatty acids became apparent (radical addition to the double bond or recombination of radicals, respectively).

Cyclopentyl and cyclohexyl fatty acid monomers are present in heated vegetable oils. Sebedio *et al.* (1989) have used GC/MS to identify the major products from heated linseed oil as a mixture of 1:1 *cis* and *trans* cyclopentyl and cyclohexyl isomers; the products of heated sunflower oil were found to be mostly cyclopentyl isomers. The major cyclopentyl isomers were the *cis* and *trans* isomers of methyl 7-(2'-hexyl cyclopentyl)-heptanoates and methyl 10-(2'-propylcyclopentyl)-decanoates. The major cyclohexyl isomers were *trans* and *cis* methyl 9-(2'-propylcyclohexyl)-nonanoates and made up 50% of the cyclic fatty acid isomers of linseed oil. Dobson *et al.* (1995) have used $AgNO_3$ and GC/MS for the structural analysis of cyclic diene fatty acids, and Mossoba *et al.* (1996) have presented confirmatory MS data for cyclic fatty acid monomers by using deuteration.

Ford et al. (1996) have applied ES/MS/MS to demonstrate that unsaturated acyl carnitines are the predominant molecular species of acylcarnitine which accumulate during myocardial ischemia (5–20 min): 18:2 (m/z 424) > 18:1 (m/z 426) > 16:0 (m/z 400) > 18:0 (m/z 428) carnitines. On prolonged global ischemia (20 min) substantial amounts of 20:1 (m/z 454) and 20:2 (m/z 452) acylcarnitine species represented 15% of the total increase in accumulation. The identity of the acylcarnitine species was confirmed by ES/MS/MS which demonstrated the production of ions at m/z 85 (carnitine) and the respective fatty acid in each case.

Hydroxy and hydroperoxy fatty acids. Free radical autoxidation of linoleic acid produces a mixture of mainly *cis, trans* conjugated diene 9 and 13-hydroperoxides (LOOH) as primary oxidation products. The hydroperoxides decompose in the presence of divalent metal ions to the corresponding alkoxy radicals, which react with a hydrogen to yield the corresponding hydroxyoctadecanoic acids or are stabilized as oxodienoic acids. Alternatively, alkoxy radicals can undergo secondary reactions, for example β-scission by homolytic cleavage between the alkoxy radical carbon and an adjacent C—C bond, producing aldehydes and secondary radicals (Esterbauer, Zollner and Schauer, 1990; Frankel, 1985; Porter, 1984; Porter, Caldwell and Mills, 1995). The hydroxy fatty acids are readily identified and the location of the site of oxygenation, determined by GC of the TMS ethers of the Me esters. Thus, Lenz et al. (1990) prepared (after triphenylphosphine reduction) the TMS ethers of hydroxy and hydroperoxy fatty acids for the GC/MS identification of the HPLC peaks obtained from the original methyl esters.

Kim and Sawazaki (1993) reported structural analysis of hydroxy fatty acids by LC/TS/MS/MS. The method provided a sensitive and convenient technique for the oxygenated products of polyunsaturates. Analysis of PFB derivatives in the negative ion (NI) mode under filament-on or discharge-on conditions generated abundant $[M - PFB]^-$ ions. These ions were further fragmented by CID with argon and detected in the NI mode. The NI fragmentation pattern was examined for various oxygenated polyunsaturated fatty acid standards as well as for their deuterated and/or hydrogenated forms. Characteristic fragmentation occurred at the oxygenated C—C bonds, allowing unambiguous determination of the sites of oxygenation. The amount of sample required was typically in the low tenths of a nanogram range. Using this method, the structures of epoxy and hydroxy derivatives of 4,7,10,13,16,19-docosahexaenoic acid (22:6n3) formed by soybean lipoxygenase were determined to be 13-hydroxy-16,17-epoxy-22:5n3 and 15-hydroxy-16,17-epoxy-22:5n3.

The newer MS techniques, such as FAB/MS, TS/MS and ES/MS, allow the analysis of non-derivatized products. The combination of these techniques with MS/MS has also facilitated the identification of eicosanoids.

Wolf et al. (1994) presented a method for analysing leukotrienes C_4, D_4 and E_4 (LTC_4, LTD_4 and LTE_4, respectively) and prostaglandin E_2 (PGE_2) based on HPLC using a C_{18} reversed-phase column eluted with methanol/water/formic acid (75:25:0.2 vol./vol./vol.) combined with positive ES and MS/MS. The spectra of the leukotrienes were dominated by the presence of $[M + H]^+$ ions, whereas PGE_2 showed an $[M + NH_4]^+$ ion. When deuterated internal standards were used, these prostanoids could be quantified down to at least 0.5 ng ml^{-1} and the standard curves were linear over two orders of magnitude. Similarly, NI/ES of leukotriene B_4 (LTB_4) and its metabolites in chloroform/methanol (10:90) with 0.5% triethylamine produced abundant $[M - H]^-$ ions (Wheelan, Murphy and Simon, 1996; Wheelan, Zirolli and Murphy, 1996).

Loidl-Stahlhofen, Hannemann and Spiteller (1994) employed ^{18}O labelling and GC/MS to study the course of peroxidation of linoleic acid and the generation of hydroxy aldehydic compounds. The autoxidation products were analysed by EI/MS of the pentafluorobenzyloxime trimethylsilyl ether methyl ester derivatives. They identified dioxygenated fatty acids as intermediates by reduction of the oxo-acids with $NaBH_4$ and Rh/H_2 and by mass spectrometric analysis of the TMS ethers of the resulting alcohols. Notably, 2-hydroxyheptanal was identified among the aldehydic secondary oxidation products. Loidl-Stahlhofen, Kern and Spiteller (1995) described a GC/MS screening procedure for unknown hydroxyaldehydic lipid peroxidation products after pentafluorobenzyloxime (PFBO) derivatization. An unsaturated hydroxyaldehyde, 6-hydroxy-2,3-undecadienal was identified as an autoxidation product of linoleic acid. Autoxidation of oleic acid yielded several 4-hydroxy-2-alkenals and 4-hydroxyalkanals. The key ions were the molecular ions, the M-15 ions, or the ions arising from α cleavage at the C—C bond adjacent to the TMS ether function and from the loss of the PFB group.

Van Rollins and Knapp (1995) have pointed out that identification of the epoxide regioisomers of arachidonic acid [epoxyeicosatrienoic acids (EETs)] produced by cytochrome P_{450} mono-oxygenases is difficult because the Me esters overlap extensively and possess similar mass spectra. Also, the four regioisomers (14,15-, 11,12-, 8,9- and 5,6-EETs) were not resolved by gas – liquid chromatography (GLC) as PFB, TMS, and tert-butyldimethylsilyl (t-BDMS) esters. However, after being hydrolysed to the dihydroxyeicosatrienoic acids (DHETs), three of the four regioisomers were resolved as (bis)-t-BDMS ether and PFB esters by capillary GC/MS. The fourth regioisomer (5,6-DHET) was resolved after being converted to Δ-lactone. These analyses could be performed in femtogramme to picogramme quantities. The PFB esters of EETs and DHETs provided diagnostic spectra when analysed by GC/NICI/MS. The mass spectral interpretation that indicated epoxide and diol positions were validated by using synthesized EET/DHET[17,17,18,18-d_4, 5,6,8,9,11,12,14,15-d_8] standards. In the past the picolinyl esters of EETs were reported to give diagnostic spectra by conventional GC/EI/MS (Harvey, 1992), but this technique required microgramme quantities of analytes.

It has been shown that NICI/FAB/MS and FAB/MS/MS can produce carboxylate anions and structural information for free fatty acids (Adams and Gross, 1987), underivatized hydroxy fatty acids (Wheelan, Zirolli and Murphy, 1993), leukotrienes and phosphatidylglycerols (PGs) (Murphy and Zirrolli, 1994). Detection limits, however, are variable and interference from the chemical matrix usually results in the loss of well-defined spectra at levels ranging from 10–100 ng of total lipid. Wheelan, Zirolli and Murphy (1996) reported the low energy CID of the carboxylate anions generated by ES of LTB_4 and 16 of its metabolites by means of ES/MS/MS. The CID spectra of the carboxylate anions revealed structurally informative ions whose formation was determined by the position of hydroxyl substituents and double bonds present in the LTB_4 metabolite. Major ions resulted from charge-remote α-hydroxy fragmentation or charge-directed α-hydroxy fragmentation. The mechanisms responsible for all major ions observed in the CID spectra were studied by means of stable isotope labelled analogues of the LTB_4 metabolites. Wheelan, Murphy and Simon (1996) have described a GC/MS study of oxo and chain-shortened LTB_4 metabolite formation in Ito cells. Their study has extended previous knowledge concerning fragmentation mechanisms for derivatized hydroxy-substituted unsaturated fatty acids. The Ito cells, which form a hepatic perisinusoidal stellate cell line, metabolized the LTB_4 by the Δ^{10} and Δ^{14} reductase pathways. GC/EI/MS analysis of the 12-oxo metabolites, derivatized as the PFB/TMS derivatives, resulted in unique fragmentation indicative of the oxo substituent and double bond position. Further metabolism of 10,11-dihydro-LTB_4 and 10,11,14,15-tetrahydro-LTB_4 by carboxy terminus β-oxidation resulted in chain-shortened monohydroxy metabolites.

The peroxidation of unsaturated fatty acids is a complex process involving reactions with molecular oxygen to yield fatty acid hydroperoxides, which in tissues are degraded to a variety of products, including alkanals, alkenals, hydroxyalkenals, ketones and alkanes, representing both the carboxyl and methyl terminals of the fatty chain. The vast majority of peroxidation processes proceed through a free radical mediated chain reaction initiated by the abstraction of a hydrogen atom from the unsaturated fatty acid by a free radical, followed by a complex sequence of propagative reactions. In contrast, singlet oxygen attack on unsaturated fatty acids is believed to give lipid hydroperoxides by a non-radical, non-chain process. The free radical peroxidation of fatty acids has been reviewed by Porter (1984, 1995) and Frankel (1985). A host of diene hydroperoxides, cyclic peroxides, bicyclic peroxides and epoxy alcohols may be formed in free fatty acid autoxidation. The primary products that lead to this mixture are the four simple hydroperoxides of linoleic acid: the two (9- and 13-) *trans, cis* dienes and the two (9- and 13-) *trans, trans* dienes.

Prostaglandins and other oxygenated fatty acids. There are 5-, 12-, and 15-lipoxygenases which insert hydroperoxy groups at these positions in an unsaturated fatty acid by singlet oxygen molecule. The cyclooxygenase pathway leads to the formation of prostanoids, which include prostaglandins and thromboxanes by bis-oxygenation of arachidonic acid via insertion of 2 mol of oxygen. The lipoxygenase and cyclooxygenase pathways of arachidonic acid metabolism have been discussed by Smith and Laneuville (1994). GC/MS has provided the most sensitive and specific method for the detection of eicosanoids. One major drawback of prostanoid determination by GC/MS is the requirement of sample purification and derivatization (Schuyl et al., 1994). However, recent improvements in GC/MS methods have increased the throughput of this analysis allowing the determination of as many as seven PGs per sample in a 15 to 30 min time interval. Thus, Schweer, Watzer and Seyberth (1994) determined seven prostanoids in 1 ml of urine by GC/NICI/MS triple stage quadrupole MS based on addition of deuterated standards and preparation of PFB esters and TMS ethers of the oximes. The GC/NICI/MS method has recently been successfully applied to the analysis of the PFBO derivatives of other prostaglandins. The method has been applied to the synthetic, biologically active, PG analogues (e.g. riprostil) which cannot form the PFB ester because they lack a carboxyl group. Schweer and Fischer (1994) reported GC/NICI/MS/MS spectra of the most abundant ion in the high-mass region of O-2,3,4,5,6-PFBO/TMS ether/Me ester derivatives of 6-oxo-prostaglandin $F_{1\alpha}$ (6-oxo-$PGF_{1\alpha}$), 6-oxo-PGE_1, 6-oxo-PGE_2 and 6-oxo-PGE_3 and 11,16-dihydroxy-9-oxo-16-methyl-prost-13-en-l-oic acid (1-carboxy-riprostil) and of the PFBO/TMS derivative of 1,11,16-trihydroxy-16-methyl-prost-13-en-9-one (riprostil). Schweer et al. (1994) have shown that by single ion monitoring (SIM) with isotope dilution, solid phase extraction, and limited TLC clean-up, it is possible to measure PGs in a sensitive manner, although handling of several hundreds of samples would not be feasible. The method was used for determination of PGE_1 and its main metabolites 15-keto-PGE_0 and PGE_0 by GC/NICI/MS/MS.

The F_2 isoprostanes are complex metabolites of arachidonic acid generated via non-enzymatic free radical oxidation and are isomeric to prostaglandin $F_{2\alpha}$ ($PGF_{2\alpha}$) enzymatically produced by prostaglandin H_2 (PGH_2) synthetase. Waugh and Murphy (1996) have identified four regioisomers of F_2 isoprostanes formed by free radical oxidation of arachidonic acid. ES/MS/MS was used to detect the elution of these isomers from the HPLC column by monitoring the characteristic loss of 44 amu (C_2H_4O) from the 1,3-diol cyclopentane ring. The type I and type IV regioisomers were the major F_2 isoprostane products, but the complexity of the isomers was too great for a simple GC/MS assay to identify precisely a particular stereoisomer within a regioisomeric family.

In view of the earlier identification of isoprostane-containing phosphatidylcholines (PCs) as indicators of oxidative stress (Morrow et al., 1992, 1994), the recent isolation and identification of a major urinary metabolite of F_2 isoprostane 8-iso-$PGF_{2\alpha}$ as 2,3-dinor-5,6-dihydro-8-iso-prostaglandin $F_{2\alpha}$ (Roberts et al., 1996) is of special interest. The urinary metabolite of 8-iso-$PGF_{2\alpha}$ was analysed by GC/NICI/MS after conversion to the PFB ester, TMS ether and catalytic hydrogenation.

Feretti and Flanagan (1994) have reported the EI/MS spectra of three piperidides and four TMS derivatives of the urinary metabolite of PGF_α and $PGF_{2\alpha}$. They proposed fragmentation pathways on the basis of substituent shifts and data from both deuterium-labelled methyloxime analogues and pyrrolidide homologues.

Mamer et al. (1994) have synthesized and used several acylating reagents to introduce quaternary phosphonium or ammonium or ternary sulphonium functions into a simple model of a peptido leukotriene. These authors demonstrated that acylation of the free amine function of LTC_4, LTD_4 and LTE_4 to produce 5-triphenylphosphoniumvaleryl-amide derivatives enhanced chemical stabilities and significantly increased responses in FAB/MS under continuous flow liquid secondary ion MS relative to native compounds. With selective ion monitoring the sensitivity of detection of LTD_4 was extended to 3 pg. Recently, a sensitive and specific LC/MS/MS with ionspray was developed and validated by Wu et al. (1996) to quantitate LTE_4 in human urine. LTE_4-d_3 was used as internal standard.

Gallon and Pryor (1993) have used negative methane chemical ionization MS and other analytical techniques to identify the allylic nitrite and nitro compounds from the reactions of low levels of nitrogen dioxide with methyl linoleate and methyl linolenate in the absence of oxygen. MS indicated the presence of a nitrite and a nitro group, confirming the molecular weight and the position of the functional group. The NICI spectra of the allylic nitrite (nitro) isomers of methyl linoleate and methyl linolenate displayed a m/z 46 ion and a m/z 62 ion, indicative of a nitro and a nitrate functional group, and m/z 339 and m/z 337 molecular ions.

Recently, the suitability of negative ES/MS and ES/MS/MS for the identification of free fatty acids has been demonstrated (Kerwin, Wiens and Ericsson, 1996; MacMillan and Murphy, 1995). MacMillan and Murphy (1995) analysed the hydroperoxides of long-chain conjugated keto acids. Kerwin, Wiens and Erricsson (1996) used ES/MS and ES/MS/MS to characterize saturated and unsaturated fatty acids, including the location of some of the double bonds. ES/MS-induced fragmentation of monounsaturated fatty acids was minimal but it increased with increasing degree of unsaturation. Although the spectra were reproducible, double and triple bonds could not be located with certainty. Kerwin and Torvik (1996) extended this MS technique to the hydroxy and hydroperoxy fatty acids, with emphasis on its utility for rapid and sensitive analysis of biological samples. NICI/ES/MS and ES/MS/MS were used to characterize saturated and unsaturated monohydroxy fatty acids and fatty acid metabolites formed following incubation with soybean lipoxygenase. Ions corresponding to $[M - H]^-$ of eicosanoids were readily observed by means of ES/MS, but double bond migration precluded the use of MS to localize double bonds or the position of hydroxyl moieties. However, by following MS analysis with negative ion ES/MS/MS of precursor ions, the position of oxygenation could be determined for picogramme quantities of underivatized monohydroxy fatty acids. Spectra of deuterated analogs supported charge-driven vinylic processes as the most common mechanism of fragmentation. Kerwin and Torvik (1996) also showed that EI/MS/MS of saturated monohydroxy fatty acids yield several unexpected product ions. The MS/MS spectrum of 2-OH stearic acid gave a prominent peak 46 amu less than the precursor ion, in contrast to the loss of 44 amu from the precursor ion of most fatty acids and oxygenated fatty acids observed during ES and FAB. The loss of 46 amu from fatty acids with hydroxyl groups on the terminal carbon could be rationalized as the loss of C_2H_5OH from the precursor ion in FAB/MS/MS spectra. A similar loss of 46 amu occurred in ESI/MS/MS spectra of 12-OH dodecanoic acid and 16-OH palmitic acid. Although the hydroperoxide derivatives of linoleic and arachidonic acids gave m/z 311 and m/z 335, respectively, prominent peaks at m/z 293 and m/z 317 were also observed, which were identified as the dehydration products

Margalit, Duffin and Isakson (1996) have developed a new method for eicosanoid assessment in biological samples by ES/MS/MS in the multiple reaction mode (MRM). MRM experiments were performed by setting the first quadrupole to pass the $[M - H]^-$ of a selected eicosanoid while simultaneously setting the third quadrupole to pass a selected fragment ion of that eicosanoid. The eicosanoids were detected and quantitated in SIM by monitoring the $[M - PFB]^-$ ion. In this study, 14 biologically significant eicosanoids were quantitated in a single sample. Limits of detection ranged from 0.5 pg for thromboxane B_2 (TXB_2) to 10 pg for 6-keto $PGF_{1\alpha}$. Eicosanoid levels determined by MS/MS were similar to those obtained by immunoassay and GC/MS. The method was applied to the analysis of eicosanoids in human blood and fluid from inflamed rat air pouch. The most abundant metabolites in lipopolysaccharide-stimulated whole blood were PGE_2, TXB_2, 6-keto $PGF_{1\alpha}$, PGE_1, PGD_2 and $PGF_{2\alpha}$, and LTB_4 and LTC_4. Yamane, Abe and Yamane (1994) used HPLC and LC/TS/MS for identifying epoxy polyunsaturated fatty acids and epoxyhydroxy polyunsaturated fatty acids from incubation mixtures of rat tissue homogenates.

Isbell and Kleiman (1996) have used GC/EI/MS to characterize the eicosanolactones, TMS ethers of the ethyl esters of 20:0 and 22:1 hydroxy ethyl esters, as well as bis TMS ether of 22:0 dihydroxy ethyl ester derived from hydrolysis of mineral-acid-catalysed condensation of meadowfoam

(*Limnanthes alba*) fatty acids into estolides. The ester linkages were found to be mainly in the original double bond positions, but some positional isomerization was also observed.

6.3.2 Sterols and steryl esters

Although sterols and steryl esters constitute some of the simplest lipid molecules, complications arise when these molecules must be isolated and identified from complex biological matrices. Both GC/MS and LC/MS have been effective (Evershed, 1994; Hoving, 1995). However, problems arise when the multitude of combinations of ring and fatty chain peroxidation products in the steryl esters must also be considered.

Natural sterols and steryl esters. Hoving (1995) has reviewed the chromatographic methods employed in the analysis of cholesterol and cholesteryl esters and has commented upon the importance of MS in confirming the identity of the chromatographic peaks. Lutjohann *et al.* (1993) used combined GC/EI/MS to evaluate the use of deuterated cholesterol and deuterated sitostanol for measurement of cholesterol absorption in humans. SIM was performed by cycling the quadrupole mass filter between different m/z values at a rate of 3.7 cycles s^{-1}. The results showed that deuterated cholesterol and deuterated sitostanol are reliable markers for measuring cholesterol absorption in humans. Faix *et al.* (1993) have applied the mass isotopomer distribution analysis technique to quantify periodicities in the biosynthesis of serum cholesterol during oral and intravenous administration of sodium [1-^{13}C]- or [2-^{13}C]-acetate. Cholesterol biosynthesis was calculated from mass isotopomer fractional abundances in free cholesterol determined by GC/MS. For TMS-cholesterol analysis, the parent M_0 fragment (m/z 368), representing underivatized free cholesterol minus [OH], was monitored as were M_1-M_4 (m/z 369–372).

Recently, LC/MS has made significant contributions in the analysis of sterols and steryl esters. Takatsu and Nishi (1993) have employed discharge-assisted LC/TS/MS for the determination of total serum cholesterol. The method incorporates stable isotope dilution using [3,4-^{13}C] cholesterol as an internal standard. $[MH - H_2O]^+$ ions were monitored by the SIM method. Satisfactory agreement between the analytical result and the certified value of the National Institute of Standards and Technology standard reference material was obtained with a relative standard deviation of 0.6%. The method does not require sterol derivatization. Yang *et al.* (1992) used FAB/MS to identify cholesteryl sulphate (m/z 465) as the $[M - H]^-$ ion recovered from the appropriate TLC fraction.

Interesting applications of GC–MS and LC–MS methods may be found in studies of the diverse marine oxylipids (Gerwick, 1996, and references cited therein).

Oxygenated sterols and steryl esters. There is extensive evidence that the autoxidation products of cholesterol and cholesteryl esters exert pronounced biological effects. Consequently, there is much interest in the identification of the autoxidation products. Soft ionization MS has proved to be highly effective for this purpose. Generally, cholesterol autoxidation proceeds by free radical processes similar to those in other lipids (Smith, 1996). However, cholesterol autoxidation yields epimeric 7-hydroxycholesterols, and 7-ketocholesterol, as well as epimeric 7-hydroperoxides, as the primary oxidation products. Alternatively, cholesterol epoxidation could yield epimeric 5,6-epoxides as the primary oxidation products and cholestane triol as the hydrated product. Some 66 cholesterol decomposition products of air-aged cholesterol have been identified. In the case of the cholesteryl esters, oxidation may involve both the steroid ring and the unsaturated fatty acid chain, yielding other primary and secondary products of cholesterol oxidation. Only a small number of these products are readily isolated and identified in tissue extracts. In a specific instance, De Fabiani *et al.* (1996) used GC/MS to identify cholesta-5,7,9(11)-trien-3β-ol as an indicator of sterol hydroperoxide formation in plasma of patients with Smith–Lemmli–Optiz syndrome.

Sevanian *et al.* (1994) applied GLC and LC/TS/MS for the analysis of plasma cholesterol-7-hydroperoxides and 7-ketocholesterol. Analysis of human and rabbit plasma identified the commonly occurring oxidation products, yet dramatic increases in 7-ketocholesterol and cholesterol-5β, 6β-epoxide were observed. The study failed to reveal the presence of cholesterol-7-hydroperoxides, which were either too unstable for isolation, metabolized or further decomposed. The principal ions of cholesterol oxides monitored by LC/TS/MS were: m/z 438 (cholestane triol); m/z 401 (cholesterol-7-hydroperoxide); m/z 401 (7-ketocholesterol); m/z 367 (7α-hydroxycholesterol); m/z 399 (cholesta-3,5-dien-7-one); and m/z 385 (cholesterol-5α,6α-epoxide). The major ions were supported by minor ions consistent with the steroid structure. Kamido *et al.* (1992a, b) synthesized the cholesteryl 5-oxovaleroyl and 9-oxononanoyl esters as stable secondary oxidation products of cholesteryl arachidonate and linoleate, respectively. These compounds were identified as the 3,5-dinitrophenylhydrazone (DNPH) derivatives by reversed-phase LC/NICI/MS. These standards were used to identify cholesteryl and 7-ketocholesteryl 5-oxovaleroyl and 9-oxononanoyl esters as major components of the cholesteryl ester core aldehydes generated by copper-catalysed peroxidation of low-density lipoprotein (LDL). In addition to 9-oxoalkanoate (major product), minor amounts of the 8, 9, 10, 11 and 12 oxo-alkanoates were also identified among the peroxidation products of cholesteryl linoleate. Peroxidation of cholesteryl arachidonate yielded the 4, 6, 7, 8, 9 and 10 oxo-alkanoates of cholesterol as minor products. The oxysterols resulting from the peroxidation of the steroid ring were mainly 7-keto, 7α-hydroxy and 7βhydroxycholesterols and the 5α-, 6α-, and 5β-, 6β-epoxy cholestanols. Subsequently, Kamido et al. (1993) used the synthetic ω-oxo-alkanoates of cholesterol to identify the above core aldehydes among the Fe^{2+}/t-butyl hydroperoxide oxidation products of liposomal LDL and synthetic cholesteryl esters. The LC/MS identification of the core aldehydes was confirmed by preparation of the DNPH derivatives. Kamido et al. (1995) have demonstrated the formation of the C5 and C9 cholesteryl ester core aldehydes in both LDL and high-density lipoprotein (HDL) isolated from normolipemic subjects, although quantitative differences were observed. More recently, Kamido et al. (1997) have succeeded in demonstrating by LC/ES/MS the presence of the cholesteryl ester core aldehydes in atheromas and xanthomas as the DNPHs (m/z 664 and m/z 720).

Kritharides et al. (1993) used ammonia CI/MS with DIC probe to detect the oxidation products of cholesterol and cholesteryl esters produced in LDL particles incubated with copper. A low ion source temperature (120°C) was used because of the thermal instability of the compounds analysed, for example cholesteryl linoleate, cholesteryl linoleate hydroperoxide and cholesteryl linoleate hydroxide, and the inability to ascertain intact molecular weights (MWs) of known standards with higher ion source temperatures. Cholesterol standard (MW 386) produced ions of mass 404 $[M + 18]^+$, 386, 369 $[MH - H_2O]^+$. Cholesteryl linoleate (MW 648) produced ions of mass 666 $[M + 18]^+$, 648, 386, 369 and 368, which was consistent with the thermal decomposition of cholesteryl linoleate to cholesterol during the analysis. Cholesteryl linoleate hydroperoxide standard (MW 680) produced major ions at mass 698 $[M + 18]^+$, 682 and 664. The production of cholesteryl linoleate hydroxide (MW 664) was thought to occur during handling and heating, leading to m/z 682 $[M + 18]^+$. Standards of cholesteryl linoleate hydroxide (9-OH and 13-OH) produced molecular ions of m/z 682, 664 and 665, supporting chromatographic identification. The formation of 7-ketocholesterol (MW 400) was confirmed by the identification of major ions of m/z 418 and 401 $[M + 18]^+$ and $[M + H]^+$, respectively. 7-Ketocholesterol and cholesterol linoleate hydroperoxide were the two major products of prolonged oxidation.

Bortolomeazzi et al. (1994) used GC/EI/MS with an ion trap to identify the thermal oxidation products of cholesteryl acetate as the 7β-hydroperoxy and 7α-hydroperoxy cholesteryl acetate, 7keto-cholesteryl acetate, the α and β isomers of 7-hydroxycholesteryl acetate, the α- and β-5,6-epoxy isomers and several derivatives arising from the loss of acetate and water. Dzeletovic et al. (1995b) have observed that saponification during sample preparation did not hydrolyse all of the oxysterol esters completely and that separation of oxysterols from cholesterol by HPLC was tedious and incomplete. They developed a stable isotope dilution GC/EI/MS SIM method for the determination of cholesterol oxidation products in human plasma. Nine oxysterols were determined by using deuterium-labelled internal standards.

Plasma from 31 healthy volunteers revealed that 27-, 24- and 7α-hydroxycholesterol were the most abundant cholesterol oxidation products. The 5,6-oxygenated products were present mainly unesterified whereas the other oxidation products were mostly in esterified form. Dzeletovic et al. (1995a) employed this isotope dilution GC/EI/MS method to determine the time course of cholesterol oxidation in LDL. Aliquots of 0.1–1 ml of oxidized LDL were withdrawn for oxysterol analysis. The same products, mainly 7- and 5-oxygenated cholesterol, were formed when incubated with cupric ions or soybean lipoxygenase. During the oxidations, esterified cholesterol was preferentially consumed, and consumption of polyunsaturated fatty acids and formation of conjugated dienes preceded the appearance of oxysterols. The formation of the initial oxidation products of cholesterol, 7α- and 7β-hydroperoxycholesterol, was quantitatively determined by analysing samples from different time points by TLC and comparing the intensity of the cholesterol hydroperoxide spot upon spraying with sulphuric acid. In the copper oxidized samples the hydroperoxide spot was barely detectable in the sample incubated for 1 h, but the intensity increased as the sample was incubated up to 8 h, after which it decreased. In the soybean lipoxygenase samples, the 7-hydroperoxides were seen over the whole time course, after first appearing in the sample incubated for 4 h. Chisolm et al. (1994) used proton and carbon NMR and CI/positive ion GC/MS to identify the major cytotoxin in oxidized LDL as 7β-hydroperoxycholest-5-en-3β-ol which was also a component of human atherosclerotic lesions. Oshima and Koizumi (1993) used GC/MS of the TMS derivatives to demonstrate the formation of cholesterol oxidation products in fish products. 7β-Hydroxycholesterol and 7keto-cholesterol were the most prominent components.

6.3.3 Neutral glycerolipids

The neutral glycerolipids have been subject to MS investigation because of the complexity of their fatty acid composition and the inability of conventional chromatographic systems to distinguish between positional isomers, reverse isomers and enantiomers of isobaric acylglycerol species. Recent interest in the metabolic role of different molecular species of MGs, DGs, structured triacylglycerols (TGs) and their oxidation products has further increased the demand for MS analyses. The initial MS studies of TGs were performed in the 1960s using direct probe EI/MS and they established the principles of TG fragmentation (reviewed by Kuksis and Myher, 1989, 1995). In addition to the DG-like fragment ions, the spectra consisted of $[M]^+$ and $[M - 18]^+$ ions at low abundance, which were used to determine the molecular weight distribution of TGs of several vegetable oils.

Monoacylglycerols and diacylglycerols. The early GC/MS methods of analysis of MGs and DGs utilized TMS ethers, which were unstable and could not be purified before analysis, or acetates, which did not yield sufficiently characteristic spectra for polyunsaturates. Myher et al. (1978) demonstrated that the t-BDMS ethers yielded abundant amounts of $[M - 57]^+$ ion sufficient to provide molecular weights both for saturated and for unsaturated species. The 1,3-isomers were identified by the $[M - \text{acyloxymethylene}]^+$ fragment ions, which are absent in the sn-1,2- and sn-2,3-DGs. The abundance ratio of the ions due to losses of the acyloxy radical $[M - RCOO]^+$ from position 1 (or 3) and position 2 indicated the proportions of the reverse isomers, for example sn-1-palmitoyl 2-stearoyl, and sn-1-stearoyl 2-palmitoylglycerols. The GC/MS method in combination with the t-BDMS ether preparation has been extensively employed in the past for the analysis of the structure of free DGs and DGs released from glycerophospholipid (GPL) by phospholipase C as well as for the analysis of MGs (Kuksis and Myher, 1990, 1995; Myher and Kuksis, 1995). Kuypers, Buetikofer and Shackleton (1991) have employed the acylglycerol benzoate derivatives for LC/TS/MS resolution, identification and quantification of the molecular species of the DG moieties released from GPLs by phospholipase C.

Ioneda and Ono (1997) have reported EI/MS spectra of the per-O-benzoyl 1-monomycoloylglycerol, which yielded $[M]^+$, $[M - \text{benzoic acid}]^+$, $[M - \text{two units of benzoic acid}]^+$, $[M-1\text{-hydroxy-2,3-dibenzoylglycerol}]^+$ and benzoic acid, $[M - 422]^+$.

Hubbard et al. (1996) have detailed a method for the measurement of femtomolar quantities of 1-stearoyl-2-arachidonoyl sn-3-glycerol generated in 10^5 human basophils via combined GC/NICI/MS. Conversion of the DG to the PFB ester conveys electron-capture properties to the DG, which undergoes limited fragmentation under NICI/MS conditions with generation of an intense molecular anion at m/z 838. Monitoring m/z 838 for detection of 18:0–20:4, and m/z 841 for detection of 18:0d_3–20:4, employed as internal standard, provided the analytical basis for GC/MS quantification of the endogenous DGs in human basophils. The GC/MS assay was highly selective and sensitive, with a detection limit of less than 0.20 pg (about 30 fmol) for endogenous 18:0–20:4n6. An additional feature of the GC/MS method detailed above was its potential application for simultaneous detection and quantification of several DG molecular species in human basophils and mast cells. The GC/MS method was not able to resolve the PFB derivatives of the reverse isomers of the DG, for example sn-1-18:0–sn-2-18:2 and sn-1-18:2–sn-2-18:0 DGs. Fluoro- and nitro-substituted compounds, which have electron-capture properties in negative CI, also confer improved sensitivity for negative mode ES. Myher et al. (1996) have shown that the dinitrophenyl urethane (DNPU) of enantiomeric DG resolved on chiral phase columns are readily detected as the $[M - H]^-$ ion by negative ES. It may be possible to determine the double bond position in the fatty acid chains associated with individual enantiomeric DG molecules by fragmenting the fatty chains of the DNPU derivatives at increased cap voltages during LC/ES/MS. (Itabashi, Kuksis and Ravandi 1997, unpublished results). In the case of DGs and MGs a distinction between the two primary positions can be made by GC/MS of the TMS or t-BDMS ethers (Myher et al., 1978) of the enantiomeric DGs derived by silolysis of the enantiomeric DGs resolved by chiral phase HPLC (Itabashi et al., 1990). A similar chiral phase HPLC resolution is possible for the sn-1- and sn-3-MGs. Superior results have been obtained with the t-BDMS ethers as these derivatives also permit the distinction of the reverse isomers of mixed 1,2-DGs. The abundance ratio of the ions due to losses of the acyloxy radical $[M - RCOO]^+$ from position 1 (or 3) and position 2 indicated the proportions of the reverse isomers, for example sn-1-palmitoyl, 2-stearoyl, and sn-1-stearoyl, 2-palmitoylglycerols. The above method of determining the reverse isomers of sn-1,2-DGs should be satisfactory for the analysis of oligoenoic but not polyenoic DGs. For the latter, the yield of the $[M - acyloxy]^+$ fragment falls off with increasing unsaturation. Kuksis and Myher (1989) have reviewed catalytic deuteration of unsaturated lipids as a means of equalizing the response and the use of the number of deuterium atoms introduced as indicators of the number of double bonds present. Itabashi, Kuksis and Myher (1997) have demonstrated the resolution of reverse isomers of mixed short-chain DGs within an enantiomer class by using a previously described chiral phase HPLC and the DNPU derivatives. Reverse isomer separations of mixed short-chain DGs were also demonstrated by GLC of the TMS ethers. The identity of the peaks was confirmed by LC/CI/MS and GC/EI/MS (Marai et al., 1992).

The structures of MGs and DGs arising during biosynthesis and lipolysis as well as during stereospecific analyses of TG have been the subject of extensive HPLC investigation, often in combination with MS. Thus, Itabashi, Marai and Kuksis (1991) have employed chiral phase LC/MS for the separation, identification and quantification of molecular species of DGs overlapping between enantiomers and within each enantiomer class. For this purpose they used positive chemical ionization MS, which provided significant $[RCO + 74]^+$ and $[M - DNPU]^+$ fragments necessary for unequivocal identification of DGs as the DNPU derivatives in the effluents of the chiral phase HPLC column. Under the direct inlet LC/MS conditions, significant intensities were also obtained for the negative ions corresponding to $[M]^-$, $[DG + Cl]^-$, $[DNPU]^-$, and $[DNPU + Cl]^-$, which did not provide additional structural information, although they increased the sensitivity of detection of the DG derivatives. Marai et al. (1992) demonstrated that the detection of the DG-DNPU derivatives as the negative chloride attachment fragment ions increased the sensitivity of analysis many-fold over positive CI/MS. The highest response was obtained for the $[DNPU]^-$ fragment ion. For this purpose a chiral column containing the R(+) naphthylethylamine polymer was eluted with an isocratic mixture of either isooctane/t-butyl methyl ether/isopropanol/acetonitrile, 80:10:5:5 (solvent A) containing 1%

dichloromethane, or with solvent A (20%) and hexane/dichloromethane/ethanol, 40:10:1 (solvent B, 80%). The method was applied to the analyses of enantiomeric diacylglycerols generated by Grignard degradation of butterfat, corn oil and palm kernel oil. Later, Lehner, Kuksis and Itabashi (1993) used chiral phase HPLC in combination with on-line NICI/MS to establish the structure of the enantiomeric DG, synthesized by purified 2-MG acyltransferase, as well as the enantiomeric DG utilized for TG synthesis by purified DG acyltransferase. Either 2-monooleoylglycerol or rac-1,2-dioleoylglycerols served as substrates for the MG- and DG acyltransferases, respectively. In both instances the DG enantiomers were resolved and identified as the DNPU derivatives, which gave the $[M]^-$ molecular ion (m/z 829) and a characteristic $[M - DNPU + 35]^-$ fragment ion (m/z 655). This corresponded to the addition of a chlorine group to the dioleoylglycerol molecule. The NICI was found to be about 100 times more sensitive than the positive CI current for the chlorine adducts of these compounds under the chiral phase LC/MS conditions (Marai et al., 1992). Kuksis et al. (1994) and Yang et al. (1995) reported the use of reversed phase LC/MS with filamenton TS ionization for the analyses of the DNPU derivatives of DG previously resolved into enantiomers by chiral phase HPLC. The molecular species were identified on the basis of the $[M - 1]^-$ ion plots extracted from the total negative ion current and the comparison of the reversed-phase HPLC retention times with those of the DNPU derivatives of standard DG. This method was used to compare the molecular species of the corresponding enantiomeric DG moieties of liver and VLDL TG. Later, Yang et al. (1996) utilized this method to identify and quantify the deuterium labelled DG enantiomers derived from the liver and very-low-density-lipoprotein (VLDL) TG following in vivo infusion of perdeuterated ethanol into rats. The labelling of the DG was estimated from the deuterium labelling of DG. The total ^2H excess in the DG molecule is the sum of the ^2H excess in glycerol and that in the fatty acids. Myher et al. (1996) used chiral phase HPLC with on-line ES/MS for the separation and identification of the molecular species of X-1,3-, sn-1,2-, and sn-2,3-DG. The HPLC system consisted of the chiral phase column and eluent hexane/dichloromethane/ethanol (40:10:1) by volume. After post-column mixing with 0.3 ml min^{-1} isopropanol/1% aqueous ammonia, mass spectra of eluting peaks were acquired by nebulizer-assisted ES/MS. The DG species were identified on the basis of the $[M - 1]^-$ ion plots extracted from the total NI current and the chiral phase HPLC retention times established for standard DG DNPU.

Triacylglycerols. The effectiveness of GC/EI/MS for the structural characterization of natural TG was demonstrated by using TG separated by GLC on packed and open tubular columns (Nakeham and Frew, 1982). The GC/EI/MS combination provided a great deal of compositional information with a quadrupole MS, even without the production of molecular ions.

Thus in the odd-carbon TG C51, the strong $[M - RCO_2]^+$ ions at m/z 563 and 565 were from 31:1 and 33:0 DG fragments. Other $[M - RCO_2]^+$ ions at m/z 537 (31:0), 551 (32:0), 577 (34:1), 591 (35:1), 593 (35:0) and 605 (36:1) corresponded to combinations of shorter and longer acyl chains on the glycerol moiety. By comparing the intensities of $[RCO]^+$ and $[RCO + 74]^+$ ions for each acyl chain length and correcting for variations in ion intensity arising from increasing chain length and unsaturation as determined for standard compounds it was possible to deduce the probable composition of the component TGs 16:0/17:0/18:1 or 16:1/17:0/18:0. The proportion of each acyl combination in the total TG mixture could be estimated from the percentage each combination represented for a given chain length in the sample. Relative abundances of chain lengths were obtained from the GLC or reconstructed ion current trace.

Myher et al. (1988) combined GLC on capillary columns containing a polarizable liquid phase with on-line EI/MS and GC/MS for the characterization of the more complex TGs in bovine milk fat. The EI spectra were recorded at 70 eV and a hydrogen carrier gas head pressure of 0.34 bar. The CI spectra were determined at 210 eV and a similar carrier gas head pressure, but the ion source pressure was increased to 0.6 T by admitting methane via a separate capillary inlet. Prior to GC/MS, the TG mixtures were segregated into long and short chain length saturates, *cis* and *trans* monoenes, dienes and trienes by TLC on silver nitrate–silica gel. The peaks were identified by carbon number and by $[M - RCOOH]^+$ fragment ions. Oshima, Yoon and Koizumi (1989) applied selective ion monitoring to the analysis of molecular species of vegetable oil TGs separated by open tubular GLC columns containing a polarizable methylphenylsilicone phase. This liquid phase separates TGs on the basis of both chain length and degree of unsaturation. It was demonstrated that peak assignments could be made by selecting certain characteristic ions with the same retention time on the SIM profile. Thus, three ions of the RCO type corresponding to the fatty acyl residues, $[R_1CO]^+$, $[R_2CO]^+$ and $[R_3CO]^+$, and the corresponding three M-acyl ions, $[M - OCOR_1]^+$, $[M - OCOR_2]^+$ and $[M - OCOR_3]^+$, could be selected instead of the molecular ion. The SIM method was specifically applied to the peak assignment of the TGs from palm, cottonseed and safflower oils and certain new fatty acid combinations were proposed. Kalo and Kemppinen (1993) employed GC/EI/MS on polarizable phenylmethylsilicone capillary columns to characterize enzymatically modified butterfat samples. In addition to $[RCO]^+$ and $[M - RCOO]^+$ fragments, the $[RCO + 74]^+$ and $[RCO + 128]^+$ fragments proved useful in the identification of molecular species of TGs. The TG composition of the modified butter oils was calculated and compared with random distribution and the composition of the original oil. CI with methane, isobutane or ammonia as a reactant gas in the positive ion mode gave less fragmentation and a greater abundance of ions consisting of intact molecule, for example $[M + H]^+$ or adduct ions. More recently, Evershed (1996) showed that NICI at an ion source block temperature of 300°C overcame the problem with interpretation of EI mass spectra during high-temperature GC/MS. Abundant $[RCO_2]^-$, $[RCO_2 - 18]^-$, and $[RCO_2 - 19]^-$ ions, believed to be produced by nucleophilic gas-phase ammonolysis, were used to identify the individual fatty acid moieties associated with peaks in TG total ion chromatograms. The polarizable stationary phase produced significantly enhanced resolution of TG molecular species compared with high-temperature stable apolar stationary phases. The resolution of complex mixtures of TG could be further improved by use of the Biller–Biemann enhancement technique to produce mass resolved chromatograms.

Kuksis, Marai and Myher (1983; Marai, Myher and Kuksis, 1983) combined NICI/MS with reversed-phase HPLC using acetonitrile and propionitrile and obtained $[M + H]^+$ and $[M - H - RCO_2H]^+$ ions from the TGs of a number of vegetable oils and animal fats. This LC/MS method was later applied to the identification of the less common isologous short-chain TGs in the most volatile 2.5% molecular distillate of butteroil (Myher, Kuksis and Marai, 1993). Both saturated and unsaturated TGs containing normal and branched-chain odd-carbon fatty acids in combination with short-chain acids were identified, and over 150 molecular species were quantified. The HPLC/MS analysis of the TG species made use of a data base consisting of MS ions $[M]^+$ and $[MH - RCOOH]^+$ as well as the corresponding elution times. This extremely powerful approach allowed identification of species that had the same elution times but different mass spectra, or species that had the same spectra but different elution times. The peak intensities and elution times for each ion mass were entered into a computer and the data were then sorted according to elution times. The molecular species were identified manually, according to previously described principles (Kuksis, Marai and Myher, 1983; Marai, Myher and Kuksis, 1983). The above reversed-phase LC/MS routine was later applied in the characterization of the uncommon TG structures generated by randomization of butteroil (Marai, Kuksis and Myher, 1994) and single-cell oils (Myher, Kuksis and Park, 1996). The uncorrected quantitative estimates showed considerable divergence from the calculated values, which, however, did not indicate non-randomness of fatty acid association. Furthermore, the LC/MS estimates differed from the estimates obtained by HPLC with light scattering. This was apparently because of the variable response of the different molecular species in the two different detector types, none of which showed a true mass or mole composition. The detection sensitivity was increased from 10-fold to 100-fold by inclusion of dichloromethane in the LC/NICI/MS with direct liquid inlet interface. This led to the exclusive detection of $[M + Cl]^-$ ions (Kuksis, Marai and Myher, 1991a, b). The m/z values of the $[M + Cl]^-$ ions defined unambiguously the number of acyl carbons and the number of double bonds present in the TG molecules of most fats and oils. The abundance of the $[M + Cl]^-$ ions determines the proportions of different molecular weight species. Comparisons of the LC/CI/MS and polar capillary GLC/FID estimates for corn oil TGs gave close agreement.

Kallio and Currie (1993a,b) demonstrated that NICI/MS provides a sensitive method for the detection of TG. Samples were introduced via a direct exposure probe into the ion source following $AgNO_3$ separation prior to MS. The TG formed deprotonated molecular ions with ammonia as the reactant gas. In addition, CID of the $[M - H]^-$ ions yielded the daughter ion spectra, which provided information on the fatty acid constituents and their distribution between the sn-2- and sn-1,3-positions of TG. This method was used by Laakso and Kallio (1993) to investigate the configurational isomers of the monoenoic fatty acyl residues in the disaturated monoenoic TGs of winter butterfat. The fatty acid compositions of the corresponding molecular weight species of disaturated *trans* and disaturated *cis* monoenoic TGs were similar. Kallio and Rua (1994) have discussed some of the problems encountered in MS assay of the fatty acid composition of the sn-2- and sn-1(3)-positions of the major TGs of human milk fat. Recently, Laakso and Kallio (1996) have reported an optimized NICI/MS analysis of TG using ammonia. Abundant $[M - H]^-$ ions were produced without the formation of $[M + 35]^-$ cluster ions which would interfere with the molecular weight region of the TG spectra. Desorption of the sample from the probe was achieved by using a heating rate of 40 mA s^{-1}, which minimized thermal degradation of unsaturated molecules and the reducing effect of double bonds on the MS response to TG. The method was useful for semi-quantitative work with mixed TGs; as the discrimination caused by differences in molecular size and unsaturation of TG with 50–55 acyl carbons was negligible and no correction factors were needed. The method was applied to analyse the TGs from blackcurrant (*Ribes nigrum*), lingonberry (*Vaccinium vitisidaea*) and raspberry (*Rubus idaeus*) grown wild in Finland. The same method was later applied to the determination of the saturated, monoenoic, dienoic and trienoic TGs in the colostrum of cows (Laakso *et al.*, 1996) and of the α- and γ-linolenic acid rich seed oils (Laakso and Voutilainen, 1996).

Spanos *et al.* (1995) employed direct probe desorption chemical ionization to examine the bovine milk fat TG fractions collected by reversed-phase HPLC. MS/MS analyses were obtained by CID of selected $[M + 1]^+$ ions followed by linked scanning at constant magnetic field strength/electrostatic analyser voltage (B/E), where the magnetic field (B) was scanned at a constant ratio to the electrostatic analyser voltage (E.) Helium was used as the CID gas.

Duffin, Henion and Shieh (1991) introduced an ES/MS/MS method for MGs, DGs and TGs based on the analysis of the $[M + NH_4]^+$ ions produced in an ion spray (pneumatically-assisted ES) source. Samples were dissolved in a solution of chloroform/methanol, 70:30 vol./vol., which was modified by the addition of alkali-metal or ammonium salts, or by inclusion of formic acid to favour the addition of cationic species to the sample molecule. Ammonium adducts were preferred for the MS/MS studies as the sodium adducts were resistant to fragmentation by CID. ES analysis of MG, DG and TG standards yielded $[M + NH_4]^+$ ions, with no fragmentation, for all species that were present at low picomole per microlitre concentrations. Acylglycerols that contained unsaturated fatty acid chains were observed to exhibit a response in the mass spectrum greater than those with saturated chains, and the response resulting from the molecular adduct ions of the acylglycerols decreased in the order MGs > DGs > TGs. CID of the $[M + NH_4]^+$ ions yielded high-quality spectra that contained an abundance of information. At 50 eV collision energy, ions corresponding to $[M - RCOO]^+$ were the main CID fragmentation products, along with some acylium ions. For TGs, it was possible to determine the nature of the fatty acids, but not their positional placement. ES/MS/MS characterization of an unknown lipid material isolated from a mammalian cell culture reactor yielded molecular adduct ions for a diverse mixture of TGs. MS/MS analysis of $[M + NH_4]^+$ ions that were produced from the lipid mixture revealed an abundance of 14:0, 16:0, 18:0 and 18:1 fatty acid components of the TGs. The MS/MS data were in good agreement with the fatty acid profile generated by GLC with flame ionization detection. However, relative amounts of individual TGs and the fatty acids that form the TG were not deduced from the ES/MS/MS data because of the demonstrated dependence of ion current abundance on analyte polarity. Myher *et al.* (1994; Myher, Kuksis and Park, 1996) have used pneumatically-assisted ES in combination with HPLC for the analysis of polyunsaturated TGs. Mixtures of TGs were separated by C-18 reversed-phase HPLC with 0.85 ml min^{-1} flow and a gradient of 20%–80% isopropanol in acetonitrile. A flow of 0.15 ml min^{-1} isopropanol containing 1% ammonia was added post-column. With lower voltages (170 V) at the capillary exit, $[M + NH_4]^+$ and $[M + Na]^+$ ions were observed with little or no fragmentation. When the voltage was increased to 250 V, abundant $[M - RCOO]^+$ ions were observed and the $[M + NH_4]^+$ ions had almost completely disappeared. However, the $[M + Na]^+$ ions, being more resistant to CID fragmentation, remained (Myher, Kuksis and Park, 1996).

Byrdwell and Emken (1995) and Neff and Byrdwell (1995) have described the application of atmospheric pressure chemical ionization (APCI)/MS to the qualitative analysis of simple homogeneous (mono-acid) TGs and genetically modified soybean oils separated by reversed-phase HPLC. The mass spectra obtained were relatively simple and consisted primarily of DG ions, $[M - RCOO]^+$ or $[DG]^+$, and protonated molecular ions, $[M + 1]^+$. The degree of unsaturation was the primary factor in determining the amount of DG ions compared with TG ions formed. DG fragments were the base peaks in spectra of TGs containing less than three or four sites of unsaturation. Mass spectra of TGs with three or four sites of unsaturation contained either DG fragments or protonated TG ions as base peaks, depending on the fatty acid distribution within the TG. No TG protonated molecular ion was observed from TGs containing only saturated fatty acids. Byrdwell et al. (1996) have reported quantitative analysis of several model and natural TG samples with d_{12}-tripalmitoylglycerol as internal standard. The samples included a mixture of 35 TGs with randomly distributed fatty acids, of randomized and normal soybean oil TGs and of randomized and normal lard. The results showed that the TG compositions obtained by APCI/MS data without application of response factors had average relative errors very similar to those obtained by LC/FID. Numerous TG species were identified by means of LC/APCI/MS which were undetected by means of LC/FID. Response factors derived from a synthetic mixture were not widely applicable to samples of disparate composition. The TG compositions obtained using response factors calculated from fatty acid composition showed less average relative error than was obtained from LC/FID data and were in good agreement with predicted compositions for the synthetic mixture and for randomized soybean oil and lard samples. These results were comparable to those obtained by other MS methods of TG quantification which reported similar effects of unsaturation and/or chain length (Kuksis, Marai and Myher, 1983; Mares, Rezanka and Novak, 1991; Myher et al., 1984; Rezanka and Mares, 1991).

Kusaka et al. (1996) have used LC/APCI/MS to show that normal plant TGs give mass spectra with peaks of $[M+H]^+$ and $[M-R_1(R_2)COOH+H]^+$ ions where $[M+H]^+$ and $R_1(R_3)$ COOH represent, respectively, the protonated molecular ion and the fatty acid at the sn-1 (or sn-3) position of the TG. It was possible to discriminate fatty acids between sn-1 (or sn-3) and sn-2 positions. Based on the difference in the mass spectra between rac-1-18:0–2-18:1–3-18:2 and rac-1,3-18:2–2-18:0 it was concluded that fatty acyl groups at the sn-1 (or sn-3) and sn-2 positions of TGs can be distinguished. LC/APCI/MS of hydroperoxidized TGs gave characteristic fragment ions $[M-H_2O_2+H]^+$, $[M-H_2O+H]^+$ and $[M-R_1(R_3)COOH-H_2O_2+H]^+$.

Schuyl et al. (1995) have reported an on-line combination of $AgNO_3$–HPLC with ES/MS. Mass spectra of TGs exhibited only abundant $[M+Na]^+$ ions without any information on fatty acid moieties. Laakso and Voutilainen (1996) used $AgNO_3$–HPLC in combination with on-line APCI/MS for the examination of TGs of seed oils rich in α and/or γ-linolenic acid moieties. Mass spectra of most TGs exhibited abundant $[M+H]^+$ and $[M-RCOO]^+$ ions, which defined the molecular weight and the molecular association of fatty acyl residues of a TG, respectively. Regioisomeric forms of TGs were not determined from the seed oil samples, although differences were measured with reference compounds for the relative abundances of $[M-RCOO]^+$ ions formed by a loss of a fatty acyl residue from the sn-2 position and the sn-1/3 positions. $AgNO_3$–HPLC/APCI/MS provided valuable information for structure elucidation of seed oil TGs: 43 molecular species were identified from cloudberry (*Rubus chamaemorus*) seed oil, 39 from evening primrose (*Oenothera biennis*) oil, 79 from borage (*Borago officinalis*) oil, 44 from alpine currant (*Ribes alpinum*) and 56 from blackcurrant (*Ribes nigrum*) seed oils. The quantification of the species, however, required further studies because the abundance of the $[M+H]^+$ and $[M-RCOO]^+$ ions were strongly affected by the structure of the molecule.

Oxygenated acylglycerols. Lipid peroxidation in biological tissues attracts much attention because of its possible contribution to the functional modulation of biomembranes and lipoproteins. It is believed to be involved in free-radical-mediated damage, carcinogenesis and ageing processes. Research requires specific, sensitive and reproducible procedures to quantify the lipid hydroperoxides in each lipid class as primary products and the alcohols and aldehydes as secondary products of the peroxidation reaction. The identification and quantification of lipid oxidation products is therefore of great practical and theoretical interest and MS has assumed a major role in these analyses as a result of the development of mild ionization techniques.

Porter, Caldwell and Mills (1995) discussed the mechanisms of free radical oxidation of unsaturated fatty acids and pointed out that whereas monoenes and dienes produce relatively simple products, the polyunsaturates yield extremely complex mixtures. In addition, polyenes such as arachidonates can yield cyclic products. Monocyclic peroxides, bicyclic peroxides and epoxy alcohols, all formed as several stereoisomers, give rise to literally hundreds of different products in the autoxidation of arachidonates. Other polyene structures may give more or less complicated product mixtures depending on the number of double bonds in the system. Morrow *et al.* (1992, 1994) isolated from natural sources the products of the peroxyl radical cyclizations and discussed the possible mechanisms involved. The oxidation may take place directly on the intact lipid molecules or the oxidation products may become subsequently esterified to the glycerolipid molecules. In either case, the potential for product diversity is greatly amplified when dealing with the peroxidation products of the intact glycerolipids. The low molecular weight products of lipid peroxidation have been examined extensively by GC/MS after pentafluorobenzyloxime (PFBO) derivatization (Loidl-Stahlhofen, Kern and Spiteller, 1995). The pentafluorobenzylhydroxylamine is widely used in GC/MS for derivatization of carbonyl compounds. These derivatives are well suited to detection by NI/MS combined with SIM. Hydroperoxides are the primary autoxidation products of all lipids and they may be oxidized further along many different routes.

In contrast to unesterified fatty acids, many more species of oxidized compounds are formed during autoxidation of even simple mixtures of unsaturated TGs. Several autoxidation products have been isolated in model TGs and have been indirectly characterized by GC/MS of the TMS ethers. For this purpose, the derivatives were either reduced with sodium borohydride or underwent hydrogenation with PtO_2 catalyst in methanol, transmethylation following on. Neff, Frankel and Miyashita (1990) found that the 9- and 13-hydroperoxide isomers of trilinoleoylglycerol occurred in tissues in the same 1:1 ratio as in autoxidized methyl linoleate, as determined by GC/MS analysis of the TMS ethers. The secondary oxidation products of trilinoleoylglycerol were identified by GC/MS of derivatives obtained after reduction with sodium borohydride or after hydrogenation. After hydrogenation, silylation and transmethylation, the mixtures of the three secondary products formed were found to contain methyl 9,13-di-OTMS stearate, 9(13)-mono-OTMS stearate, and stearate in a ratio of 1:1:1; methyl 9,13-di-OTMS stearate, epoxystearate, and stearate in a ratio of 1:1:1; and methyl 9,13-di-OTMS stearate and stearate in a ratio of 2:1, respectively. This indicated that the three secondary oxidation products were: bis-(dioxygenated linoleoyl)(mono-oxygenated linoleoyl) monolinoleoylglycerol; bis-(dioxygenated linoleoyl)(mono-epoxyenelinoleoyl) monolinoleoylglycerol; and bis-(dioxygenated linoleoyl) monolinoleoylglycerol, respectively. Since the GC/MS was performed on the fatty acid methyl esters no information was obtained regarding their positional distribution in the TG molecules. Frankel, Neff and Miyashita (1990) have used GC/MS to identify the oxidized fatty acids recovered from oxidized TGs and isolated by preparative reversed-phase HPLC as both major and minor oxidation products of trilinolenoylglycerol. Monohydroperoxides (22.1%) and epidioxides (8.8%) were found to be major products. Minor oxidation products included tris-hydroperoxides (0.9%), bis-hydroperoxides (1.7%), mono-9,12-, 13-16-, and 9,16-dihydroxyperoxides (1.6%) and unidentified compounds (0.7%). GC/MS analysis was done using suitable GC derivatives. GC/MS analyses showed that the mono-, bis- and tris-hydroperoxides contained a mixture of one, two and three 9-, 12-, 13-, and 16-hydroperoxide isomers. After hydrogenation, transmethylation and silylation, GC/MS of the monohydroperoxy epidioxides showed the presence of one hydroperoxy epidioxide group per trilinolenoylglycerol molecule. These fatty acids were identified as a mixture of methyl 9,10,12- and 13,15,16-tri-OTMS stearate derivatives corresponding to the 9-hydroperoxy-10,12- and 16-hydroperoxy-13,15-epidioxides in autoxidized methyl linoleate. Similar studies with synthetic 18:3/18:3/18:2 TGs showed that the molecules with 18:3 in the 1,2 position were slightly less stable to oxidation than those with 18:3 in the 1,3 position, while synthetic 18:2/18:2/18:3 were less stable with 18:2 in the 1,3 position than in the 1,2 position (Miyashita et al., 1990).

Kuksis et al. (1993), Sjovall et al. (1995) and Sjovall and Kuksis (1995) have reported the direct identification of the hydroperoxides and core aldehydes of selected synthetic and natural TGs. Mixed hydroperoxide and core aldehyde derivatives were obtained following treatment with tert-butyl hydroperoxide and Fe^{2+} ions. The oxidation products were identified by normal-phase TLC and reversed-phase LC/MS with on-line TS/MS. The major components of a 2–4 h peroxidation of 18:2/18:1/16:0 were resolved by normal-phase TLC. The core aldehydes were converted into the dinitrophenylhydrazone (DNPH) derivatives which were then resolved in the neutral lipid solvent system. A total of nine bands indicative of the presence of hydrazones were obtained which, however, overlapped with various hydroperoxy derivatives of the TGs, as demonstrated by LC/MS examination of the individual TLC bands. Among the hydroperoxides identified were mono- (m/z 947) and di- (m/z 979) hydroperoxides. The mass of the tri-hydroperoxy derivative (m/z 1011) was outside the range of the instrument and was not identified. However, TLC suggested its presence. The mixed hydroperoxide and core aldehyde (ALD) derivatives were identified as minor components, for example 16:0/18:1 (OOH)/9:0 ALD (m/z 959) and 16:0/18:1 (OOH)/9:1 ALD (m/z 957). The major core aldehydes were the 16:0/18:1/9:0 ALD (m/z 927) and 16:0/18:1/8:0 ALD (m/z 913). Characteristically, the various ions were found in more than one LC/MS peak, which suggested resolution of positional and geometric isomers as well as of different isobaric triacylglycerols. The isolation and identification of triacylglycerol hydroperoxides and core aldehydes in various synthetic and natural TG mixtures exposed to autoxidation and chemical oxidation has been pursued systematically by Sjovall and Kuksis (1997a, b) using reversed-phase HPLC with on-line ES/MS. The mild ionization technique permitted high recoveries of the hydroperoxides and the mixed hydroperoxide–core aldehyde derivatives. Specifically, Sjovall and Kuksis (1997a, b) identified the mono- and di-hydroperoxide, core aldehyde and epoxide derivatives of some 20 synthetic TGs and utilized this knowledge of the chromatographic and mass spectrometric properties to establish the presence of these oxidation products in autoxidized natural vegetable oils. The hydroperoxides and core aldehydes of the DGs derived from peroxidized glycerophospholipids by phospholipase C digestion have been investigated by means of NICI GC/EI/MS (Kamido et al., 1992a, b) and LC/TS/MS (Kamido et al., 1995). The 16:0(18:0)/5:0 ALD and 16:0(18:0)/9:0 ALD were isolated as the DNPH derivatives and identified as major components among tert-butylhydroperoxide and copper ion peroxidation products of egg yolk and LDL phosphatidylcholine (PC) (Kamido et al., 1992a, b, 1995). Tokumura et al. (1996) have recently isolated the corresponding carboxy-acid esters rather than the core aldehyde esters as the major products of tert-butylhydroperoxide oxidation on the basis of GC/MS of the t-BDMS ether/Me esters of the oxo-diacylglycerols released from the oxidized PC by phospholipase C.

6.3.4 Neutral glycerophospholipids

The neutral glycerophospholipids, PC and phosphatidylethanolamine (PE), make up the bulk of the complex lipids derived from glycerophosphoric acid. Owing to their zwitterionic structure and amphipathic character, special efforts are necessary for MS analysis. The more successful approaches have involved the soft ionization methods. Of the early soft ionization methods, ammonia CI has been the method of choice because it favours the formation of abundant ions containing molecular weight information. Other soft ionization methods for glycerophospholipids involve FAB desorption, TS, and ES ionization, all of which yield abundant $[M + H]^+$ ions in the positive ion mode and $[M - H]^-$ ions in the negative ion mode. Significant progress was realized by the introduction of the tandem or MS/MS system of analysis which permitted the identification of the fragmentation products from individual molecular ions. This approach allowed the identification of many species of natural glycerophospholipids, except the isobaric species which require preliminary chromatographic resolution. Limited applications of MS to the determination of molecular species of glycerophospholipids have been reported during discussion of specific MS techniques (Jensen and Gross, 1988; Murphy, 1993; Myher and Kuksis, 1995) and in previous reviews on the application of MS to analyses of phospholipids (Kim and Salem, 1993) and other lipids (Evershed, 1994, 1996; Kuksis and Myher, 1989, 1995).

Choline glycerophospholipids. Crawford and Plattner (1983) and Jungalwala, Evans and McCluer (1984) showed that positive ion ammonia CI/MS of PC yielded $[M + H]^+$ and other prominent ions in the spectra, identifying the fatty acid composition. However, Crawford and Plattner (1984) concluded that isobutane CI/MS was better suited for quantification of underivatized phospholipids (PLs). When isobutane is used as the reagent gas, the $[M + H]^+$ ions are less intense than those seen in the corresponding ammonia CI/MS, and there are no ions between the $[M + H]^+$ and the $[M - 183]^+$ ion that would significantly interfere with quantification of any other molecular species. In ammonia CI/MS the disadvantage of ion interference is minimized when the technique is combined on-line with HPLC. Thus, in an early application, Jungalwala, Evans and McCluer (1984) used a moving belt interface to couple normal-phase HPLC with ammonia or methane CI/MS. Specific ions for individual PL bases were identified and used in single ion monitoring of the PL peaks. CI/MS of each PL also provided extensive information on the molecular species of the individual class of PL. Jungalwala, Evans and McCluer (1984) give the major characteristic fragment ions in the positive and negative ammonia and methane CI/MS of various PLs. The following ions were recognized in positive ion ammonia CI/MS of PC: $[M + 1]^+$, $[M - 41]\{[M + NH_4]^+ - N(CH_3)_3\}$, $[[M]^+ - 182]$ {i.e. M^+ – phosphocholine) and $[M]^+ - 147([M + N_2H_7]^+ - 182)$, DG adduct}. From the relative intensity of the DG-related ions one can calculate directly the percentage composition of the molecular species present in each PL.

Besides the above-mentioned DG-related ions in the positive CI/MS, the GPL base related ions were identified. In the case of PC, m/z 142, 156, 172, 184, 186 and 196 were identified. The ion at m/z 142 was the most characteristic ion for choline-containing PL and was used for monitoring these PLs after HPLC. The negative ion ammonia CI/MS of PC differed considerably from the positive ion spectra. These spectra contained $[M-1]^-$, $[M]^- - 13$, $[M]^- - 33$, $[M]^- - 42$, $[M]^- - 60$, and $[M]^- - 184$ as the major ions in the high mass range. The choline base was represented by m/z 182 and 123. The ion, m/z 123, presumably resulted from m/z 182 by loss of trimethylamine. The negative ion methane CI/MS of PC contained $[M]^- - 1$, $[M]^- - 13$, $[M]^- - 24$, $[M]^- - 33$, $[M]^- - 42$, $[M]^- - 60$, $[M]^- - 130$, $[M]^- - 148$, and $[M]^- - 184$ in the high mass range. The rat brain PC peak also split into two major peaks; the earlier peak contained mostly long-chain fatty-acid-containing species, whereas the latter contained short-chain fatty-acid species.

In one of the more successful early FAB/MS studies of molecular species of PC, Gross (1984) used positive ion analysis of intact PL from canine sarcolemma. The PC consisted predominantly of species with protonated parent ion $[MH]^+$ molecular masses of m/z 760 (16:0/18:1) and m/z 758 (16:0/18:2). Other major species identified possessed protonated parent ion masses $[MH]^+$ of m/z 744 (16:0/18:1 plasmalogen); 766 (16:0/20:4 plasmalogen), and 782 (16:0/20:4). Even small amounts of parent ions with masses of m/z 742, 780, 786, 788, 792, 794 and 808 were easily identified. A high plasmalogen content of sarcolemmal choline glycerophospholipid was supported by the finding of the predominant fragment ion peak at m/z 480, which resulted mainly from the loss of the 18:2 ketene from the 16:0/18:2 (m/z 742 plasmalogen), the loss of the 18:1 ketene from the 16:0/18:1 (m/z 744 plasmalogen) and the loss of the 20:4 ketene from the 16:0/20:4 (m/z 766 plasmalogen). Jensen, Tomer and Gross (1986) have shown that PC does not yield $[M-H]^-$ ions, but instead produces three characteristic high mass ions, $[M-CH_3]^-$, $[M-HN(CH_3)_3]^-$ and $[M-HN(CH_3)_3-C_2H_2]^-$, along with two low mass ions arising from the fatty acid portions. For the 16:0/18:1 PC species, these ions are m/z 744, 699, 673, 255 and 281. In the CID spectrum, ions of m/z 435 and 409 arise from the losses of R_1CHCO and R_2CHCO, respectively, and ions of m/z 417 and 391 arise from losses of R_1CH_2COOH and R_2CH_2COOH, respectively. The losses of R_2CHCO and R_2CH_2COOH occur more readily than losses of R_1CHCO and R_1CH_2COOH, and the RCH_2COOH loss occurs about twice as readily as the RCHCO loss. The three high mass ions of m/z 506, 461 and 435 result from losses of various portions of the choline moiety. Pyrenyl PC, where a fatty acid substituent is replaced with a pyrene functionality, is a common fluorescent biological probe. On FAB/MS it yielded an m/z corresponding to $[M-CH_3^+]^-$ which may be attributed to the loss of one of the methyl groups of the choline.

Gasser et al. (1991) used both positive and negative ion FAB/MS for the characterization of PC in rabbit lung lavage fluid. A chloroform/methanol extraction followed by a simple HPLC separation was superior to TLC separation to obtain stable, long-lasting protonated molecular ions and diagnostic fragment ions, which permitted the identification of the polar head-group. In combination with 3-nitrobenzyl alcohol as liquid matrix, they established a procedure that yielded a fast sample preparation method, a good signal-to-noise ratio for determining minor species and reduced formation of $[M + H - 2H]^+$ ion species.

Kayganich-Harrison and Murphy (1994a) employed negative ion FAB/MS/MS together with stable isotope labelling to assess directly events of biosynthesis and metabolism of arachidonic acid-containing PL molecular species by cells in culture. Mast cells, cultured with $[^{13}C]$ linoleic acid, converted it into arachidonic acid which was then incorporated into cellular PLs. Over a 24 h period, the extent of label enrichment in each arachidonate-containing PL molecular species was monitored. Specific incorporation of $[^{13}C_{17}]$-labelled arachidonate was determined from the ratio of the carboxylate anions at m/z 320 and 303, which correspond to $[^{13}C_{17}]$ arachidonate and unlabelled arachidonate, respectively. The carboxylate anions were produced by CID of each specific molecular anion.

Kim and Salem (1986) developed a technique for rapid and detailed molecular species analysis of PL with use of reversed-phase HPLC with on-line TS/MS. In conjunction with a hexane/methanol/0.2 M ammonium acetate mixture as mobile phase, the technique was generalized for natural mixtures of PC and PE. The positive ion spectra of PC gave fragments similar to those of ammonia CI, but the TS produced much simpler spectra, with extremely low background. As an example, 16:0/18:1 GPC gave the DG ion at m/z 578, the result of a loss of the phosphocholine group. The ions derived from phosphocholine were detected at m/z 142 and 184, with the peak at m/z 142 usually more intense than that of m/z 184. The molecular ion was present as a protonated form at m/z 761. The m/z 142 ion was monitored for the detection of PC. Coupling of reversed-phase HPLC with MS detection allowed an extensive separation of the molecular species of egg yolk PC. The DG ion peaks, which are predominant in the TS spectra of PC, allowed an easy reconstruction of the 10 major molecular species of PC.

Kerwin, Tuininga and Ericsson (1994) described the use of positive and negative ion ES/MS and ES/MS/MS to identify PL head-groups and their alkyl, alkenyl and acyl constituents of PC and PE. The spectra were acquired by ES/MS/MS by using a flow injection of chloroform extracts of biological samples without prior chromatographic separation of the GPL. The ES spectra contained mainly $[M + H]^+$ and $[M - H]^-$ ions, in the positive and negative modes, respectively. The composition of molecular species could be determined from the carboxylate ions that were produced by CID of the $[M + H]^+$ ions. Likewise, Han and Gross (1994) reported that ES/MS facilitates the structural determination and quantitative analyses of individual molecular species of synthetic and naturally derived PL with a sensitivity two to three orders of magnitude greater than that previously achieved by FAB/MS. This method allowed quantification of PL, with coefficients of determination of better than 0.99 and accuracies better than 95%. A positive ion ES spectrum of erythrocyte plasma membrane PL extract containing 13.5 pmol of PC in 25 nl of whole blood showed 14 molecular species of PC and 4 molecular species of sphingomyelin (SPH). ES/MS also facilitated the quantification of structurally diverse lipids differing in their surface interactions. Such direct quantification by FAB/MS has not been possible. An equimolar mixture of the plasmenylcholine (m/z 767), plasmanylcholine (m/z 769) and PC (m/z 783) resulted in three sodiated molecular ion peaks of equal intensity (within 5%).

Myher *et al.* (1994) and Kim, Wang and Ma (1994) reported improved ES/MS analysis of molecular species of PL by combining this instrumentation with HPLC. Myher *et al.* (1994) combined normal-phase HPLC with ES/MS for the analysis of total lipid extracts of PLs, including the hydroperoxides of intact PLs. The PLs were resolved on a silica column by elution with a gradient of chloroform/methanol/water/ammonia. As already noted, $[M + H]^+$ and $[M - H]^-$ ions were produced exclusively by positive and negative ES, respectively, at 150 V. At higher voltages increased intensities were seen for the DG-type of ions in the positive ion mode. Interestingly, dimers of PC were also seen. In these experiments, the column effluent was split 1:100 before admission to the ES source.

Kim, Wang and Ma (1994) used 0.5% ammonium hydroxide in a water/methanol/hexane mixture and a C18 column to resolve complex mixtures of molecular species of PLs, which were detected as protonated and sodiated molecular species. Samples were injected onto a C18 HPLC column (5 μm, 2.1 mm × 15 cm) and separated by means of the aforementioned mobile phase changing linearly from 12:88:0 to 0:88:12 in 17 min after holding at the initial composition for 3 min. The column flow rate was 0.5 ml min^{-1} and the effluent was split 1:100. The capillary exit voltage was generally set at 200 V and raised up to 300 V for the detection of DG ions.

Platelet activating factor. The determination of the platelet activating factor (PAF) occupied a special place among early analyses of PLs by MS. MS methods such as GC/MS and FAB/MS were crucial for the demonstration of the chemical heterogeneity of PAF molecules. Murphy (1993) reviewed the use of FAB/MS/MS to identify the glyceryl ether GPC species (1-O-alkyl-2-acyl) from the production of a single carboxylate anion following CID of $[M - 15]^-$ and a very abundant neutral loss of arachidonic acid from $[M - 86]^-$. Diacyl species with an odd number of carbon atoms in one alkyl chain would be isobaric with such glyceryl ether lipid species, but

CID of [M − 15]⁻ resulted in the production of two carboxylate anions, one of which would have an odd number of carbon atoms in the alkyl chain. Mild acid hydrolysis, followed by re-analysis of the precursors of m/z 303, was necessary to differentiate the isobaric 1-O-alkyl-2-acyl species and the 1-O-alk-1-enyl-2-acyl (plasmalogen) species of PE and PC. 1-O-alkyl-2-acyl molecular species with n double bonds in the $sn - 1$ hydrocarbon chain are isobaric with plasmalogen molecular species having $n - 1$ additional double bonds in the $sn - 1$ hydrocarbon chain. The mass fragmentation pattern observed by Kim and Salem (1987) for beef heart PAF molecules was similar to that of other PLs in that the DG fragment was the base peak. However, the system employed in this case gave poor separation for 15:0 alkyl, 16:0 acyl and 16:0 alkyl species, although 18:0 acyl and 18:0 alkyl species were well separated. Effective separations of alkyl chain homologues were later obtained by Weintraub, Lear and Pinckard (1993) by means of differential electron capture mass spectrometry of the PFB derivatives of the diradylglycerols.

HPLC systems interfaced with an ES/MS/MS have been successfully used for structural analysis and chemical characterization of PAF extracted from clinical samples. Silvestro *et al.* (1993) developed an improved LC/MS technique, with an ES interface, for the determination of PAF and lyso-PAF in biological samples. HPLC separations were performed using a reversed phase column. The mass spectra showed an intense [M + H]⁺ ion. CID of protonated molecular ions gave characteristic daughter ions corresponding to the phosphocholine group. By selective ion monitoring, a detection limit of 0.3 ng was obtained for all molecules; by multiple ion monitoring the sensitivity for the lyso-PAF limit was decreased to 3 ng. Borgeat *et al.* (1994) then demonstrated that MS/MS daughter ion monitoring (m/z 524 to 184) could be used to detect PAF in the low picogramme range. Rizea Savu *et al.* (1996) combined normal-phase HPLC separation and MS/MS detection, using an ion-spray LC/MS interface, for the quantitative determination of acyl-PAF, PAF and related GPL classes. MS of positive ions [M + H]⁺ and CID of protonated molecular ions gave characteristic daughter ions corresponding to the polar head-group. Detection limits of 0.1–0.3 ng sample^{-1} were obtained by multiple reaction monitoring. The method was applied in the analysis of acyl-PAF and PAF levels in treated human endothelial cells. Woodard *et al.* (1995) have described a direct derivatization of PAF extracts with PFB anhydride. Individual species of PFB-derivatized PAF were separated by GLC prior to mass spectral analyses. Quantitative estimates of six different species were obtained. The electron capture mass spectra and specific ions for two internal standards and six species of PAF were recorded with negative and positive ion detection. The method was applied to the analysis of PAF in saliva. The predominant PAF in human saliva was 1-O hexadecyl-2-acetyl-*sn*-glycero-3-phosphocholine or 16:0-alkyl-PAF, representing only 30% of the total

PAF. Substantial amounts of 18:1 and 18:0-alkyl PAF and 16:0-acyl-PAF were also identified.

Ethanolamine glycerophospholipids. Jungalwala, Evans and McCluer (1984) reported that in the positive ion ammonia CI/MS of PE, $[M]^+ - 140([M]^+ -$ phosphoethanolamine) and $[M+35]^+ - 140$ were the major ions in the high mass range. The characteristic ion for ethanolamine base was found at m/z 141 and had the greatest intensity in the spectrum. The presence of this peak was useful for monitoring PE after HPLC. The $[M-140]^+$ and $[M-105]^+$ ions for the alk-1-enyl/acyl glycerophosphoethanolamine (GPE) were distinguished from the corresponding ions derived from diacyl GPE in that the former had 14 units less mass than had the latter type. However, they cannot be distinguished from the corresponding odd chain fatty-acid-containing diacyl GPEs. The positive ion methane CI/MS of PE was similar to that of positive ion ammonia CI/MS in the high mass range in having $[M-140]^+$ as the major ion, but $[M-105]^+$ and $[M-123]^+$ were only of low intensity. In negative ion methane CI/MS of PE, the major ion in the high mass range was $[M-88]^-$, and the characteristic ion for ethanolamine GPL was found at m/z 191. Bovine brain PE contained both diacyl and alkylacyl GPE, the majority being alk-enyl/acyl GPE. Based on the intensity of $[M-140]^+$ ions, the composition of alkenylacyl and alkylacyl GPE was calculated to be 69%, whereas diacyl GPE was 31% of the total PE. The major diacyl and alkenylacyl and alkylacyl GPE species were identified and listed, and about 15 other molecular species of PE were identified but not listed. The positive ion methane CI/MS of PE was similar to that of positive ion ammonia CI/MS, except that the characteristic ion for ethanolamine PL was at m/z 124 in these spectra. The ethanolamine-containing PLs were identified by m/z 141. The m/z 142 ion is known to be specific for choline-containing PLs. The m/z 198 ion is found to be associated only with inositol-containing PLs. The HPLC/MS of PLs also showed that individual PLs were chromatographically resolved to some extent based upon molecular species. The long-chain fatty-acid-containing species were eluted before the short-chain fatty-acid-containing species. The rat brain ethanolamine-containing GPLs were partially resolved into two peaks. Mass spectral analysis of the first emerging peak showed that it contained mostly alkenylacyl GPE of m/z 561, 587 and 589, corresponding to plasmalogen species 16:1/18:1, 18:1/18:2 and 18:2/18:2, respectively. The front peak, however, also contained some diacyl GPE of m/z 623, 627, and 651, corresponding to 16:0/22:6, 18:0/22:4 and 18:0/22:6, respectively. The second peak contained mostly diacyl GPE 18:0/18:1 (m/z 605) and 18:0/18:2 (m/z 603).

Gross (1984) described an early application of FAB/MS to the characterization of tissue PL including PE. FAB/MS of canine sarcolemmal ethanolamine PL showed major ion peaks with protonated molecular ions at m/z 752 and 750 and lower-intensity ions at m/z 724 and 768. In positive ion mode, ethanolamine, in contrast to choline GPL, underwent a facile loss of ethanolamine phosphate, particularly from the diacyl species, m/z 768 − 141 = 627. The facile loss of phosphoethanolamine in diacyl, but not plasmalogen ethanolamine PL may account for the overrepresentation of plasmalogens in parent ion peaks when compared with quantitative GLC analysis of the products. Jensen, Tomer and Gross (1987) extended systematically the previous work on FAB/MS of PE by using synthetic standards. Thus, the FAB/MS of 16:0/16:0 GPE in triethanolamine matrix was characterized by ions of m/z 732, 716, 690, 647 and 255. The latter three ions were thought to correspond to the $[M - H]^-$ ion, the $[M - H - C_2H_5N]^-$ ion and the carboxylate anion, respectively. The two higher mass ions appeared to be adducts of the ethanolamine head-group and the matrix. The FAB/MS of 16:0/16:0 N,N-dimethyl PE in triethanolamine matrix was characterized by ions of m/z 718, 647 and 255, corresponding to $[M - H]^-$, $[M - H - C_4H_9N]^-$ and the carboxylate anion, respectively. No high mass adduct ions corresponding to those of m/z 732 and m/z 716 were seen for the N,N-dimethyl compound. The CID of the $[M - H]^-$ ion of 16:0/16:0 GPE (m/z 690) yielded the following daughter ions: m/z 452, $[M - H - RCHCO]^-$; m/z 434, $[M - H - RCH_2COOH]^-$; m/z 255, (the carboxylate anion); and a series of other high mass ions. Jensen and Gross (1988) discussed the fragmentation mechanism for PLs in the negative ion mode following FAB high-energy CID and noted a higher abundance in the product ion mass spectra of the sn-2-carboxylate anion. However, NI/MS of two isomeric PEs, 1-16:0/2-18:1 and 1-18:1/2-16:0, gave markedly different ion intensities. For both PEs examined, the intensity of the carboxylate anion in the 2 position was approximately twice that of the carboxylate in the sn-1 position. The low-intensity ions resulting from the elimination of RCH_2COOH and RCHO were found at m/z 434 and 452, respectively, in the product ion mass spectra of 1-16:0/2-18:1 GPE; and at m/z 460 and 478, respectively, in the mass spectra of the other isomer, indicating that the formation of these ions also involved the preferential loss of the acyl group from the sn-2 position.

Kayganich-Harrison and Murphy (1992) used FAB/MS/MS for the identification of molecular species of diacyl-, alkylacyl-, and alk-l-enylacyl GPE in human polymorphonuclear leukocytes. For natural mixtures, however, a preliminary chromatographic resolution is desirable (see below).

Sandoval, Huang and Garrett (1995) used FAB/MS and FAB/MS/MS to characterize N-acyl PE from dry and soaked seeds of cotton. The major species were identified as 16:0/18:2 GPE (N-palmitoyl), 16:0/18:2 GPE (N-linoleoyl) and 18:2/18:2 GPE (N-palmitoyl).

Kim and Salem (1986) described a reversed-phase LC/TS/MS technique for the detailed analysis of molecular species of PLs, including PE. An analysis of standard diacyl GPE showed that instead of the peaks at m/z

142 and 184 for phosphocholine, phosphoethanolamine produced peaks at m/z 124 and 141 resulting from the dehydration of both the protonated phosphoethanolamine and the ammonium adduct of phosphoethanolamine, respectively. A similar application of the technique to the analysis of egg yolk PEs yielded nine major molecular species, which again were readily identified on the basis of the DG ions. At the same time, direct quantification of each molecular species was possible since their TS/MS responses were found to be comparable and reproducible. Under optimum conditions of the analytical system, 50 ng sample^{-1} was near the lower limit for the simultaneous detection of all fragments, including the molecular ion. The deviation in peak areas between replicate analyses was less than 10% even for a 50 ng injection.

Kerwin, Tuininga and Ericsson (1994) reported the identification of molecular species of PE by using ES/MS and ES/MS/MS. Positive and negative ion ES/MS for molecular species of commercial preparations of PE produced similar profiles. Positive ion ES/MS/MS provided information about the nature of the head-group. The identity of alkyl or alkenyl substituents in PE molecular species could be established from residual ions following loss of ethanolamine plus loss of acyl moiety in the sn-2 position, and cyclization of a phosphate oxygen with C-2 of glycerol. ES/MS did not provide information on the position (sn-1 or sn-2) of fatty acids, and was not capable in all instances of differentiating between alkylacyl and alkenylacyl substituents without prior separation of these lipid substances. With positive and negative ion ES/MS, over 50 molecular species were identified from bovine brain PEs, where at least 60% alkylacyl or alkenylacyl species were present.

Han and Gross (1994) applied ES/MS and ES/MS/MS to the structural characterization and quantification of individual molecular species of PE of human erythrocyte plasma membrane. A negative ion ES spectrum of a total lipid extract of plasma membrane GPLs showed over 25 molecular species of PE. The negative ion ES/MS spectra of PE were far more sensitive than were positive ion spectra owing to the case of loss of a proton from the ammonium ion in ethanolamine PLs. The human erythrocyte plasma membrane ethanolamine PLs are predominantly composed of plasmalogen molecular species, and this was evident in the ES spectrum with, 1-(O)-Z-octadec-1'-enyl-2-eicosatetra-5', 8', 11', 14'-enoyl-sn-glycero-3-phosphoethanolamine (i.e. 18:0/20:4 plasmenylethanolamine, m/z 751) representing the major individual molecular species present (about 20 mol%). Other plasmenylethanolamines, containing 20:4, 22:4, 22:5 and 22:6 fatty acids at the sn-2 position, were also identified. The ES/MS/MS analysis of isobaric ethanolamine GPLs, for example m/z 777 after NI/ES ionization, produced two carboxylic anions (m/z 329 and 331) corresponding to 22:5 and 22:4 fatty acids, respectively. Also, several ethanolamine lyso PL-type ions were identified facilitating the assignment of these species as 18:0–22:5 and 18:1–22:4

plasmenylethanolamines in a 2:1 molar ratio. Han et al. (1996) have exploited the analytical power and sensitivity of ES/MS both to identify plasmenylethanolamines as the largest source of arachidonic acid mass released during thrombin stimulation and to demonstrate the presence of multiple novel molecular species of plasmenylethanolamines in human platelets.

Kim, Wang and Ma (1994) have improved ES/MS analysis of the molecular species of PE by linking the spectrometer to HPLC and increasing the exit voltage. The capillary exit voltage was generally set at 200 V and raised up to 300 V for the detection of positive DG-type ions. The nitrogen drying gas flow was approximately $3 \, l \, min^{-1}$, and the temperature was 220 °C. As a result of the increased exit voltage, there was an increase in the production of 18:0/22:6 DG (m/z 651) from 18:0/22:6 GPE, which at 200 V gave $[M + H]^+$ as a maximum.

Myher et al. (1994) and Ravandi et al. (1995a) combined normal-phase HPLC with ES/MS for the analysis of total lipid extracts containing PE and glucosylated PE. In negative ion mode the normal and glycated PE showed intense peaks for the $[M - H]^-$ ions, which were identified, quantified and compared with the molecular species composition of PE determined by independent methods. The uncorrected averaged mass spectra taken over the red-cell PE classes agreed rather closely with reported values for molecular species obtained by much more elaborate analytical systems (Myher, Kuksis and Marai, 1993).

6.3.5 Acidic glycerophospholipids

The acidic PLs, such as phosphatidylserine (PS), phosphatidylinositol (PI), phosphatidylglycerol (PG), diphosphatidylglycerol (DPG) and phosphatidic acid (PA), produce primarily negative ions with the soft ionization techniques. The positive ions, when formed, are detected at much lower intensities. The principles of FAB/MS (Murphy, 1993) and TS/MS (Kim and Salem, 1993) of the acidic PLs have been reviewed. The following discussion emphasizes the combination of chromatographic and MS methods and summarizes the more recent applications.

Serine glycerophospholipids. FAB/MS of PS is quite similar to that discussed previously for PC and PE, showing a significant $[M + 1]^+$ pseudo-molecular ion along with DG-type and MG-type fragment ions. In the negative mode, the $[M - 1]^-$ pseudo-molecular ion, $[M - 88]^-$ and the abundant carboxylate ions from each fatty acyl group predominate (Murphy, 1993).

Jungalwala, Evans and McCluer (1984) were the first to combine HPLC with on-line FAB/MS for the analysis of glycerophospholipids. Separation of PL was achieved on a Brownlee silica gel cartridge column with ammonia-containing solvent as described for neutral PLs (subsection on choline

GPLs in Section 6.3.4). Under these conditions fairly good reproducibility was achieved, but PS tailed to some extent into the lyso PI peak. In the positive ion CI/MS with ammonia or methane the major ions arising from PS resulted from the loss of phosphoserine, $[M - 184]^+$ and $[(M + 35) - 184]^+$, in the high mass range. The characteristic ion of all serine-containing PLs was located at m/z 105 and was attributed to the serine head-group. In the negative ion methane CI/MS of PS, m/z 139 and 173 were the major ions in the low mass range, whereas $[M - 132]^-$ was the major ion in the high mass range. The $[M + H]^+$ ions and other high mass ions corresponding to the DG moieties allowed the identification of the molecular species resolved by the HPLC column. The major species of PS from bovine brain were identified by positive ion ammonia CI/MS as: 18:0/18:1 (54%, m/z 605 and 640); 18:0/18:2 (7%, m/z 603 and 638); 18:0/20:1 (17%, m/z 633 and 668); and 18:0/20:2 (6.3%, m/z 631 and 666). Small amounts of alkenylacyl glycerophosphoserine (GPS) were also identified, as 18:1/22:6 (7%, m/z 607 and 642) and 18:1/18:0 (2%, m/z 591 and 626).

Jensen, Tomer and Gross (1986) used FAB/MS/MS to examine commercial samples of PS. It was shown that 18:0/18:1 GPS could be satisfactorily desorbed from triethanolamine matrix as the $[M - H]^-$ ion (m/z 788). CID of $[M - H]^-$ caused loss of serine, which led to abundant amounts of the m/z 701 ion. Ions attributable to the carboxylate of 18:0 (m/z 283) and 18:1 (m/z 281) were also seen. Satisfactory desorption of any PS by using positive ion FAB/MS was not achieved.

Chen (1994) has since reported partial characterization of the molecular species of PS from human plasma by HPLC and FAB/MS. Ions at m/z 788 and 810 originated from the deprotonated molecules of 36:1 GPS and 38:4 GPS, respectively. Loss of fatty acids from the ions at m/z 788 and 810 gave peaks at m/z 524 and 506. The two molecular species within the human plasma PS were therefore identified as 1-stearoyl-2-oleoyl and 1-stearoyl-2-arachidonoyl-*sn*-3-phosphoserine. The absence of the palmitoyl species of PS was surprising. Chen (1997) has subsequently reviewed the analysis of aminophospholipids by FAB/MS in relation to the ES/MS procedures and has pointed out the advantages of each method.

Kim and Salem (1987) extended the filament-on LC/TS/MS technique to a general determination of the molecular species of the major PL classes. Positive ion TS of PI, PS, and PAF showed fragmentation patterns similar to those of PC and PE (Kim and Salem, 1986) as they produced DG, MG, head-group and molecular ion species. PS produced intense MG ions as did PI, although the DG ion was still the base peak in most of these spectra. The head-group of PS yielded m/z 105 attributable to serine. Reversed-phase LC/TS/MS yielded six molecular species of bovine (liver) PS: 18:0/22:6; 18:1/18:1; 18:0/18:1; 18:/22:3; 18:0/20:1; and 20:1/20:1. Although quantification of molecular species was possible on the basis of the DG ion intensity further method development was necessary in order to achieve greater precision.

In a parallel study Hullin, Kim and Salem (1989) have described the reversed-phase separation of individual molecular species of PE and PS (rat brain and human red blood cells) as the trinitrobenzenesulphonic acid (TNP) derivatives. This allowed excellent resolution of the diacyl and alkenylacyl species of PE. The identity of the resolved species was confirmed by TS/MS. The TNP derivatives fragmented in a manner similar to other PLs, with DGs and MGs providing the base peaks. Under reversed-phase HPLC conditions, the retention times of the TPN derivatives depended both on the polar head-group and on the fatty acid composition of the PL. The TNP derivatives of PS eluted at a lower methanol concentration than the corresponding species of PE. The brain PS was resolved into 11 components, the identities of which were confirmed by LC/TS/MS (Kim and Salem, 1987).

Kerwin, Tuininga and Ericsson (1994) used flow injection ES/MS to characterize the molecular species of PS recovered as the Na^+ salts from HPLC columns. In positive ES/MS, pseudo-molecular ions were seen for both $[M + H]^+$ and $[M + Na]^+$, whereas in negative ES/MS, the pseudo-molecular ion was $[M - H]^-$. Positive ion MS/MS of several PS molecular ions confirmed the presence of sodium adducts and provided confirmation of head-group composition. MS/MS of m/z 834, which corresponds to the sodium salt of 18:0/20:4 diacyl PS, showed major peaks at m/z 747 (loss of CH—C(COOH)H—NH_2 from the head-group); m/z 651 (loss of serine); m/z 627 (loss of serine plus sodium); and m/z 208 attributable to the cation of phosphoserine plus sodium. Negative ion MS/MS of m/z 810 confirmed the presence of $RCOO^-$ ions corresponding to 18:0 and 20:4. Similar fragmentation was found after MS/MS of peaks corresponding to 18:0/18:1 and 18:0/22:6 diacyl PS. The negative ion spectrum, lacking the sodium adducts, was more easily interpreted and permitted the recognition of over 30 different (primarily diacyl) PS molecular species in bovine brain PS.

Myher *et al.* (1994) subjected normal and oxygenated GPLs to normal-phase HPLC and on-line ES/MS. The solvent system used was the gradient of chloroform/methanol/ammonium hydroxide employed earlier for neutral PL (subsection on choline GPLs in Section 6.3.4). It completely resolved the common PLs but resulted in a partial overlap between the corresponding natural and oxidized PLs. In the positive ion mode, PS yielded a very weak signal. In the negative ion mode, intense $[M - H]^-$ ions were observed for bovine brain PS and PI. No fragment ions were apparently formed at an exit voltage of 150 V. Ravandi *et al.* (1995a) used this system to characterize synthetic glucosylated PS, which gave a $[M - H]^-$ ion at m/z 897 (16:0/16:0 GPS). DG-type ions could be obtained for PS in the positive ion mode by increasing the exit voltage from 150 V to 250 V.

Kim, Wang and Ma (1994) reported an improved technique for molecular species analysis of PLs including acidic PLs, with use of HPLC with an ES interface. By using 0.5% ammonium hydroxide in a water/methanol/hexane mixture and a C18 column, they separated complex mixtures of PL classes along with the molecular species. Under these conditions, PS eluted first followed by PI, PA, PE and PC in a successive manner for PLs containing a given fatty acyl composition. In the positive ion mode, protonated or natriated molecular species of PS were detected. Formation of sodium adducts was most prominent with the acidic PLs. As the capillary exit voltage increased from 200 V to 300 V the intensity of the DG fragment increased and peaked at approximately 300 V. For all 18:0/18:1, 18:0/20:4 and 18:0/22:6 species, the sensitivity of PS, as monitored by $[M + H]^+$, was approximately 5–20 times less than that for PC. In a special application of this method, Kim, Wang and Ma (1994) monitored the incorporation of 22:6n3 into glioma cell PLs. The DG fragments were monitored for 22:6n3 containing PL species. It was apparent from the results that 22:6n3 is taken up into PA, PI and PC first, then remodelled to PE, and eventually retained as PE plasmalogens in the C-6 glioma cells. PS accumulated during the first 24 h did not appear to undergo significant change. In comparison with the existing LC/MS techniques, marked improvement in sensitivity was observed. It was estimated that the limit of quantification was approximately 0.5 pmol before effluent stream splitting (5 fmol after 1/100 split) when sodiated adducts were monitored instead of protonated molecules. In general, the ES technique was approximately 20–50 times more sensitive than the existing methodologies which use LC/TS/MS, which generate DG fragments as the major ions (Kim and Salem, 1986, 1987).

Ma and Kim (1995) developed an on-line LC/TS/MS method for the analysis of PL molecular species in rat brain. After total lipid extraction, the extract was subjected to analysis with on-line reversed-phase HPLC and filament-on TS/MS. By using non-conventional HPLC conditions, partial separation of individual PL classes (PS, PI, PE and PC) and partial separation of molecular species within each class were achieved. By monitoring the retention time and the characteristic fragment ions (DG ions) formed in the filament-on TS process, individual molecular species in each PL class could be identified. Although non-linear calibration curves were observed for all DG ions monitored, even in the presence of an internal standard, semiquantitative and quantitative results could still be obtained for a mixture of PLs.

Inositol glycerophospholipids. Jungalwala, Evans and McCluer (1984) showed that positive ion ammonia CI/MS of 18:0/20:4 PI had $[M - 259]^+ ([M - \text{phosphoinositol}]^+)$ and $[M + 35]^+ - 259$ as the major ions in the high mass range. The characteristic ions of PI were *m/z* 198 and 180. The major molecular species of bovine brain PI were identified as: 18:0/20:4 (40%, *m/z* 627 and 662); 18:0/18:2 (10%, *m/z* 603 and 638); 18:0/18:1 (14%, *m/z* 605 and 640); and 18:0/20:3 (11%, *m/z* 629 and 664). Nine other minor molecular species were also identified. In the negative ion methane CI/MS of PI, $[M - 207]^-$ was the major ion. The amount of PI in rat brain is very small. Nevertheless, the specific ion *m/z* 198 clearly identified the appropriate HPLC peaks as PI, with the major species being 18:0–20:4 (m/z 662 and 627). The brain PS was also resolved into two peaks. The initial peak was associated mostly with SPH with C18 sphingenine and 22:0 (m/z 586 and 604). The last peak in the chromatogram was attributable to PS (m/z 105).

Jensen, Tomer and Gross (1987) used FAB/MS/MS to examine the PI of soybeans. In the negative ion mode, $[M - H]^-$ ions were obtained for the principal components at m/z 833 and 857, with minor components at m/z 831, 859 and 861. CID of the $[M - H]^-$ ion, m/z 833, gave m/z 255 and 279, corresponding to the carboxylate anion of 16:0 and 18:2, respectively. The MG-type ions resulting from loss of 16:0 (m/z 577) and loss of 18:2 (m/z 553) were also seen. In a similar manner, the ions of m/z 831 and 857 in the mixture were characterized as PI species, 16:0/18:3 and 18:2/18:2, respectively. Another series of ions in the CID mode arose from the remote charge site fragmentation of the $[M - H]^-$ ion, which provides for highly specific structural characterization of a single alkyl chain. This fragmentation, however, is of diminished analytical value when it involves the fragmentation of the one alkyl chain in competition with the parallel fragmentation of the other. According to Kim and Salem (1987), LC/TS/MS of PI gave rather intense MG fragment ions compared with those of PC or PE. The peak at m/z 198 corresponded to an ammoniated inositol fragment and was characteristic of the PI class.

Kerwin, Tuininga and Ericsson (1994) identified over 30 molecular species of bovine liver PIs using flow injection ES/MS in either positive or negative ion mode. However, the negative ES spectra were easier to interpret. Negative ion MS/MS of the parent ion of 16:0/18:1 (m/z 835) produced ions at m/z 255 and 281 attributable to $RCOO^-$ from 16:0 and 18:1, respectively, thus showing the presence of only the 16:0/18:1 molecular species. In the positive ion mode, MS/MS of m/z 859, which corresponds to the Na^+ adduct of 16:0/18:1 and 18:0/16:1 GPI, produced ions at m/z 599 and 577, corresponding to the loss of H^+ and Na^+ adducts, respectively, of phosphoinositol. An ion at m/z 282 was attributed to the Na^+ adduct of phosphoinositol. An alkylacyl GPI species was recognized at m/z 873 as a Na^+ adduct. The positive ES MS/MS spectrum of the Na^+ adduct at m/z 873 produced two major peaks corresponding to the loss of the phosphoinositol head-group as described above. The MS/MS spectrum in the negative ion mode produced m/z 241 from the PI head-group and four peaks from the acyl and/or alkyl constituents: m/z 267, 269, 281 and 283. As m/z 267 and 269 could be attributable to the alkyl $[RCH_2O^-]$ substituents of 18:1 and 18:0, it was concluded that there are 18:0/18:1 and 18:1/18:0 alkylacyl species present. In PI species, an RCO^- ion is formed during negative ion MS/MS because of steric interference from the bulky phosphoinositol head-group which prevents cyclization and stabilization of the ion, as described for the PE species.

Kerwin et al. (1995) employed flow ES/MS and flow ES/MS/MS to examine PC, PE and PI species from fungal mycelium and nuclei grown in defined medium with and without isoprenoids which induce reproduction. A very large percentage of the PC (69–80 mol%) and PI (74%–79 mol%) molecular species from mycelia and nuclei were found to contain ether linkages. PE species had only 13–20 mol% of ether-containing moieties.

Other acidic glycerophospholipids. Selected samples of PG, cardiolipin (CL), PA and their lysoderivatives have also been subject to examination by MS and MS/MS. However, no systematic studies on natural mixtures have been reported.

Jensen, Tomer and Gross (1987) extended the FAB/MS characterization of PLs to PA, PG and CL. PA is the simplest PL. FAB/MS of 18:0/18:0 GP gave $[M - H]^-$ (m/z 703), $[M - H - RCHCO]^-$ (m/z 437) and a carboxylate anion (m/z 283) as daughter ions. The corresponding ions for the 16:0/16:0 GP were m/z 647, 409 and 255. CID confirmed the origin of the fragment ions in the corresponding parent molecules. Kim, Wang and Ma (1994) used reversed-phase LC/ES/MS to analyse the molecular species of PLs, including PA. Using a gradient of ammonia/water/methanol/hexane as eluant, PS eluted first (17 min) followed by PI, PA, PE and PC in successive order for PLs containing the same fatty acid composition. This technique was applied to monitor the incorporation of 22:6n3 into acidic PL. Incorporation was most prominent in PA.

According to Jensen, Tomer and Gross (1987) the negative ion spectrum of PG was similar to that of PI. The 16:0/16:0 GPG species gave a $[M - H]^-$ ion at m/z 721, and a carboxylate fragment at m/z 255 corresponding to 16:0. CID of $[M - H]^-$ yielded several informative daughter ions. A fragment at m/z 647 was formed by loss of $C_3H_6O_2$, presumably involving the loss of the unsubstituted glycerol portion of the molecule with hydrogen-transfer to the phosphate oxygen. Another ion, at m/z 465, was formed by loss of RCH_2COOH, and an ion at m/z 483 was formed by loss of RCHCO; yet another ion, at m/z 255, was attributable to the carboxylate anion of 16:0. Itabashi and Kuksis (1997) have reported the chiral phase LC/ES/MS separation and identification of the enantiomeric and diastereomeric PG isomers as the DNPU derivatives. In the negative ion mode, the dioleoyl GPG gave $[M - H]^-$ at m/z 1192, $[M - H - DNPU]^-$ (m/z 982) and $[M - bis - DNPU^-]$ (m/z 773). The fragment ions were produced in increased amounts following increases in the exit voltage from 150 V to 250 V. In the positive ion mode, the $[M + H]^+$ and $[M + Na]^+$ ions were observed without fragmentation by using an exit voltage of 200 V.

Jensen, Tomer and Gross (1987) have also used FAB/MS to examine a sample of bovine heart CL. In the negative ion mode the major ions were formed at m/z 1448, 833, 695, 415 and 279. CID showed that the $[M - H]^-$ ion, m/z 1448, fragmented to yield m/z 833, 695, 415 and 279. The fragment at m/z 833 was attributed to the loss of one acyl glycerol portion from one PA group, thereby yielding PG phosphate. Ions at m/z 695 and 415 could be accounted for as PA and (PA − RCH$_2$COOH) ions, respectively. The desorption of the [M − H]$^-$ ion to produce m/z 279 indicated that the CL was made up of four residues of 18:2, for example tetralinoleoyl GPG. CID of the m/z 695 and 415 ions provided further confirmation of the proposed structure of the major CL species of bovine brain.

Robinson (1990) used FAB/MS with NI detection to examine not only the structures of the unmodified CL but also the structures of acetylated CL (Ac$_n$ CL), succinylated CL (Suc CL), and monolyso CL (MLCL) and dilyso CL (DLCL). The specific chemical modifications yielded the expected substitutions of the 2- and/or 2'-hydroxyl(s). The observed [M − H]$^-$ values and the expected [M − H]$^-$ values for the synthetic tetrahydropyranyl derivative of cardiolipin were: CL, m/z 1448; MLCL, m/z 1186; DLCL, m/z 924; AcCL, m/z 1490; Ac$_2$MLCL, m/z 1270; Ac$_3$DLCL, m/z 1050; and Suc CL, m/z 1548.

Kates et al. (1993) reported positive FAB/MS of hydrogenated *Escherichia coli* (*E.coli*) CL which showed major ion peaks at m/z 523, 551, 563, 579, 591 and 607, corresponding, respectively, to 14:0/14:0, 16:0/16:0, 16:0/cy-17:0, 16:0/18:0, 16:0/cy-19:0 and 18:0/18:0 molecular species of DGs, consistent with the fatty acid analytical data. Negative ion FAB/MS did not yield any diagnostic ion peaks. Kates et al. (1993) also examined the mass spectrum of 2'-deoxy CL (diphosphatidyl-1,3-propanediol). In negative FAB/MS the dipalmitoyl species gave m/z 811, an unidentified, but possibly [M-dipalmitoylglycerol+ OH + Na − H]$^-$ or [phosphatidylpropanediol phosphate +Na − H]$^-$) ion; m/z 705, base peak, [M − H − dipalmitoyl PA + H$_2$O]$^-$ or [phosphatidylpropanediol − H]$^-$; and m/z 647, [M−phosphatidylpropanediol +OH]$^-$ or [diphosphatidyl PA−H]$^-$. Positive FAB/MS gave m/z 707 [M − dipalmitoyl PA + H$_3$O]$^+$; and m/z 551 [M − phosphatidylpropanediol phosphate + H]$^+$ or [dipalmitoylglycerol − OH]$^+$. No parent ion peak for 16:0 deoxy CL was detected by negative or positive FAB/MS.

Smith, Snyder and Harden (1995) have used negative ion ES/MS/MS to analyse the PLs of four *Escherichia* and *Bacillus* species. The principal bacterial PLs detected by this technique were PC and CL, accompanied by small amounts of PE. The negative ion spectrum for a chloroform/methanol extract of *B. licheniformis* showed three prominent peaks, at m/z 693, 707 and 721, accompanied by several minor peaks in the PL region of the mass spectra. The low-intensity peaks at m/z 647, 654, 661, 668 and 675 revealed that the M + 1 and M + 2 isotope peaks differed by half-unit mass increments rather than by unit mass increments, which indicates that these compounds were doubly charged. The peak at m/z 647 was of sufficient intensity to provide an informative product ion mass spectrum and had the correct molecular weight (1296) for a CL with four 15:0 fatty acids. Higher mass peaks in the series were CLs with replacement of one or more of the 15:0 fatty acids with longer chain fatty acids. The product ion mass spectrum for the $[M - H]^-$ ion at m/z 693 showed a prominent ion at m/z 241, and as no other ions were present in the fatty acyl $[RCH_2COO]^-$ region of the mass spectrum, it implied that a 15:0 fatty acid is the acyl substituent both at the sn-1 and the sn-2 positions. Subtraction of the mass of two 15:0 fatty acids from the $[M - H]^-$ ion at m/z 693 yields m/z 209, which is indicative of PG. The negative ion mass spectra for chloroform/methanol extracts of *B. stearothermophilus* were similar to those obtained for *B. licheniformis* as they also showed doubly charged CL at m/z 647–683, and PGs were observed at m/z 693, 707, 721, 735 and 749. In contrast, the mass spectrum for *B. thuringiensis* did not contain any CLs, and the PGs were found at lower masses (m/z 665–721), with the ion at m/z 693 being the most prominent. The ES/MS/MS method was used to determine the differences in the PG composition among *B. licheniformis*, *B. stearothermophilus*, *B. thuringiensis* and *E. coli* extracts. The differences were apparent in the molecular weight distributions and in the fatty acid composition of isobaric PGs. ES/MS has also been employed for the determination of the GPL of *Lactobacillus* spp. (Drucker *et al.*, 1995). Prominent anions were consistent with the presence of PG (37:2; 36:2, 35:1, 34:1 and 33:1).

Bergqvist and Kuksis (1997) reported negative ion LC/ES/MS of bovine heart CL (tetralinoleoyl species) as m/z 1448 and m/z 724, corresponding to the singly and doubly charged molecules, respectively. The CL of the heart of a rabbit receiving fish oil in the diet demonstrated a much wider range of CL species under similar experimental conditions.

Wissing and Behrbohm (1993) used mass spectrometry and NMR analyses to identify a DG pyrophosphate (DGPP), a hitherto unknown PL. The MS of DGPP was obtained by FAB/MS with negative ionization with use of triethylamine as matrix. The mass spectrum of the dioleoyl derivative of DGPP showed a pseudo-molecular ion $[M - H]^-$ for DGPP at m/z 779. Fragment ions at m/z 699 (PA), m/z 515 (lyso-DGPP), m/z 497 (dehydro-lyso-DGPP), m/z 435 (lyso PA), m/z 281 (oleic acid), m/z 177 (pyrophosphate), m/z 153 (dehydroglycerol phosphate), m/z 97 ($H_2PO_4^-$) and m/z 79 (phosphoryl radical) further confirmed the presence of the pyrophosphate of 18:0/18:0 GP.

Wolucka, Rozenberg and de Hoffmann (1996) applied desorption NICI/MS/MS to the analysis of nanomole quantities of semi-synthetic polyisoprenyl phosphates, the chain length of which ranged from 7 to 20 isoprene units. The DCI spectrum of all the compounds tested showed the presence of independently generated ions $[M - HPO_3 - H]^-$, $[M - H_3PO_2 - H]^-$ and $[M - H_3PO_4 - H]^-$ resulting from the loss of a part of or the entire phosphate group of polyisoprenyl phosphate. In MS/MS, the $[M - H_3PO_4 - H]^-$ fragment produced a series of ions 68 amu apart, indicative of the polyisoprenoid nature of the compound. The mechanism of the fragmentation was investigated with deuterated and α-saturated derivatives. Saturated dolichyl phosphates could be distinguished easily from the corresponding polyprenyl phosphates on the basis of a 2 amu shift and the α-series of fragments and because of the presence of an additional (A + 14) series of ions, 14 amu heavier than fragments resulting from the allylic cleavages of α-saturated polyisoprenoid chain.

6.3.6 Oxygenated glycerophospholipids

Early work identified the oxygenated fatty acid constituents of soybean PC as: 15,16-epoxy-9,12-octadecadienoates, 12,13-epoxy-9-octadecenoate, both with double bonds and epoxide groups predominantly of *cis* configuration; 13-oxo-9,11- and 9-oxo-10,12-octadecadienoates; 9,10,13-trihydroxy-11-and 9,12,12-trihydroxy-10-octadecenoates. Porter (1984) has pointed out that autoxidation of dilinoleoyl GPC yields two series of products, one series containing monoxidation products, the other, compounds resulting from dioxidation. The first series contains products of autoxidation of dilinoleoyl GPC in which one acyl chain has been oxidized (at either C-1 or C-2 of the glycerol moiety). A total of 16 different oxidation products are formed from dilinoleoyl GPC. Eight of the products derive from oxidation at the C-1 acyl chain and eight from the C-2 acyl linoleate. Four of the eight products in each of the acyl chains are *trans,cis*-diene hydroperoxide products, whereas the other four products have *trans,trans*-diene stereochemistry. Within the set of *trans,cis* or *trans,trans* products, half are formed at C-9 of linoleate with a 10,12-conjugated diene unit, the other half are formed with hydroperoxide substitution at C-13 with a 9,11-diene functionality. Finally, hydroperoxide at C-9 or C-13 may have either an *R* or *S* configuration. Although these stereoisomers would be enantiomers in autoxidation of simple linoleate esters, this is not the case in the oxidation of phospholipid esters. Since natural GPCs have only an *R* configuration at the C-2 glycerol carbon, the possibility of having *R* and *S* configurations in a particular oxidation product leads to a mixture of diastereomers, not enantiomers. General mechanisms of free-radical peroxidation of unsaturated lipids have been discussed by Frankel (1985) and Porter, Caldwell and Mills (1995).

Morrow *et al.* (1992) reported that the free-radical-catalysed peroxidation of arachidonic acid proceeds via the intermediate formation of F_2-isoprostane-containing PLs. As such species are believed to lead to remarkably distorted molecules, which would be anticipated to alter membrane properties dramatically, it is likely that these species would be immediately expelled from the red cell lipid bilayer or become subject to lipolysis. Negative FAB/MS gave the M-15, M-60 and M-86 ions for the stearoyl and palmitoyl F_2-prostanoates, along with m/z 353, representing the carboxylate anion of F_2-isoprostane from the *sn*-2-position. On normal-phase HPLC, the PC F_2-prostanoates were retained much longer than the common PC species. Morrow *et al.* (1994) have shown the major urinary metabolite, representing 29% of the total extractable recovered radioactivity in urine, to be structurally identified by GC/MS as 2,3-dinor-5,6-dihydro-8-iso-prostaglandin $F_{2\alpha}$. The compound was identified as the TMS ether of the methyl ester derivative. This means that F_2 isoprostanes should be found in the PL of red blood cells. Despite the apparent presence of the di-hydroperoxy and trihydroperoxy derivatives of the arachidonoyl PC in the PLs of human red cell lipids, the ions anticipated for the F_2 esters of GPC and GPE could not be demonstrated (Kuksis et al., 1996).

Zhang et al. (1994) reported the analysis of intact hydroxyeicosatetraenoic (HETE) acid-containing PCs by using negative liquid secondary ion (LSI) MS/MS. GPLs containing esterified HETE were initially separated by means of normal-phase HPLC, then further purified by reversed-phase HPLC. For MS, the individual HETEs and the HETE PCs were placed on the LSI probe. CID of the $[M - H]^-$ ion from the HETE acid regioisomers (m/z 319) resulted in loss of water and loss of CO_2, yielding m/z 301 $[M - H - H_2O]^-$ and m/z 257 $[M - H - (H_2O + CO_2)]^-$. Negative LSI/MS of purified HETE PC species gave an ion at m/z 810, $[M - CH_3]^-$. CID of the m/z 810 ion gave product ions at m/z 283 and m/z 319, corresponding to stearate at the sn-1 position and HETE at the sn-2 position, respectively. From CID of the negative ion at m/z 319 and examination of the product ion spectra, the HETE regioisomer present in PC could be identified. The HETE-containing PC were produced enzymatically by incubation of 5(S)-, 12(S)- and 15(S)-HETE to stearoyl lyso-PC in the presence of coenzyme A, adenosine triphosphate (ATP) and rat liver microsomes. However, Zhang et al. cite references to extensive studies indicating that 5-, 12- and 15-HETEs are incorporated into GPLs of a wide variety of cell-types.

Kim and Salem (1987, 1989) reported a LC/TS/MS assay of GPL hydroperoxyl derivatives of docosahexaenoic acid. Zhang et al. (1995) examined by LC/TS/MS the hydroperoxides obtained by two commonly employed PC hydroperoxide (PCOOH) generation methods. The results revealed that over 90% of the hydroperoxides generated by the photooxidation–methylene blue method were mono- and di-PCOOH, whereas over 70%–95% of the hydroperoxides generated by the azo compound [2,2′-azobis-2,4-dimethylvaleronitrile (AMUN)] incubation were not PCOOH but rather AMVN-derived hydroperoxides. Spectra were obtained in the positive ion mode, with discharge off, using a source temperature of approximately 200°C and a repeller voltage of 90 V. Two species of PC were employed in the peroxidation studies: di-18:2 GPC, which could form a mono- or di-PCOOH, and 18:0/18:2 GPC, which could form only the mono-PCOOH. The peak resulting from photooxidation of di-18:2 GPC contained ions at m/z 783, 815 and 847, which are equivalent to $[M + H]^+$ of di-18:2 GPC (MW 782), its mono-PCOOH (MW 814) and di-PCOOH (MW 846), respectively. The m/z 783 ion was probably produced from the loss of the hydroperoxide moiety from the PCOOH. The peak from the peroxidation of 18:0/18:2 GPC contained ions of m/z 787 and 819, which were equivalent to $[M + H]^+$ of 18:0/18:2 GPC (MW 786) and its mono-PCOOH (MW 818). No m/z equal to the di-PCOOH (MW 850) was found, indicating that there was mono-PCOOH but no di-PCOOH of 18:0/18:2 GPC present, as expected. In contrast, the PCOOH generated by AMVN incubation contained less than 30% authentic PCOOH. The other 70% was AMVN-derived or some other hydroperoxides. The present findings that over 70%–95% of the hydroperoxides generated by AMVN incubation with PC are not PCOOH is alarming as AMVN incubation is a widely used method to generate hydroperoxides. The successful synthesis of PCOOH by the photooxidation–methylene-blue method makes it possible to examine other hydroperoxide standards such as PEOOH, cholesterol OOH, cholesteryl ester OOH, and others, and to establish a common basis for comparing experimental data, which has not been possible in the past.

Beckman *et al.* (1994) used normal-phase HPLC of constituent hydroxy fatty acids followed by GC/MS analysis to reveal that the oxidized GPL in the skin of CD_1 mice, following application of the tumour-promoter phorbol esters, were oxidized derivatives of linoleic acid, including 9- and 13-hydroxyoctadecadienoic acids (9- and 13-HODE). Sodium borohydride reduction increased product yield by approximately 50%, suggesting the additional presence of GPL hydroperoxides in the oxidized lipids.

Guo *et al.* (1995) used oxidation with 2,2′-azobis(2-amidinopropane)dihydrochloride (AAPH) to produce GPL hydroperoxides, which, after reduction and saponification, were analysed by GC/MS as the hydroxy fatty acid methyl ester TMS ethers. The hydroxy and hydroperoxy moieties in PC were identified as 9-hydroxy and 13-hydroxy octadecanoic acids, derived from 18:2. Aldehydic GPE was the most prominent aldehydic GPL; negligible amounts of aldehydic GPC were formed. This study demonstrated that the process of oxidation for the individual PLs clearly differs among PLs and depends on the structure of each PL. The fragmentation of the oxygenated polyunsaturated esters leads to the formation of oxidized PLs that contain aldehyde moieties (Esterbauer, Zollner and Schauer, 1990). Itabe, Kobayashi and Inoue (1988) identified oxidized PC with a saturated dicarboxylic acid (azelaoyl group) at the 2-position during the oxyhaemoglobin-induced peroxidation of linoleoyl GPC. Species with short-chain dicarboxylic acid moieties were detected among PCs that were oxidized by an Fe^{2+}/ascorbate/EDTA system and similar oxidized PCs were found in bovine brain tissues (Tanaka *et al.*, 1993). Tokumura *et al.* (1996) demonstrated that short-chain dicarboxylate- and monocarboxylate-containing GPCs are generated by peroxidation of human plasma and egg yolk LDL with Cu^{2+}. Surprisingly, the core aldehydes of PC were not produced. Potential aldehyde precursors of the dicarboxylic acid esters have been found in oxidized biomembranes and plasma lipoproteins (Kamido *et al.*, 1995) by means of reversed-phase LC/MS with direct liquid inlet interface. For this demonstration, PLs containing the aldehyde ester groups were isolated as the DNPH derivatives and identified by LC/MS as the DNPH derivatives of DGs released from the PC by phospholipase C. Subsequently, Kamido *et al.* (1997) used similar methods to isolate and identify the PL core aldehydes from human atheromas and xanthomas. Interestingly, Itabe *et al.* (1996) have obtained immunochemical evidence for the plausible formation of 1-palmitoyl-2-(9-oxononanoyl) GPC during LDL peroxidation. By using a monoclonal antibody raised against peroxidized LDL these workers were able to demonstrate that the antibody reacts with several peroxidation products of PC, which covalently modify polypeptides. Among these products were the C_8 and C_9 core aldehydes of PC.

The chain-shortened oxidized PC lipids have also been characterized by Kayganich-Harrison and Murphy (1994b) by FAB/MS and FAB/MS/MS. They detailed the FAB and CID behaviour of 1- hexadecanoyl-2-(5-oxopentanoyl)-GPC and 1-hexadecanoyl-2-pentanedioyl-GPC. There are two prototypic oxidized PC lipids produced by hydroxyl-radical-initiated peroxidation of 1-hexadecanoyl-2-arachidonoyl-GPC. The identities of these two products were established by initially using GC/ES/MS to analyse the corresponding DG tert-butyldimethylsilyl ether derivatives. The aldehyde moiety of 1-hexadecanoyl-2-(5-oxopentanoyl)-GPC was found to react with FAB amine-containing liquid matrices (e.g. diethanolamine) to form an oxazolidine derivative (87 amu increase in mass). No reaction was observed with matrices such as glycerol, thioglycerol, and 3-nitrobenzoyl alcohol. This derivative can be used to distinguish an aldehyde-containing oxidized GPL from isobaric species not containing an aldehyde moiety. Results also showed that 1-hexadecanoyl-2-pentanedioyl-GPC formed primarily $[M - H]^-$ ions rather than $[M - 15]^-$, $[M - 60]^-$, and $[M - 86]^-$ negative ions expected for the PC species. The ionized Ω-carboxyl group of the 2-substituent allowed the phosphocholine moiety to remain as a zwitterion. Following CID, an *N*-methyl group from the choline was transferred to the Ω-carboxyl of the *sn*-2 substituent prior to generation of the carboxylate anions. This resulted in a 14 amu increase in the observed *sn*-2-carboxylate anion. The methyl transfer mechanism was confirmed with a trideuterio methyl group from PC containing [$^2H_9$]choline.

Ponchaut *et al.* (1996) employed FAB/MS/MS to demonstrate that the PC isolated from heart mitochondrial fractions contain molecular species with monohydroxylated fatty acyl moieties. Detailed analysis of the negative ion mass spectrum of the PC revealed the presence of six carboxylate ions at *m/z* 267, 269, 271, 295, 297 and 299, which did not correspond to known substituents of mammalian GPLs. CID of these carboxylate anions showed that the ion at *m/z* 297 was attributable to 12-hydroxyoctadecamonoenoic acid (12-HOME). Analysis of the fragmentation pattern obtained by CID of the anions at *m/z* 295 and 299 indicated that these ions only differ from 12-HOME in the number of double bonds; thus they correspond to hydroxyoctadecadienoate and hydroxyoctadecanoate, respectively. The ions at m/z 267, 269 and 271 have masses that correspond to hydroxyhexadecadienoate, hexadecenoate and hydroxyhexadecanoate. The precursor spectrum of the ion at m/z 297 displayed high mass ions at m/z 690, 715, 737, 763, 784 and 808. Analysis of the spectra obtained by CID of these high mass ions allowed the determination of their fatty acid composition. This analysis revealed that 12-HOME may be accompanied in the PC molecule by arachidonate or eicosatrienoate, as shown by the presence of ions at m/z 808 and 810 at m/z 763 and 765 and at m/z 737 and 739, which represent the M-15, M-60 and M-86 ions, respectively, of PC species with 12-HOME and arachidonate or eicosatrienoate. Similar analyses of the precursor ion spectra of the other five monohydroxylated fatty acids indicated that they were incorporated in PC molecules and were accompanied by the same fatty acyl groups.

Synthetic 16:0(18:0)/18:2 GPC containing ozonide, hydroperoxide and core aldehyde groups were characterized by flow injection and LC/MS in positive ion ES/MS by Myher et al. (1994) and by normal-phase HPLC with on-line negative ion ES/MS by Ravandi et al. (1995b). Characteristic $[M + 1]^+$ ions were obtained at a 200 V exit voltage without noticeable fragmentation. Subsequently, negative ion ES/MS was utilized on-line with a reversed-phase HPLC column for the resolution and identification of the ozonides of CL (Bergqvist and Kuksis, 1997). Mono-, di-, tri-, tetra-, penta-, hexa-, hepta- and octa-ozonides were observed for the tetralinoleoyl species of CL. Application of this method to the analysis of rabbit heart CLs after exposure to ozone resulted in a large variety of ozonized species owing to the more complex nature of the starting material. In addition, ω-oxo aldehydes were found to be produced from the various ozonides. Harrison and Murphy (1996) performed a systematic study of the positive and negative ions produced by ozonized PCs of various degrees of unsaturation. Characteristic ions were recorded for the tri- and tetra-ozonides of 18:0/20:4 GPC, mono- and di-ozonides of 16:0/18:2 GPC, and the mono-ozonides of 16:0/18:1 GPC. In addition, the core aldehyde mono-, di- and tri-ozonides were observed with the 18:0/20:4 GPC.

The amounts of aldehydic GPL should be closely related to the proportion of polyunsaturated fatty acids (PUFAs), such as arachidonic acid, because the main accumulation of aldehydic residues was observed in highly PUFA-rich GPE, whereas generation of aldehydes even in PUFA GPC was rather limited.

6.3.7 Glycosylated glycerophospholipids

The glycosylated GPLs comprise the well-established PI lipid anchors and the newly discovered glucosylation products of the aminophospholipids. Despite the glycation these PLs retain their lipid properties, which is essential for their function.

Glycosylated phosphatidylinositol. The structure, biosynthesis and function of glycosylated PI in parasitic protozoa and higher eukaryotes was reviewed by McConville and Ferguson (1993). This PI derivative has received special MS attention as it serves as an anchor for membrane proteins. It possesses a unique structure, which has been subject to extensive chromatographic and enzymatic investigation and MS. Although small amounts of the glycoinositol PI may occur in tissues in the free form, the bound form has received the most interest. Acetylcholinesterase (AchE) on the surface of human erythrocytes is an amphipathic globular dimeric protein that is anchored in the plasma by covalently attached glycoinositol GPL (Roberts, Kim and Rosenberry, 1987; Roberts *et al.*, 1988a). Similar glycolipid anchors have been identified in many proteins, the best characterized of which are the tryponosome variant surface glycoproteins (a review is given in Ferguson and Williams, 1988). Roberts *et al.* (1988b) established the structure of the anchor of human erythrocyte AchE by FAB/MS with negative ion monitoring and by the complementary technique of CID. The MS/MS studies revealed molecular and daughter ions which indicated a plasmanylinositol with a palmitoyl group on an inositol hydroxyl. Positive and negative FAB/MS of the major products isolated by HPLC indicated the complete glycoinositol PL anchor. Deeg *et al.* (1992) examined the glycan components in the glycoinositol PL anchor of human erythrocyte AchE. Four glycans denoted α, β, γ and δ were resolved by anion exchange HPLC after sequential treatment with proteinase K, methanolic KOH and a PI-specific phospholipase C. Glycans α and β were analysed by ES/MS, and respective parent ions of m/z 1266 and 1477 were observed. The fragmentation pattern produced by CID of these parent ions was consistent with a common linear core glycan sequence prior to radiomethylation of ethanolamine–phosphate–mannose–mannose–mannose–glucosamine–inositol. The new fragments involving an intact radiomethylated glucosamine–inositol bond were proposed as new diagnostic indicators in the search for minor glycoinositol PL in cells and tissues. Haas *et al.* (1996) used ES/MS to examine the glycoinositol PL anchor and protein C terminus of bovine erythrocyte AchE. The lipid portion of the AchE glycoinositol GPL anchor and a C-terminal tryptic fragment that contained the residual glycerophosphoinositol (GPI) glycan, were isolated by HPLC. Analysis by ES/MS revealed a parent ion at m/z 3798. The fragmentation patterns produced by CID of the +4 and +5 states of the parent ion indicated a 23-amino acid peptide in amide linkage to ethanolamine-PO_4-Hex-Hex-Hex-(PO_4-ethanolamine) (Hex*N*Ac)-Hex-*N*(Me)$_2$-inositol phosphate. The glycan structure was completely consistent with that obtained previously for the PI anchor of human erythrocyte AchE except for the addition of the Hex*N*Ac substituent.

Redman *et al.* (1994) described the application of ES/MS and CID to study intact glycosyl-PI-peptides from a *Tryponosoma brucei* variant surface glycoprotein. CID of the $[M + 4H]^{4+}$ pseudo-molecular ions of two glycosyl-

PI-peptide glycoforms produced easily interpretable daughter ion spectra from which detailed information on the lipid moiety, carbohydrate sequence and site of peptide attachment could be obtained. All of the CID-induced cleavage events occurred in the glycosyl-PI portion of the glycosyl-PI-peptide. This technique supplied complementary data to the highly sensitive oligosaccharide sequencing procedures and should greatly assist glycosyl-PI anchor structure–function studies. Subsequently, Heise, de Almeida and Ferguson (1995) determined the complete structure of the lipid component of the glycosyl-PI anchor of *Tryponosoma cruzi* $1G_7$-antigen by ES/MS, GC/MS, phospholipase C sensitivity and HPTLC of the DG component after benzoylation. These analyses showed that the lipid moiety of the $1G_7$-antigen is composed essentially of 1-O-hexadecyl-2-O-hexadecanoyl-PI and 1-O-hexadecyl-2-O-octadecanoyl-PI. The ES/MS also revealed the presence of small amounts of putative inositol–phosphoceramide structures and confirmed the absence of inositol–acylated species.

Glycosylated aminophospholipids. Ravandi *et al.* (1995a) employed normal-phase HPLC with on-line ES/MS for the identification of glucosylated aminophospholipids formed upon exposure of the liposomal PE and PS to 5–100 mM glucose. Although the formation of the glycated aminophospholipids had been postulated, previous attempts to demonstrate their presence in hyperglycaemia or in chemical reaction mixtures had not been successful. The mild ES ionization permitted the detection of the molecular ions of glycated PE and PS following incubation of egg yolk PE and bovine brain PS with glucose. The mass spectra averaged over the residual PE, after incubation with the glucose, and over the Gly PE HPLC peaks indicated close agreement between the molecular species. Thus, all species of the red blood cell PE were glucosylated in their mass proportions. Likewise, good agreement was observed between the residual PS and GlyPS species. In both instances, the glycated aminophospholipids possessed the same pattern of molecular species as the residual aminophospholipids. This indicated that all molecular species were readily and equally accessible to glycation in both GPLs, alkylacyl and alkenylacyl PE species inclusive. Thus the ions corresponding to the molecular species were present in identical proportions, except that the masses differed by 162 amu. The glycation products were Schiff bases, which could be stabilized by reduction with sodium cyanoborohydride.

Ravandi *et al.* (1996) were subsequently able to demonstrate the presence of glucosylated PE in the red blood cells of diabetic subjects in both hyperglycemia and during normalization of plasma glucose. Surprisingly, there was no glucosylation of the PS and plasmalogenic PE species. In still another application of the method, it was demonstrated that the polyunsaturated species of glycated PE were more susceptible to peroxidation than were non-glycated PE species (Kuksis *et al.*, 1996) and that the hydroperoxidized species of glycated PE were subject to preferential attack by phospholipase A_2. Still other studies have utilized normal-phase LC/ES/MS to demonstrate that the glycation and peroxidation of the PE species leads to an increase in oxidation of non-glycated phospholipid species in the cell membrane and in the surface monolayer of the lipoproteins. Eventually there is increased hydroperoxidation of the cholesteryl ester interior. Upon increasing the capillary outlet voltage, the glycated PE is fragmented to yield glucose, the parent PE and the derived DG species (Ravandi *et al.*, 1997).

6.3.8 Sphingomyelins, sphingolipids and gangliosides

Lipids with a common sphingosine or related long-chain base are widely distributed in nature. Sphingosine is the most abundant but accounts for only 70–75% of the long-chain bases in sphingolipids isolated from cellular membranes (Sweeley, 1991). Kim and Salem (1993) have reviewed the MS of ceramides and sphingomyelin (SPH), and Murphy (1993) has discussed the principles of the MS of sphingolipids in general. Kuksis and Myher (1995; Myher and Kuksis, 1995) have updated the subject with an emphasis on LC/MS applications.

Ceramides. Ceramides are amides of long-chain bases with fatty acids and constitute the lipophilic tail of SPH and sphingolipids. Free ceramides occur naturally to a limited extent but are released from SPH and sphingolipids during metabolism. There is currently much interest in ceramides because of their postulated role as modulators of cell growth and differentiation (Hannun and Obeid, 1995). Their characterization is usually performed by hydrolysis and separate identification of the fatty acid and long-chain base components, but intact ceramides can also be readily determined by GLC (Myher *et al.*, 1981) or HPLC (Yang *et al.*, 1992), both techniques permitting on-line combination with MS. However, complex mixtures of ceramides can also be characterized by direct-probe FAB/MS and FAB/MS/MS. Thus, FAB/MS of precursors and fragment ions has allowed the characterization of the fatty acid composition of a complex ceramide mixture prepared from beef brain lipids. The use of a triple quadrupole yielded far better results than did linked scanning on a sector mass spectrometer, as it was possible to achieve unit mass resolution of the precursor ions, allowing discrimination of saturated and unsaturated fatty acids with the same number of carbon atoms (Rubino, Zecca and Sonnino, 1994). The structure of the *N*-linked fatty acid could be derived from an analysis of the array of ions generated by CID charge remote fragmentation of its carbon backbone. This technique allows the determination of the position of the double bonds and of other structural motifs such as methyl branching and cyclopropane rings on polymethylene chains such as those of fatty acids and alcohols. By this method, the location of the double bond in 24:1 acid was established to be at C_{15}. The characterization of the long-chain bases by fragmentation analysis was not successful as the CID spectrum did not yield any charge remote fragmentation. Sawabe *et al.* (1994) were able to locate the double bonds in the long-chain base of a cerebroside isolated from mushrooms by using the B/E constant linked scan method. The structure of cerebrosides having the long-chain base of 9-methyl-C_{18}-sphinga-4,8-dienine could be determined in general by the presence of the characteristic fragment ions of $[C_{19}$-sphingadienine $+ H - HOH]^+$ at m/z 276, and $[C_{19}$-sphingadienine$+ H]^+$ at m/z 294. The fatty acid carbon number could be calculated from the characteristic fragment ions of [ceramide $- 180]^+$ ($[MH - GlcOH - 180]^+$) in positive mode FAB/MS.

Sphingomyelins. Conjugation of the hydroxyl group of ceramides with phosphocholine yields SPHs which are major components of cell membranes. The first detailed analyses of the molecular species of SPH were obtained by Jungalwala, Evans and McCluer (1984b) by means of positive ion ammonia CI/MS. The major high molecular weight ions for bovine brain SPH were the ceramide ions $[M - 182]^+$ and $[(M + 18) - 182]^+$. The ceramide ion was easily distinguished from the DG ions as its mass number was even (because of the nitrogen) in contrast to odd for the DG ion. The major molecular species of bovine brain SPH were: (sphingenine) d18:0–18:0 (48%, m/z 530 and 548); d18:0–16:0 (4%, m/z 502 and 520); d18:0–20:0 (5%, m/z 558 and 576); d18:0–22:0 (6%, m/z 586 and 604); d18:0–24:1 (7%, m/z 612 and 630); and d18:0–24:0 (7%, m/z 614 and 632). C_{20}-sphingenine with a 16:0 fatty acid cannot be resolved from C_{18}-sphinganine with an 18:0 fatty acid. The bovine brain contained 7% and 10%, respectively, of the C_{18}-sphinganine and C_{20}-sphingenine (Jungalwala *et al.*, 1984). Upon LC/MS analysis with ammonia, rat brain SPH was resolved into two major peaks. The initial peak consisted mainly of SPH with (sphingenine) d18:0–22:0 (4%, m/z 586 and 604); d18:0–24:1 (12%, m/z 612 and 630); and d18:0–24:0 (6%, m/z 614 and 632). The last peak described contained SPH with d18:0–16:0 (5%, m/z 502 and 520); d18:0–18:0 (54%, m/z 530 and 548); d18:0–20:0 (5%, m/z 558 and 576) and d18:0 (sphinganine) with 16:0 (1%, m/z 522), and d18:0 (sphinganine) with 18:0 (10%, m/z 532 and 550). The SPH were well resolved from the other PL classes by normal-phase HPLC. Kim and Salem (1987) described the separation of the molecular species of bovine brain SPH by means of reversed-phase LC/MS with filament-on TS ionization. A total of nine molecular species were separated and the structures were confirmed by MS analysis. Ceramide ions were the base peaks in the spectra. Most natural SPHs have C_{18}-sphingenine backbones, but 18:0 or 20:1 sphingoid moieties may also be present, which complicate the assignment of chromatographic peaks. Thus the 22:0 and 24:1 SPH species co-elute, but TS/MS provide sufficient information for their identification. The ceramide peaks were observed at m/z 605 and 631 as base peaks for each species, whereas the fatty acyl amide moiety $[RC(=O)NH_3]^+$ was seen at m/z 340 and 366 for 22:0 and 24:1 SPH, respectively. Although the peak at m/z 631 could represent the ceramide ion of either the 24:1 species with the C_{18}-sphingenine or the 22:1 species with the C_{20}-sphingenine structure. However, the peak at m/z 366 which corresponds to the 24:1-amide ion excludes the latter possibility. PL head-group information was obtained from the peak at m/z 142, which represents the loss of trimethylamine from the ammonium adduct of phosphocholine.

Ann and Adams (1993) have explored the use of alkali metal ions as adducts to improve structural determination of other types of sphingolipids. CID spectra of $[M + Cat]^+$ ions (Cat^+ = alkali metal ions) of ceramides and neutral glycosphingolipids provide immediate information about the lengths of the sphingolipid base and N-acyl chains and, most importantly, the locations of substituents on the N-acyl chain.

Valeur et al. (1994) have determined the composition of the SPH obtained from bovine brain, chicken egg yolk and bovine milk by means of LC/MS with discharge-assisted TS. Positive ion TS/MS exhibited prominent ions for ceramides and other ions related to the amine base structure as well as fragments which can be utilized for identification of molecular species. No molecular ions were observed under the conditions used. In agreement with earlier results, the bovine brain SPH possessed the 18:0 and 24:1 species as major components, whereas the SPH of egg yolk was largely the 16:0 species. In contrast, the bovine milk SPH contained the 22:0, 23:0 and 24:0 species as major components in nearly equal proportions. Quantitative aspects of discharge-assisted TS of SPHs were examined by determining the dose–response curve for d18:1–18:1 selectively monitored at the ion m/z 546.7, which forms the base peak in the spectrum. Over the dose range studied, a slight deviation from linearity (0–800 ng) was observed, indicating that a precise direct dose comparison with an external standard should be made only within a restricted dose range. The detection limit was defined at a signal-to-noise ratio of 3 and was found to be about 10 ng of d18:1–18:1, a value typically found in TS for many other complex lipids.

Kerwin, Tuininga and Ericsson (1994) have reported the positive and negative ion ES/MS and ES/MS/MS for commercial preparations of SPH. The samples were analysed by flow ES/MS. Molecular ion adducts were the primary products formed by positive ionization, for example $[M + H]^+$, $[M + Na]^+$ and others. The negative spectra contained the negative ions corresponding to the loss of these groups. Bovine brain SPH yielded 17 molecular species. In positive ion MS, molecular ions were found at $[M]^+$. In negative ion MS, molecular ion adducts were found as $[M + \text{acetate} - CH_4]^-$, that is $[M + 44]^-$ arising from the loss of one methyl group from the quaternary nitrogen, and acetate complexation to the choline head-group. Positive ion MS/MS of $[M + H]^+$ ions yielded a strong peak at m/z 184, corresponding to the phosphocholine head-group. In negative ion MS/

MS spectra, the major ion found was the result of a loss of acetate from the parent ion described above. Less intense peaks were attributable to the loss of acetate and an additional CH_3 from the head-group. Another peak of low intensity was seen at m/z 168 arising from phosphocholine head-group minus one methyl moiety. Attempts to increase fragmentation to generate peaks characteristic of the N-acyl and sphingenine/sphinganine constituents by using 1% methanolic formate were not successful. Han and Gross (1994) used flow ES/MS and ES/MS/MS for the determination of molecular species of human erythrocyte plasma membrane SPH. The sensitivity of ES/MS was two to three orders of magnitude greater than that achievable with FAB/MS. Four molecular species of SPH were demonstrated, including species at m/z 726, 754 and 782, corresponding to the 16:0, 18:0 and 20:0 amides of sphingosine.

Kwon *et al.* (1996) used ES/MS to characterize the content and composition of SPH isolated from pancreatic islets. More than four molecular species were identified with the most abundant of which containing sphingosine as the long-chain base and palmitic acid as the acyl chain, but stearoyl and arachidoyl species were also found.

Glycosphingolipids. The neutral glycosphingolipids or cerebrosides are ceramide monohexosides, lactosides and higher sugar glycosides. The great complexity and number of new glycosphingolipid components being reported challenge the best contemporary methods of characterization. These lipids have been investigated by means of EI/MS, CI/MS and FAB/MS with or without prior chromatographic resolution and the results have been reviewed (Murphy, 1993). Recent examples are provided by the systematic analysis of glycosphingolipids from the human gastrointestinal tract by Natomi *et al.* (1993), who reported enrichment of sulphatides with hydroxylated long-chain fatty acids in the gastric and duodenal mucosa based on the analyses by GC and negative ion FAB/MS. The sulphatides were isolated from the acidic glycosphingolipid fraction by column chromatography on Iatrobeads and the homogeneity of the isolated fractions examined by TLC by using standard sulphoglycosphingolipids. Intense molecular ions, $[M - H]^-$, were observed at m/z 778, 794, 878, 890, 906 and 934, corresponding to the sulphatides (I^3SO_3 – GalCer) with 4-sphingenine and the fatty acids 16:0, 16h:0, 22h:0, 24:0 24h:0 and 26h:0, respectively. The relative intensities of the ions were in good agreement with the fatty acid composition of the sulphatides isolated from the fundic mucosa and determined by GC. In addition, the fragment ions of the terminal sulphate group at m/z 97, $[HSO_4]^-$, were clearly identified. High-energy CID of the ceramide fragments from underivatized glycosphingolipids provided little information on the location of substituents such as double bonds and hydroxyl groups in the sphingoid and N-acyl chains. Ju Wei and Her (1994) showed that permethylation can overcome this problem. The product ion mass spectra of the ceramide fragments produced via high-energy CID provide detailed information regarding the structure of the ceramides. Because the ceramide fragments, and not the pseudo-molecular ions, were selected as the precursor ions, the size of glycosphingolipids had little effect on the quality of the product ion spectra.

ES/MS offers several advantages over FAB/MS for polysaccharide analysis, and this technique has been utilized for the characterization of the structural heterogeneity of native glycolipid samples (Gibson et al., 1993). Reinhold et al. (1994) reported that application of ES/MS to methylated glycosphingolipid samples provides a sensitive molecular mass profile with no detectable fragmentation and little matrix background. Structural details of the major components within a bovine brain preparation were obtained by utilizing low-energy CID tandem MS and periodate oxidation. Glycosidic fragments defined details that allowed the determination of structural isomers, and specific fragments of the ceramide moiety differentiated sphingosine from N-acyl heterogeneity. In contrast to high-energy (8 keV) MS/MS, low-energy CID of multiply charged molecular ions provided abundant structurally diagnostic fragments. These results served as a basis for analysis of more complex and higher molecular mass preparations.

Taki et al. (1995) and Isobe et al. (1996) reported the application of TLC/MS. Glycosphingolipids developed by TLC were transferred to a polyvinylidene difluoride membrane by TLC blotting, after which the glycosphingolipid band on the membrane was excised and placed on a direct probe tip and a few microlitres of triethanolamine was added as the matrix. The sample was analysed by secondary ion MS. The glycosphingolipids on the membrane were bombarded with a 20 kV caesium beam.

Gangliosides. The sphingolipids known as gangliosides contain several sugar residues and one or more N-acetylneuraminic acid (sialic acid) residues attached to the C-1 hydroxyl groups of ceramide. Other acid sphingolipids contain ceramides with multiple carbohydrate moieties and sulphonic acid, known as sulphatides. Gangliosides are sialic-acid-containing glycosphingolipids predominantly located in the outer leaflet of the plasma membrane. They are believed to have an important function in cell–cell interaction and become altered during malignant transformation. The gangliosides exhibit striking variability in the oligosaccharide moiety, their long-chain sphingosine base and fatty acid composition (Sweeley, 1991). A comprehensive review of the structural diversity of this class of lipids has been presented by Kates (1990).

The general methods of MS analysis of the glycosphingolipids have been summarized by Murphy (1993) and by Peter-Katalinic (1994) who have suggested critical experimental sequences with FAB and derivatization procedures for efficient structural characterization of an unknown glycosphingolipid. Early attempts to analyse complex glycosphingolipids involved direct probe EI/MS, CI/MS and DCI of progressively higher molecular weight derivatives and eventually intact underivatized glycosphingolipids. Introduction of the FAB/MS substantially advanced the capability to analyse glycosphingolipids by permitting the abundant formation of molecular ion species and structurally specific fragment ions from relatively small amounts of derivatized and underivatized samples. Still further advancement came from the FAB/MS/MS approach, which allowed further fragmentation of specific high mass ions selected in the initial mass spectrometric analysis. These analyses were subsequently preceded by an HPLC fractionation of the glycosphingolipids. Peter-Katalinic (1994) has reviewed the analysis of the glycoconjugates by FAB/MS and related MS techniques. Negative ion FAB/MS was found to be a very convenient established method for quick sequencing of neutral and acidic native glycosphingolipids because of the clear-cut fragmentation from the lipid part of the molecule, yielding alcoholate-type fragment ions. Information on the size of the sugar chain, the number and type of monosaccharides involved, the presence of other functional groups, the degree of linearity compared with branching and the type of ceramide portion present could also be obtained in a single experiment. More recently, the glycosphingolipids have been analysed by ES/MS and ES/MS/MS with or without prior chromatographic resolution.

Stroud et al. (1996a, b) have employed both FAB/MS and ES/MS to characterize monosialogangliosides of human myelogenous leukemia HL60 cells and normal human leukocytes. The gangliosides were extracted and subjected to extensive segregation and examination of the E- and P-selectin binding ability of each fraction (Stroud et al., 1996a). Fractions resolved by HPLC were reanalysed by means of HPTLC and the pure components subjected to negative ion and positive ion FAB/MS. The tentative identification of pseudo-molecular ion species in positive ion mode was confirmed by the addition of NaOAc to the FAB matrix, which produced a mass shift of 22 amu for each $[M + H]^+$ ion. One group of subfractions gave pseudo-molecular ($[M - 1]^-$ ions (nominal m/z 1516, 1626 and 1628), which were consistent with the sugar composition NeuAc–Hex$_3$–HexNAc plus ceramide and consisted of the nitrogenous base/fatty acid combinations d18:1/16:0, d18:1/24:1 and d18:1/24:0. Ceramide ions were found at m/z 536, 646 and 648, respectively. Ceramide fragments established the linear sequence NeuAc–Hex–HexNAc–Hex–Hex. Six other subfractions that were examined similarly yielded FAB/MS spectra which indicated mainly variations in the sugar composition with minimal alterations in the ceramide species. ES/MS analysis was performed on methylated samples which were dissolved in methanol/water solutions containing 0.25 mM NaOH. For CID studies, multiply charged precursor ions were selectively transmitted by the first mass analyser and directed into the argon collision cell. The ES/MS spectrum of the permethylated monosialogangliosides from HL60 cells was dominated by three major ions formed by the adduction of 2, 3 and 4 sodium ions to a single component, m/z 1683.5^{2+}, 1129.8^{3+}, and 853.3^{4+}, respectively. These data indicated a molecular weight of 3321.7. The triply charged ion was selected for CID, which yielded major fragments on the reducing side of each HexNAc residue and produced the sequence Neu$_5$Ac–LacN–(deoxyHex)LacN–LacN–LacN–Hex–cer from both the reducing terminus and the non-reducing terminus. A facile HexNAc rupture positioned the fucosyl moiety on the penultimate lactosamine residue. The results revealed that only monosialogangliosides having a polylactosamine core with more than 10 monosaccharide units showed E-selectin binding. In a parallel paper, Stroud et al. (1996b) isolated the E-selectin binding gangliosides from myelogenous leukemia HL60 cells and compared them with the monosialogangliosides isolated from HL60 cells having unbranched polylactose structures showing clear E-selectin binding. The binding fractions were identified as monosialogangliosides having a series of unbranched polylactosamine cores. FAB/MS and ES/MS with CID of permethylated fractions were again used along with chromatographic and NMR methods. It was shown that the E-selectin binding epitope in HL60 cells was carried by unbranched terminally α2-3 sialylated polylactosamine having at least 10 monosaccharide units (4 N-acetyllactosamine units) with internal multiple fucosylation at GlcNAc. Monosialogangliosides from normal human neutrophils showed an essentially identical pattern of gangliosides with selectin binding properties.

Wegener, Kobbe and Stoffel (1996) described a useful and very efficient microbore HPLC method for the resolution of minute amounts of gangliosides (e.g. serum or tumour specimens). One of the greatest advantages of the HPLC procedure described here over that of HPTLC is that native non-derivatized gangliosides are separated and can be analysed further by reversed-phase HPLC, GLC and MS.

6.3.9 Lipid A and other lipopolysaccharides

The lipopolysaccharides are found uniquely on the surface of Gram-negative bacterial cells where, among other functions, they play an important role as a permeation barrier, making bacteria resistant to antibiotics. These compounds exhibit great structural variability and cause symptoms associated with organic dust diseases. The lipid A subgroup within the lipopolysaccharides is an acylated and phosphorylated disaccharide of glucosamine moiety of a lipopolysaccharide as found for the outer membrane of *E. coli*. Similar lipid structures have been identified in other bacterial outer membranes, and mass spectrometry has played a crucial role in the elucidation of their structures.

Lipid A has been analysed by ES/MS (Chan and Reinhold, 1994; Gibson et al., 1993; Harrata, Domelsmith and Cole, 1993). As a result of facile deprotonation, monophosphoryl lipid A is readily analysed by negative ion

ES. Dephosphorylated lipid A molecules easily formed adducts with Na^+ (Harrata, Domelsmith and Cole, 1993). The mass spectra of lipid A of *Enterobacter agglomerans* revealed the presence of lipid A ions which differed in the nature of attached fatty acid side chains (Harrata, Domelsmith and Cole, 1993). At least two heptadecyl forms of lipid A were detected, one of which had a structure which appeared to be the same as the structure of heptaacyl lipid A produced by *Salmonella minnesota*. The second structure differed only by the nature of the side chain at position 3' of the disaccharide backbone where the hydroxymyristoyl group replaced the myristoyloxymyristoyl substituent. CID prior to mass analysis enabled the identification of fragment ions which could be distinguished from at least eight intact deprotonated molecules present in crude lipid A. Wang and Cole (1996) have used ES/MS/MS to monitor the acid and base hydrolysis products of lipid A from *Enterobacter agglomerans* in relation to detoxification mechanisms.

Gibson et al. (1993) have determined the precise molecular weights of the lipooligosaccharide (LOS) and lipopolysaccharide (LPS) preparations from various *Haemophilus*, *Neisseria* and *Salmonella* species by ES/MS. The LOS or LPS were first O-deacylated under mild hydrazine conditions to remove O-linked esters primarily from the lipid A portion. Under negative-ion conditions the O-deacylated LOS yielded abundant multiply deprotonated molecular ion, $(M - nH)_n^-$, where n refers to the number of protons removed and therefore determines the absolute charge state ($n = z$). The identification of sialic acid in the LOS of *Haemophilus* and *Neisseria* species and the variable phosphorylation of the core of *S. typhimurium* LPS have afforded insights into the biosynthetic pathways used by these organisms. Further structural studies (Kim, Phillips and Gibson, 1994) indicated that the neisserial LOS were composed of an oligosaccharide portion with a phosphorylated diheptose core attached to the toxic lipid A moiety. A conserved meningococcal LOS epitope, defined by the monoclonal antibody D6A, was expressed on group A and many group B and C meningococci of different LOS serotypes. Chan and Reinhold (1994) reported detailed structural characterization of lipid A by ES/MS/MS. Monophosphoryl lipid A (MLA) preparations of *Salmonella minnesota* Re595 were compared with *Shigella flexneri* for sample type and component distribution. Each sample was composed of in excess of 20 individual structures which differed considerably in abundance but little in composition. Component heterogeneity could be directly related to alkane chain length, lipid X-type analogues, and variations in esterification of the core 2-amino-2-deoxydisaccharide, $GlcNH_2$ β(1–6)$GlcNH_2$. The previously defined heptaacyl structure in *S. minnesota* Re595 was identified at only a 15% level with the most abundant species a hexaacyl analogue. Profiles of *S. flexneri* MLA showed an absence of any heptaacyl analogue. In this case, the pentaacyl component was the most abundant. Since the position of each acyl group had been previously established in *S. minnesota* Re595 MLA, ions in the CID spectrum of the *S. flexneri* sample could be structurally assigned.

The chemical structure of lipid A of lipopolysaccharide isolated from *Comamonas testosteroni* was recently determined by Iida *et al.* (1996) by means of methylation analysis, mass spectrometry and NMR. The lipid A backbone was found to consist of 6-O-(2-deoxy-2-amino-β-D-glucopyranosyl)-2-deoxy-2-amino-alpha-D-glucose which was phosphorylated in positions 1 and 4'. Hydroxyl groups at positions 4 and 6' were unsubstituted, and position 6' of the reducing terminal residue was identified as the attachment site of the polysaccharide group. Fatty acid distribution analysis and ES/MS of lipid A showed that positions 2, 2', 3 and 3' of the sugar backbone were N-acylated or O-acylated by R-3-hydroxydecanoic acid and that the hydroxyl groups of the amide-linked residues attached to positions 2 and 2' were further O-acylated by tetradecanoic and dodecanoic acids, respectively.

Bhat, Forsberg and Carlson (1994) have investigated the structure of the lipid A component of *Rhizobium leguminosarum* bv. phaseoli lipopolysaccharide by alkylation analysis, and NMR and by electrospray and FAB mass spectrometry of the de-O-acylated lipid A. The lipid A carbohydrate backbone was shown to be a trisaccharide containing galactouronic acid, glucosamine and the unique sugar 2-amino-2-deoxygluconic acid previously unreported in lipopolysaccharides. The fatty acids of the *R. leguminosarum* lipid were attached both as O- and N-acyl substituents to glucosamine and 2-aminogluconate. All fatty acids were hydroxylated and consisted of 3-hydroxymyristate, 3-hydroxypentadecanoate, 3-hydroxypalmitate, 3-hydroxystearate, and 27-hydroxyoctacosanoate in the approximate ratio of 3:0.2:1:0.6:1. Unlike lipid A from enteric bacteria, the *R. leguminosarum* lipid A lacked 3-acyloxyacyl substituents. However, the long-chain 27-hydroxy fatty acid carried ester-linked β-hydroxybutyrate at the 27-hydroxy position. FAB/MS of the de-O-acylated lipid A demonstrated the presence of two molecular species that differed by 28 amu owing to the fatty acid heterogeneity at the two amide linkages. One species carried amide-linked 3-hydroxymyristate and 3-hydroxypalmitate whereas the second species carried 3-hydroxymyristate and 3-hydroxystearate. Each molecular species also existed as the aldolactone, yielding molecular ions at $([M + H]^+) - 18$. The heterogeneity in the amide-linked fatty acids further distinguished the *Rhizobium* lipid A from enteric lipid A.

6.4 Summary and conclusions

The mid-1990s have seen the emergence of ES/MS as the method of choice for the efficient ionization of complex lipids. Since ES/MS yields the molecular ion with minimal fragmentation, ES/MS/MS with CID is essential for structural characterization of the molecular ions. MS/MS is less important in

LC/MS identification of the common glycerolipid molecules which are extensively resolved into lipid classes by normal-phase HPLC and into lipid classes and molecular species by reversed-phase HPLC. On-line LC/ES/MS allows a pseudo-ES/MS/MS by repeat analysis of the same sample at non-fragmenting and fragmenting ionization conditions. Although this approach has proven to be surprisingly effective, LC/ES/MS cannot replace ES/MS/MS in the identification of complex sphingolipid and lipopolysaccharide molecules as their species are not easily resolved by HPLC. The ES/MS and ES/MS/MS systems have been shown to be capable of handling total lipid extracts of progressively lower concentration and more complex nature. The greatest advantages of the ES/MS method, especially when combined with on-line chromatography, are the elimination of the need for derivatization of the sample and the ability to provide meaningful mass spectra from crude total lipid extracts. Both LC/MS and flow/MS systems may be further automated for lipid class and molecular species analysis by combining chromatography with computerized data acquisition and analysis.

The introduction of ES/MS and LC/ES/MS, however, has not solved the problem of quantification since the yields of these entities vary with experimental conditions, lipid class and frequently with the type of molecular species present. Quantification within certain lipid classes over limited ranges of concentration has been claimed by several laboratories, but systematic quantitative studies have yet to be reported. In view of the obvious advantages of ES/MS for the sensitive detection and identification of complex lipid molecules it would appear that further efforts directed at improving the quantification capability of this method would be well justified.

Acknowledgements

The studies by the author and his collaborators referred to in this review were supported by the Heart and Stroke Foundation of Ontario, Toronto, Ontario and by the Medical Research Council of Canada, Ottawa.

References

Adams, J. and Gross, M. L. (1987) Tandem mass spectrometry for collisional activation of alkali metal-cationized fatty acids: a method for determining double bond location. *Anal. Chem.*, **59**, 1576–82.

Ann, Q. and Adams, J. (1993) Collision-induced decomposition of sphingomyelins for structural elucidation. *Biol. Mass Spectrom.*, **22**, 285–94.

Beckman, J. K., Bagheri, F., Ji, C., Blair, I. A. and Marnett, L. J. (1994) Phospholipid peroxidation in tumor promoter-exposed mouse skin. *Carcinogenesis*, **15**, 2937–44.

Bergqvist, M. and Kuksis, A. (1997) Liquid chromatography with on-line electrospray mass spectrometry of oxidized diphosphatidylglycerol, in *New Techniques and Applications for Lipid Analysis* (eds R. McDonald and M. Mossoba), AOCS Press, Champaign, IL, pp. 81–99.

Bhat, U. R., Forsberg, L. S. and Carlson, R. W. (1994) Structure of lipid A component of *Rhozobium leguminosarum* bv. Phaseoli lipopolysaccharide. Unique nonphosphorylated lipid A containing 2-amino-2-deoxygluconate, galacturonate and glucosamine. *J. Biol. Chem.*, **269**, 14402–10.

Borgeat, P., Picard, S., Braquet, P. *et al.* (1994) LC–MS–MS with ion spray: a promising approach for analysis of underivatized platelet activating factor (PAF). *J. Lipid Mediators Cell Signal.*, **10**, 11–12.

Bortolomeazzi, R., Pizzale, L., Conte, L. S. and Lercker, G. (1994) Identification of thermal oxidation products of cholesteryl acetate. *J. Chromatogr. A*, **683**, 75–85.

Byrdwell, W. C. and Emken, E. A. (1995) Analysis of triglycerides using atmospheric pressure chemical ionisation mass spectrometry. *Lipids*, **30**, 173–5.

Byrdwell, W. C., Emken, E. A., Neff, W. E. and Adlof, R. O. (1996) Quantitative analysis of triglycerides using atmospheric pressure chemical ionisation mass spectrometry. *Lipids*, **31**, 919–35.

Chan, S. and Reinhold, V. N. (1994) Detailed structural characterization of lipid A: electrospray ionization coupled with tandem mass spectrometry. *Anal. Biochem.*, **218**, 63–73.

Chen, S. (1994) Partial characterization of the molecular species of phosphatidylserine from human plasma by high performance liquid chromatography and fast atom bombardment mass spectrometry. *J. Chromatogr. B*, **661**, 1–5.

Chen, S. (1997) Tandem mass spectrometric approach for determining structure of molecular species of aminophospholipids. *Lipids*, **32**, 85–100.

Chisholm, G. M., Ma, G., Irwin, K. C. *et al.* (1994) β-Hydroperoxycholest-5-en-3β-ol, a component of human atherosclerotic lesions, is the primary cytotoxin of oxidized human low density lipoprotein. *Natl. Acad. Sci. USA* **91**, 11452–6.

Couderc, F. (1995) Gas chromatography/tandem mass spectrometry as an analytical tool for the identification of fatty acids. *Lipids*, **30**, 691–9.

Crawford, C. G. and Plattner, R. D. (1983) Ammonia chemical ionization mass spectrometry of intact diacylphosphatidylcholine. *J. Lipid Res.* **24**, 456–60.

Crawford, C. G. and Plattner, R. D. (1984) Phospholipid molecular species quantitation from mass spectra of underivatized lipids. *J. Lipid Res.*, **25**, 518–22.

Deeg, M. A., Humphrey, D. R., Yang, S. H. *et al.* (1992) Glycan components in the glycoinositol phospholipid anchor of human erythrocyte acetylcholinesterase. Novel fragments produced by trifluoroacetic acid. *J. Biol. Chem.*, **267**, 18573–80.

De Fabiani, E., Caruso, D., Cavaleri, M. *et al.* (1996) Cholesta-5,7,9(11)-trien-3β-ol found in plasma of patients with Smith–Lemli–Opitz syndrome indicates formation of sterol hydroperoxide. *J. Lipid Res.* **37**, 2280–7.

Dobson, G., Christie, W. W., Brechany, E. Y. *et al.* (1995) Silver ion chromatography and gas chromatography–mass spectrometry in the structural analysis of cyclic dienoic acids formed in frying oils. *Chem. Phys. Lipids*, **75**, 171–82.

Drucker, D. B., Megson, G., Harty, D. W. *et al.* (1995) Phospholipids of lactobacillus spp. *J. Bacteriol.*, **177**, 6304–8.

Duffin, K. L., Henion, J. D. and Shieh, J. J. (1991) Electrospray and tandem mass spectrometric characterization of acylglycerol mixtures that are dissolved in non-polar solvents. *Anal. Chem.*, **63**, 1781–8.

Dzeletovic, S., Babiker, A., Lund, E. and Diczfalusy, U. (1995a) Time course of oxysterol formation during in vitro oxidation of low density lipoprotein. *Chem. Phys. Lipids*, **78**, 119–28.

Dzeletovic, S., Breuer, O., Lund, E. and Diczfalusy, U. (1995b) Determination of cholesterol oxidation products in human plasma by isotope dilution–mass spectrometry. *Analyt. Biochem.*, **225**, 73–89.

Esterbauer, H., Zollner, H. and Schauer, R. J. (1990) Aldehydes formed by lipid peroxidation: mechanisms of formation, occurrence, and determination, in *Membrane Lipid Oxidation*, vol. 1 (ed. C. Vigo-Pelfrey), CRC Press, Boca Raton, pp. 239–68.

Evershed, R. P. (1994) Application of modern mass spectrometric techniques to the analysis of lipids, in *Developments in the Analysis of Lipids* (eds J. H. P. Tyman and M. H. Gordon), Royal Society of Chemistry, Cambridge, pp. 123–60.

Evershed, R. P. (1996) High-resolution triacylglycerol mixture analysis using high-temperature gas chromatography/mass spectrometry with a polarizable stationary phase, negative ion chemical ionization, and mass-resolved chromatography. *J. Am. Soc. Mass Spectrom.*, **7**, 350–61.

Faix, D., Neese, R., Kletke, C. et al. (1993) Quantification of menstrual and diurnal periodicities in rates of cholesterol and fat synthesis in humans. *J. Lipid Res.*, **34**, 2063–75.

Ferguson, M. A. J. and Williams, A. F. (1988) Cell surface anchoring of proteins via glycosylphosphatidylinositol structures. *Ann. Rev. Biochem.*, **57**, 285–320.

Ferretti, A. and Flanagan, V. P. (1994) Mass spectra of piperidine and trimethylsilyl ester derivatives of the major metabolite prostaglandin F. *Chem. Phys. Lipids*, **74**, 65–72.

Ford, D. A., Han, X., Horner, C. C. and Gross, R. W. (1996) Accumulation of unsaturated acylcarnitine molecular species during acute myocardial ischemia: metabolic compartmentalization of products of fatty acyl chain elongation in the acylcarnitine pool. *Biochemistry*, **35**, 7903–9.

Frankel, E. N. (1985) Chemistry of free radical and singlet oxidation of lipids. *Progr. Lipid Res.*, **23**, 197–221.

Frankel, E. N., Neff, W. E. and Miyashita, K. (1990) Autoxidation of polyunsaturated triacylglycerols. II. Trilinolenoylglycerol. *Lipids*, **25**, 40–7.

Gallon, A. A. and Pryor, W. A. (1993) The identification of the allylic nitrite and nitro derivatives of methyl linoleate and methyl linolenate by negative chemical ionisation mass spectrometry. *Lipids*, **28**, 125–33.

Gasser, H., Strohmaier, W., Schlag, G. et al. (1991) Characterization of phosphatidylcholines in rabbit lung lavage fluid by positive and negative ion fast-atom bombardment mass spectrometry. *J. Chromatogr. Biomed. Applic.*, **562**, 257–66.

Gerwick, W. H. (1996) Epoxy allylic carbocations as conceptual intermediates in the biogenesis of diverse marine oxylipins. *Lipids*, **31**, 1215–31.

Gibson, B. W., Melaugh, W., Phillips, N. J. et al. (1993) Investigation of the structural heterogeneity of lipolysaccharides from pathogenic *Haemophilus* and *Neisseria* species and of R-type lipolysaccharides from *Salmonella typhimurium* by electrospray mass spectrometry. *J. Bacteriol.*, **175**, 2702–12.

Gross, R. W. (1984) High plasmalogen and arachidonic acid content of canine myocardial sarcolemma: a fast atom bombardment mass spectroscopic and gas chromatography–mass spectroscopic characterization. *Biochemistry*, **23**, 158–65.

Guo, L., Ogamo, A., Ou, Z. et al. (1995) Preferential formation of the hydroperoxide of linoleic acid in choline glycerophospholipids in human erythrocytes membrane during peroxidation with an azo initiator. *Free Rad. Biol. Med.*, **18**, 1003–12.

Haas, R., Jackson, B. C., Reinhold, B. and Foster, J. D. (1996) Glycoinositol phospholipid anchor and protein C-terminus of bovine erythrocyte acetylcholinesterase: analysis by mass spectrometry and by protein and DNA sequencing. *Biochem. J.*, **314**, 817–25.

Han, X. and Gross, R. W. (1994) Electrospray ionisation mass spectroscopic analysis of human erythrocyte plasma membrane phospholipids. *Proc. Natl. Acad. Sci. USA*, **91**, 10635–9.

Han, X., Gubitosi-Klug, R. A., Collins, B. J. and Gross, R. W. (1996) Alterations in individual molecular species of human platelet phospholipids during thrombin stimulation: electrospray ionisation mass spectrometry-facilitated identification of the boundary conditions for the magnitude and selectivity of thrombin-induced platelet phospholipid hydrolysis. *Biochemistry*, **35**, 5822–32.

Hannun, Y. A. and Obeid, L. M. (1995) Ceramide: an intracellular signal for apoptosis. *TIBS*, **20**, 72–4.

Harrata, A. K., Domelsmith, L. N. and Cole, R. B. (1993) Electrospray mass spectrometry for characterization of lipid A from *Enterobacter agglomerans*. *Biol. Mass Spectrom.*, **22**, 59–67.

Harrison, K. A. and Murphy, R. C. (1996) Direct mass spectrometric analysis of ozonides: application to unsaturated glycerophosphocholine lipids. *Anal. Chem.*, **68**, 3224–30.

Harvey, D. J. (1991) Nicotinylidene derivatives for the structural elucidation of glyceryl monoethers and mono-esters by gas chromatography/mass spectrometry. *Biol. Mass Spectrom.*, **20**, 87–93.

Harvey, D. J. (1992) Mass spectrometry of picolinyl and other nitrogen-containing derivatives of lipids, in *Advances in Lipid Methodology – One* (ed. W. W. Christie), The Oily Press, Ayr, pp. 19–80.

Heise, N., de Almeida, M. L. and Ferguson, M. A. (1995) Characterization of the lipid moiety of the glycosylphosphatidylinositol anchor of *Tryponosoma cruzi* 1G7-antigen. *Mol. Biochem. Parasit.*, **70**, 71–84.

Hoving, E. B. (1995) Chromatographic methods in the analysis of cholesterol and related lipids. *J. Chromatogr. B*, **671**, 342–62.

Hubbard, W. C., Hundley, T. R., Oriente, A. and MacGlashan, D. W. Jr (1996) Quantitation of 1-stearoyl-2-arachidonoyl-*sn*-3-glycerol in human basophils via gas chromatography–negative ion chemical ionisation mass spectrometry. *Anal. Biochem.*, **236**, 309–21.

Hullin, F., Kim, H.-Y. and Salem, N. Jr (1989) Analysis of aminophospholipid molecular species by high performance liquid chromatography. *J. Lipid Res.*, **30**, 1963–75.

Iida, T., Haishima, Y., Tanaka, A. *et al.* (1996) Chemical structure of lipid A isolated from *Comamonas testosteroni* lipopolysaccharide. *Eur. J. Biochem.*, **237**, 468–75.

Ioneda, T. and Ono, S. S. (1997) Chromatographic and mass spectrometric analyses of 1-monomycoloyl glycerol fraction from *Rhodococcus lentifragmentus* as per-O-benzoyl derivatives. *Chem. Phys. Lipids* (in press).

Isbell, T. A. and Kleiman, R. (1996) Mineral acid-catalysed condensation of meadowfoam fatty acids into estolides. *J. Am. Oil Chem. Soc.*, **73**, 1097–107.

Isobe, T., Naiki, M., Handa, S. and Taki, T. (1996) A simple assay method for bacterial binding to glycosphingolipids on a polyvinylidene difluoride membrane after thin-layer chromatography blotting and in situ mass spectrometric analysis of the ligands. *Anal. Biochem.*, **236**, 35–40.

Itabashi, Y. and Kuksis, A. (1997) Determination of stereochemical configuration of phosphatidylglycerols by chiral phase high performance liquid chromatography with electrospray mass spectrometry. *Analyt, Biochem.* (in press).

Itabashi, Y., Kuksis, A. and Myher, J. J. (1997) Chromatographic resolution of reverse isomers of *sn*-1,2-diacylglycerols. *J. Chromatogr. A*, to be submitted.

Itabashi, Y., Marai, L. and Kuksis, A. (1991) Identification of natural diacylglycerols as the 3,5-dinitrophenylurethanes by chiral phase liquid chromatography with mass spectrometry. *Lipids*, **26**, 951–6.

Itabashi, Y., Myher, J. J. and Kuksis, A. (1993) Determination of positional distribution of short-chain fatty acids in bovine milk fat on chiral columns. *J. Am. Oil Chem. Soc.*, **70**, 177–81.

Itabashi, Y., Kuksis, A., Marai, L. and Takagi, T. (1990) Chiral phase high performance liquid chromatographic resolution of diacylglycerol moieties of natural triacylglycerols on R-(+)-1-(1-naphthylethylamine). *J. Lipid Res.*, **31**, 1711–17.

Itabe, H., Kobayashi, T. and Inoue, K. (1988) Generation of toxic phospholipid(s) during oxyhemoglobin-induced peroxidation of phosphatidylcholines. *Biochim. Biophys. Acta*, **961**, 13–21.

Itabe, H., Yamamoto, H., Suzuki, M. *et al.* (1996) Oxidised phosphatidylcholines that modify proteins. *J. Biol. Chem.* **271**, 33208–17.

Jensen, N. J. and Gross, M. L. (1988) A comparison of mass spectrometry methods for structural determination of and analysis of phospholipids. *Mass Spectrom. Rev.*, **7**, 41–

Jensen, N. J., Tomer, K. B. and Gross, M. L. (1986) Fast atom bombardment and tandem mass spectrometry of phosphatidylserine and phosphatidylcholine. *Lipids*, **21**, 580–8.

Jensen, N. J., Tomer, K. B. and Gross, M. L. (1987) FAB MS/MS for phosphatidylinositol, -glycerol, -ethanolamine and other complex phospholipids. *Lipids*, **22**, 480–9.

Ju, D. D., Wei, G. J. and Her, G. R. (1994) High-energy collision-induced dissociation of ceramide ions from permethylated glycosphingolipids. *J. Am. Soc. Mass Spectrom.*, **5**, 558–63.

Jungalwala, F. B., Evans, J. E. and McCluer, R. H. (1984) Compositional and molecular species analysis of phospholipids by high performance liquid chromatography coupled with chemical ionisation mass spectrometry. *J. Lipid Res.*, **25**, 738–49.

Jungalwala, F. B., Evans, J. E., Kadowaki, H. and McCluer, R. H. (1984) High performance liquid chromatography–chemical ionisation mass spectrometry of sphingoid bases using moving-belt interface. *J. Lipid Res.*, **25**, 209–16.

Kallio, H. and Currie, G. (1993a) Analysis of low erucic acid turnip rapeseed oil (*Brassica campesteris*) by negative ion chemical ionisation tandem mass spectrometry. A method giving information on the fatty acid composition in positions *sn*-2 and *sn*-1/3 of triacylglycerols. *Lipids*, **28**, 207–15.

Kallio, H. and Currie, G. (1993b) Analysis of natural fats and oils by ammonia negative ion tandem mass spectrometry – triacylglycerols and positional distribution of their acyl groups, in *CRC Handbook of Chromatography, Analysis of Lipids* (eds K. Mukherjee, N. Weber and J. Sherma), CRC Press, Boca Raton, FL, pp. 435–58.

Kallio, H. and Rua, P. (1994) Distribution of the major fatty acids of human milk between sn-2- and sn-1,3-positions of triacylglycerols. *J. Am. Oil Chem. Soc.*, **71**, 985–92.

Kalo, P. and Kemppinen, A. (1993) Mass spectrometric identification of triacylglycerols of enzymatically modified butterfat separated on a polarizable phenylmethylsilicone column. *J. Am. Oil Chem. Soc.*, **70**, 1209–17.

Kamido, H., Kuksis, A., Marai, L. and Myher, J. J. (1992a) Identification of cholesterol-bound aldehydes in copper-oxidized low density lipoprotein. *FEBS Letters*, **304**, 269–72.

Kamido, H., Kuksis, A., Marai, L. et al. (1992b) Preparation, chromatography and mass spectrometry of cholesteryl ester and glycerolipid-bound aldehydes. *Lipids*, **27**, 645–50.

Kamido, H., Kuksis, A., Marai, L. and Myher, J. J. (1993) Identification of core aldehydes among in vitro peroxidation products of cholesteryl esters. *Lipids*, **28**, 331–6.

Kamido, H., Kuksis, A., Marai, L. and Myher, J. J. (1995) Lipid ester-bound aldehydes among copper-catalysed peroxidation products of human plasma lipoproteins. *J. Lipid Res.*, **36**, 1876–86.

Kamido, H., Nonaka, K., Yamana, K. et al. (1997) Cholesteryl ester core aldehydes (components of extensively oxidized low density lipoproteins) exist in human atherosclerotic lesions and are cytotoxic to cultured human endothelial cells. *Arteriosclerosis and Thrombosis*, submitted.

Kates, M. (ed.) (1990) Glycolipids, phosphoglycolipids and sulphoglycolipids. *Handbook of Lipid Research*, vol. 6. Plenum Press, New York.

Kates, M., Syz, J.-Y., Gosser, D. and Haines, T. H. (1993) pH–Dissociation characteristics of cardiolipin and its 2'-deoxy analogue. *Lipids*, **28**, 877–82.

Kayganich-Harrison, K. A. and Murphy, R. C. (1994a) Fast-atom bombardment tandem mass spectrometry of [^{13}C]arachidonic acid labeled phospholipid molecular species. *J. Am. Soc. Mass Spectrom.*, **5**, 144–50.

Kayganich-Harrison, K. A. and Murphy, R. C. (1994b) Characterization of chain-shortened oxidised glycerophosphocholine lipids using fast atom bombardment and tandem mass spectrometry. *Analyt. Biochem.*, **221**, 16–24.

Kayganich-Harrison, K. A. and Murphy, R. C. (1992) Fast atom bombardment tandem mass spectrometric identification of diacyl-, alkylacyl-, and alk-1-enylacyl molecular species of glycerophospho ethanolamine in human polymorphonuclear leukocytes. *Anal. Chem.*, **64**, 2965–71.

Kerwin, J. L. and Torvik, J. J. (1996) Identification of monohydroxy fatty acids by electrospray mass spectrometry and tandem mass spectrometry. *Analyt. Biochem.*, **237**, 56–64.

Kerwin, J. L., Tuininga, A. R. and Ericsson, L. H. (1994) Identification of molecular species of glycerophospholipids and sphingomyelin using electrospray mass spectrometry. *J. Lipid Res.*, **35**, 1102–14.

Kerwin, J. L., Wiens, A. M. and Ericsson, L. H. (1996) Identification of fatty acids by electrospray mass spectrometry and tandem mass spectrometry. *J. Mass Spectrom.*, **31**, 184–92.

Kerwin, J. L., Tuininga, A. R., Wiens, A. M. et al. (1995) Isoprenoid-mediated changes in the glycerophospholipid molecular species of the sterol auxotropic fungus *Lagenidium gaganteum*. *Microbiology*, **141**, 399–410.

Kim, H.-Y. and Salem, N. Jr (1986) Phospholipid molecular species analysis by thermospray liquid chromatography/mass spectrometry. *Anal. Chem.*, **58**, 9–14.

Kim, H.-Y. and Salem, N. Jr (1987) Application of thermospray high-performance liquid chromatography/mass spectrometry for the determination of phospholipids and related compounds. *Anal. Chem.*, **59**, 722–6.

Kim, H.-Y. and Salem, N. Jr (1989) Preparation and structural determination of hydroperoxy derivatives of docosahexaenoic acid and other polyunsaturates by thermospray LC/MS. *Prostaglandins*, **37**, 105–19.

Kim, H.-Y. and Salem, N. Jr (1993) Liquid chromatography–mass spectrometry of lipids. *Progr. Lipid Res.*, **32**, 221–45.

Kim, H.-Y. and Sawazaki, S. (1993) Structural analysis of hydroxy fatty acids by thermospray liquid chromatography/tandem mass spectrometry. *Biol. Mass Spectrom.*, **22**, 302–10.

Kim, H.-Y., Wang, T.-C. and Ma, Y.-C. (1994) Liquid chromatography/mass spectrometry of phospholipids using electrospray ionisation. *Anal. Chem.*, **66**, 3977–82.

Kim, J. J., Phillips, N. J. and Gibson, B. W. (1994) Meningococcal group A lipooligosaccharides (LOS): preliminary structural studies and characterization of serotype-associated and conserved LOS epitopes. *Infection and Immunity*, **62**, 1566–75.

Kritharides, L., Jessup, W., Gifford, J. and Dean, R. T. (1993) A method for defining the stages of low-density lipoprotein oxidation by the separation of cholesterol- and cholesteryl ester-oxidation products using HPLC. *Anal. Biochem.* **213**, 79–89.

Kuksis, A. and Myher, J. J. (1989) Lipids, in *Clinical Biochemistry. Principles. Methods. Applications: Mass Spectrometry*, vol. 1. (ed. A. M. Lawson), Walter de Gruyter, Berlin, pp. 265–351.

Kuksis, A. and Myher, J. J. (1990) Mass analysis of molecular species of diradylglycerols, in *Methods in Inositide Research* (ed. R. F. Irvine), Raven Press, New York, pp. 187–216.

Kuksis, A. and Myher, J. J. (1995) Application of tandem mass spectrometry for the analysis of long chain carboxylic acids. *J. Chromatogr. B*, **671**, 35–70.

Kuksis, A., Marai, L. and Myher, J. J. (1983) Strategy of glycerolipid separation and quantitation by complementary analytical techniques. *J. Chromatogr.*, **273**, 43–66.

Kuksis, A., Marai, L. and Myher, J. J. (1991a) Reversed phase liquid chromatography–mass spectrometry of complex mixtures of natural triacylglycerols with chloride attachment negative chemical ionisation. *J. Chromatogr.*, **588**, 73–87.

Kuksis, A., Marai, L. and Myher, J. J. (1991b) Plasma lipid profiling by liquid chromatography with chloride attachment mass spectrometry. *Lipids*, **26**, 240–6.

Kuksis, A., Myher, J. J., Marai, L. and Geher, K. (1993) Analyses of hydroperoxides and core aldehydes of triacylglycerols, in *Proceedings, 17th Nordic Lipid Symposium* (ed. Y. Malkki), Lipid Forum, Imatra, pp. 230–8.

Kuksis, A., Myher, J. J., Yang, L.-Y. and Steiner, G. (1994) Glycerolipid metabolism with deuterated tracers, in *Biological Mass Spectrometry: Present and Future* (eds T. Matsuo, R. M. Caprioli, M. L. Gross and Y. Sayama), John Wiley, Chichester, Sussex pp. 481–93.

Kuksis, A., Ravandi, A., Marai, L. and Myher, J. J. (1996) Identification of lipid ester-bound aldehydes from plasma lipoproteins and red cell membranes of diabetic subjects, in *Oils–Fats–Lipids 1995: The Proceedings, 21st World Congress, International Society for Fat Research (ISF)* (ed. W. A. M. Castenmiller), P. G. Barnes, Bridgwater, pp. 279–84.

Kusaka, T., Ishihara, S., Sakaida, M. *et al.* (1996) Composition analysis of normal plant triacylglycerols and hydroperoxidised rac-1-stearoyl-2-oleoyl-3-linoleoyl-sn-glycerols by liquid chromatography–atmospheric pressure chemical ionisation mass spectrometry. *J. Chromatogr. A*, **730**, 1–7.

Kuypers, F. A., Buetikofer, P. and Shackleton, C. L. (1991) Application of liquid chromatography–thermospray mass spectrometry in the analysis of glycerophospholipid molecular species. *J. Chromatogr. Biomed. Applic.*, **562**, 191–206.

Kwon, G., Bohrer, A., Han, X.L. *et al.* (1996) Characterization of the sphingomyelin content of isolated pancreatic-islets: evaluation of the role of sphingomyelin hydrolysis in the action of interleukin-1 to induce islet overproduction of nitric-oxide. *Biochim. Biophys. Acta*, **300**, 63–72.

Laakso, P. and Kallio, H. (1993) Triacylglycerols of winter butterfat containing configurational isomers of monoenoic fatty acyl residues. I. Disaturated monoenoic triacylglycerols. *J. Am. Oil Chem. Soc.*, **70**, 1161–71.

Laakso, P. and Kallio, H. (1996) Optimization of the mass spectrometric analysis of triacylglycerols using negative ion chemical ionisation with ammonia. *Lipids*, **31**, 33–42.

Laakso, P. and Voutilainen, P. (1996) Analysis of triacylglycerols by silver ion high-performance atmospheric pressure chemical ionisation mass spectrometry. *Lipids*, **31**, 1311–22.

Laakso, P., Manninen, P., Makinen, J. and Kallio, H. (1996) Post- parturition changes in the triacylglycerols of cow colostrum. *Lipids*, **31**, 937–43.

Lehner, R., Kuksis, A. and Itabashi, Y. (1993) Chiral phase HPLC analysis of the stereospecificity of mono- and diacylglycerol acyltransferases from rat intestine. *Lipids*, **28**, 29–34.

Lenz, M. L., Hughes, H., Mitchell, J. R. *et al.* (1990) Lipid hydroperoxy and hydroxy derivatives in copper-catalyzed oxidation of low density lipoprotein. *J. Lipid Res.*, **31**, 1043–50.

Le Quere, J.-L. (1993) Tandem mass spectrometry in the structural analysis of lipids, in *Advances in Lipid methodology – Two* (ed. W. W. Christie), The Oily Press, Ayr, pp. 215–45.

Loidl-Stahlhofen, A., Hannemann, K. and Spiteller, G. (1994) Generation of alpha-hydroxyaldehydic compounds in the course of lipid peroxidation. *Biochim. Biophys. Acta*, **1213**, 140–8.

Loidl-Stahlhofen, A., Kern, W. and Spiteller, G. (1995) Gas chromatographic–electron impact mass spectrometric screening procedure for unknown hydroxyaldehydic lipid peroxidation products after pentafluorobezyloxime derivatization. *J. Chromatogr. B*, **673**, 1–14.

Lutjohann, D., Meese, C. O., Crouse, J. R. III and von Bergmann, K. (1993) Evaluation of deuterated cholesterol and deuterated sitostanol for measurement of cholesterol absorption in humans. *J. Lipid Res.*, **34**, 1039–46.

Ma, Y. C. and Kim, H. Y. (1995) Development of the on-line high-performance liquid chromatography/thermospray mass spectrometry method for the analysis of phospholipid molecular species in rat brain. *Analyt. Biochem.*, **226**, 293–301.

McConville, M. J. and Ferguson, M. A. J. (1993) The structure, biosynthesis and function of glycosylated phosphatidylinositols in the parasitic protozoa and higher eukaryotes. *Biochem. J.*, **294**, 305–24.

MacMillan, D. K. and Murphy, R. C. (1995) Analysis of lipid hydroperoxides and long-chain conjugated keto acids by negative-ion electrospray mass-spectrometry. *J. Am. Soc. Mass Spectrom.*, **6**, 1190–1201.

Mamer, O. A., Just, G., Li, C.-S. *et al.* (1994) Enhancement of mass spectrometric detection of LTC_4, LTD_4 and LTE_4 by derivatization. *J. Am. Soc. Mass Spectrom.*, **5**, 292–98.

Marai, L., Kuksis, A. and Myher, J. J. (1994) Reversed-phase liquid chromatography–mass spectrometry of the uncommon triacylglycerol structures generated by randomization of butteroil. *J. Chromatogr. A*, **672**, 87–99.

Marai, L., Myher, J. J. and Kuksis, A. (1983) Analysis of triacylglycerols by reversed phase high performance liquid chromatography with direct liquid inlet mass spectrometry. *Can. J. Biochem. Cell Biol.*, **61**, 840–59.

Marai, L., Kuksis, A., Myher, J. J. and Itabashi, Y. (1992) Liquid chromatography chloride attachment negative chemical ionization mass spectrometry of diacylglycerol dinitrophenylurethanes. *Biol. Mass Spectrom.*, **21**, 541–7.

Mares, P., Rezanka, T. and Novak, M. (1991) Analysis of human blood plasma triacylglycerols using capillary gas chromatography, silver ion thin-layer chromatographic fractionation and desorption chemical ionisation mass spectrometry. *J. Chromatogr.*, **542**, 145–59.

Margalit, A., Duffin, K. L. and Isakson, P. C. (1996) Rapid quantitation of a large scope of eicosanoids in two models of inflammation: development of an electrospray and tandem mass spectrometry method and application to biological studies. *Analyt. Biochem.*, **235**, 73–81.

Miyashita, K., Frankel, E. N., Neff, W. E. and Awl, R. A. (1990) Autoxidation of polyunsaturated triacylglycerols. III. Synthetic triacylglycerols containing linoleate and linolenate. *Lipids*, **25**, 48–53.

Morrow, J. D., Awad, J. A., Boss, H. J. *et al.* (1992) Non- cyclooxygenase-derived prostanoids (F_2-isoprostanes) as formed in situ on phospholipids. *Proc. Natl. Acad. Sci. USA*, **89**, 10721–5.

Morrow, J. D., Minton, T. A., Mukundan, C. R. *et al.*, (1994) Free radical-induced generation of isoprostanes in vivo. Evidence for the formation of D-ring and E-ring isoprostanes. *J. Biol. Chem.*, **269**, 4317–26.

Mossoba, M. M., Yurawecz, P. M., Roach, J. A. G. *et al.* (1996) Confirmatory mass-spectra data for cyclic fatty acid monomers. *J. Am. Oil Chem. Soc.*, **73**, 1317–21.

Muizebelt, W. J. and Nielsen, M. W. F. (1996) Oxidative crosslinking of unsaturated fatty acids studied with mass spectrometry. *J. Mass Spectrom*, **31**, 545–54.

Murphy, R. C. (1993) Mass spectrometry of lipids, in *Handbook of Lipid Research*, vol. 7 (ed. R. C. Murphy), Plenum Press, New York, pp. 189–211.

Murphy, R. C. and Harrison, K. A. (1994) Fast atom-bombardment mass spectrometry of phospholipids. *Mass Spectrom. Rev.*, **13**, 57–75.

Murphy, R. C. and Zirrolli, J. A. (1994) Lipid mediators, eicosanoids and fatty acids, in *Biological Mass Spectrometry: Present and Future* (eds T. Matsuo, R. M. Caprioli, M. L. Gross and Y. Seyama) John Wiley, Chichester, Sussex, pp. 463–79.

Myher, J. J. and Kuksis, A. (1995) Electrospray – MS for lipid identification. *INFORM*, **6**, 1068–72.

Myher, J. J., Kuksis, A. and Marai, L. (1993) Identification of the less common isologous short-chain triacylglycerols in the most volatile 2.5% molecular distillate of butteroil. *J. Am. Oil Chem. Soc.*, **70**, 1183–91.

Myher, J. J., Kuksis, A. and Park, P. W. (1996) Stereospecific analysis of TAG oils rich in long chain polyunsaturates, in *Abstracts, 87th AOCS Annual Meeting, Indianapolis, IN*, AOCS Press, Champaign, IL, p. 2.

Myher, J. J., Kuksis, A., Breckenridge, W. C. and Little, J. A. (1981) Differential distribution of sphingomyelins among plasma lipoprotein classes. *Can. J. Biochem.*, **59**, 626–36.

Myher, J. J., Kuksis, A., Geher, K. *et al.*, (1996) Stereospecific analysis of triacylglycerols rich in long-chain polyunsaturated fatty acids. *Lipids*, **31**, 207–15.

Myher, J. J., Kuksis, A., Marai, L. and Manganaro, F. (1984) Quantitation of natural triacylglycerols by reverse-phase liquid chromatography with direct liquid inlet mass spectrometry. *J. Chromatogr.*, **283**, 289–301.

Myher, J. J., Kuksis, A., Marai, L. and Sandra, P. (1988) Identification of the more complex triacylglycerols in bovine milk fat by gas chromatography–mass spectrometry using polar capillary columns. *J. Chromatogr.*, **452**, 93–118.

Myher, J. J., Kuksis, A., Marai, L. and Yeung, S. K. F. (1978) Microdetermination of molecular species of oligo- and polyunsaturated diacylglycerols by gas chromatography–mass spectrometry of their tert-butyl dimethylsilyl ethers. *Anal. Chem.*, **50**, 557–61.

Myher, J. J., Kuksis, A., Ravandi, A. and Cocks, N. (1994) Normal phase liquid chromatography/mass spectrometry with electrospray for sensitive detection of oxygenated glycerophospholipids. *INFORM*, **5**, 478–9. Abs. 13E.

Natomi, H., Saitoh, T., Sugano, K. *et al.* (1993) Systematic analysis of glycosphingolipids in the human gastrointestinal tract: enrichment of sulfatides with hydroxylated longer-chain fatty acids in the gastric and duodenal mucosa. *Lipids*, **28**, 737–42.

Neff, W. E. and Byrdwell, W. C. (1995) Soybean triacylglycerol analysis by reverse-phase high performance liquid chromatography coupled with atmospheric pressure chemical ionization mass spectrometry. *J. Am. Oil Chem. Soc.*, **72**, 1185–91.

Neff, W. E., Frankel, E. N. and Miyashita, K. (1990) Autoxidation of polyunsaturated triacylglycerols. I. Trilinoleoylglycerol. *Lipids*, **25**, 33–9.

Oshima, T., Li, N. and Koizumi, C. (1993) Oxidative decomposition of cholesterol in fish products. *J. Am. Oil Chem. Soc.*, **70**, 595–600.

Oshima, T., Yoon, H.-S. and Koizumi, C. (1989) Application of selective ion monitoring to the analysis of molecular species of vegetable oil triacylglycerols separated by open-tubular column GLC on a methylphenylsilicone phase at high temperature. *Lipids*, **24**, 535–44.

Peter-Katalinic, J. (1994) Analysis of glycoconjugates by fast atom bombardment mass spectrometry and related MS techniques. *Mass Spectrom. Revs*, **13**, 77–98.

Ponchaut, S., Veitch, K., Libert, R. *et al.* (1996) Analysis by fast-atom bombardment tandem mass spectrometry of phosphatidylcholine isolated from heart mitochondrial fractions: evidence of incorporation of monohydroxylated fatty acyl moieties. *J. Am. Soc. Mass Spectrom.*, **7**, 50–8.

Porter, N. A. (1984) Chemistry of lipid peroxidation. *Methods Enzymol.*, **105**, 273–82.

Porter, N. A., Caldwell, S. E. and Mills, K. A. (1995) Mechanisms of free radical oxidation of unsaturated lipids. *Lipids*, **30**, 277–91.

Ravandi, A., Kuksis, A., Marai, L. and Myher, J. J. (1995a) Preparation and characterization of glucosylated aminophospholipids. *Lipids*, **30**, 885–91.

Ravandi, A., Kuksis, A., Marai, L. *et al.* (1996) Isolation and identification of glycated aminophospholipids from red cells and plasma of diabetic blood. *FEBS*, **381**, 77–81.

Ravandi, A., Kuksis, A., Myher, J. J. and Marai, L. (1995b) Determination of lipid ester ozonides and core aldehydes by high performance liquid chromatography with on-line mass spectrometry. *J. Biochem. Biophys. Methods*, **30**, 271–85.

Ravandi, A., Kuksis, A., Shaikh, N. A. and Jackowski, G. (1997) Schiff base formation between amino glycerophospholipids, amino acids and polypeptides and the core aldehydes of phosphatidylcholine. *Lipids* (in press).

Redman, C. A., Green, B. N., Thomas-Oates, J. E. and Reinhold, V. N. (1994) Analysis of glycophosphatidylinositol membrane anchors by electrospray ionisation–mass spectrometry and collision induced dissociation. *Glycoconjugate J.*, **11**, 187–93.

Reinhold, B. B., Chan, S. Y., Chan, S. and Reinhold, V. N. (1994) Profiling glycosphingolipid structural detail–periodate oxidation, electrospray, collision-induced dissociation and tandem-mass spectrometry. *Org. Mass Spectrom.*, **29**, 736–46.

Rezanka, T. and Mares, P. (1991) Determination of plant triacylglycerols using capillary gas chromatography, high performance liquid chromatography and mass spectrometry. *J. Chromatogr.*, **542**, 145–59.

Rizea Savu, S., Silvestro, L., Sorgel, F. et al. (1996) Determination of 1-O-acyl-2-acetyl-sn-glyceryl-3- phosphocholine, platelet-activating factor and related phospholipids in biological samples by high-performance liquid chromatography–tandem mass spectrometry. *J. Chromatogr. B*, **682**, 35–45.

Roberts, L. J., Moore, K. P., Zackert, W. E. et al. (1996) Identification of the major urinary metabolite of the F_2-isoprostane 8-iso-prostaglandin $F_{2\alpha}$ in humans. *J. Biol. Chem.*, **271**, 20617–20.

Roberts, W. L., Kim, B. H. and Rosenberry, T. L. (1987) Differences in the glycolipid membrane anchors of bovine and human erythrocyte acetylcholinesterases. *Proc. Natl. Acad. Sci. USA*, **84**, 7817–21.

Roberts, W. L., Myher, J. J., Kuksis, A. et al. (1988a) Lipid analysis of the glycoinositol phospholipid membrane anchor of human erythrocyte acetylcholinesterase. *J. Biol. Chem.*, **263**, 18766–75.

Roberts, W. L., Santikarn, S., Reinhold, V. N. and Rosenberry, T. L. (1988b) Structural characterization of the glycoinositol phospholipid membrane anchor of human erythrocyte acetylcholinesterase by fast atom bombardment mass spectrometry. *J. Biol. Chem.*, **263**, 18776–84.

Robinson, N. C. (1990) Silicic acid HPLC of cardiolipin, mono- and dilysocardiolipin, and several of their chemical derivatives. *J. Lipid Res.*, **31**, 1513–16.

Rubino, F. M., Zecca, L. and Sonnino, S. (1994) Characterization of a complex mixture of ceramides by fast atom bombardment and precursor fragment analysis tandem mass spectrometry. *Biol. Mass Spectrom.*, **23**, 82–90.

Sandoval, J. A., Huang, Z. H. and Garrett, D. C. (1995) N-acylphosphatidylethanolamine in dry and imbibing cottonseeds. Amounts, molecular species, and enzymatic synthesis. *Plant Physiol.*, **109**, 269–75.

Sawabe, A., Morita, M., Okamoto, T. and Ouchi, S. (1994) The location of double bonds in a cerebroside from edible fungi (mushroom) estimated by B/E linked scan fast atom bombardment mass spectrometry. *Biol. Mass Spectrom.*, **23**, 660–4.

Schuyl, P. J. W., de Joode, T., Duchateau, G. S. M. J. E. et al. (1994) Negative chemical ionization and collisionally activated decomposition mass spectra of O-2,3,4,5,6-pentafluorobenzyloxime derivatives of prostaglandins. *Biol. Mass Spectrom.*, **23**, 47–50.

Schuyl, P. J. W., de Joode, T., Duchateau, G. S. M. J. E. and Vaconcellos, M. A. (1995) HPLC-MS analysis of triglycerides, in *Proceedings of the 43rd/ASMS Conference on Mass Spectrometry and Allied Topics*, Atlanta, GA, 21–26 May, American Society for Mass Spectrometry, p. 1143.

Schweer, H. and Fischer, S. (1994) Negative ion chemical ionization and collisionally activated decomposition mass spectra of O-2,3,4,5,6,-pentafluorobenzyloxime derivatives of prostaglandins. *Biol. Mass Spectrom.*, **23**, 47–56.

Schweer, H., Watzer, B. and Seyberth, H. W. (1994) Determination of seven prostanoids in 1 ml of urine by gas chromatography–negative ion chemical ionisation triple stage quadrupole mass spectrometry. *J. Chromatogr. B*, **652**, 221–7.

Schweer, H., Meese, C. O., Watzer, B. and Seyberth, H. W. (1994) Determination of prostaglandin F_1 and its main plasma metabolites 15-keto-prostaglandin E_0 and prostaglandin E_0 by gas chromatography/negative ion chemical ionisation triple-stage quadrupole mass spectrometry. *Biol. Mass Spectrom.*, **23**, 165–70.

Sebedio, J. L., LeQuere, J. L., Morin, O. et al. (1989) Heat treatment of vegetable oils. III. GC–MS characterization of cyclic fatty acid monomers in heated sunflower and linseed oils after total hydrogenation. *J. Am. Oil Chem. Soc.*, **66**, 704–9.

Sevanian, A., Seraglia, R., Traldi, P. et al. (1994) Analysis of plasma cholesterol oxidation products using gas- and high-performance liquid chromatography/mass spectrometry. *Free Rad. Biol. Med.*, **17**, 397–409.

Silvestro, L., Dacol, R., Scappaticci, E. et al. (1993) Development of a HPLC–MS technique, with an ionspray interface, for the determination of platelet-activating factor (PAF) and lyso-PAF in biological samples. *J. Chromatogr.*, **647**, 261–9.

Sjovall, O. and Kuksis, A. (1995) Detection and quantitation of lipid ester core aldehydes in edible oils using HPLC and LC/MS with electrospray, in *Abstracts, 21st World Congress and Exhibition of the Intenational Society for Fat Research (ISF)*, 1–6 October, The Hague, Netherlands, International Society for Fat Research, pp. 52–53.

Sjovall, O. and Kuksis, A. (1997a) Analysis of oxidized triacylglycerols by LC/MS with electrospray ionization. I. Hydroperoxides, hydroxides, and epoxides. *Lipids*, submitted.

Sjovall, O. and Kuksis, A. (1997b) Analysis of oxidized triacylglycerols by LC/MS with electrospray ionization. II. Core aldehydes. *Lipids*, submitted.

Sjovall, O. and Kuksis, A., Marai, L. and Myher, J. J. (1995) Incremental elution factors as an aid in identification of oxo triacylglycerols during reverse-phase HPLC and LC/MS. *INFORM*, **6**, 508. Abs. F.

Smith, L. L. (1996) Review of progress in sterol oxidations: 1987–1995. *Lipids*, **31**, 453–87.

Smith, P. B. W., Snyder, A. P. and Harden, C. S. (1995) Characterization of bacterial phospholipids by electrospray ionisation tandem mass spectrometry. *Anal. Chem.*, **67**, 1824–30.

Smith, W. L. and Laneuville, O. (1994) Cyclooxygenase and lipoxygenase pathways of arachidonic acid metabolism, in *Prostaglandin Inhibitors in Tumor Immunology and Immunotherapy* (ed.), CRC Press, Boca Raton, FL, pp. 1–39.

Spanos, G. A., Schwartz, S. J., van Breemen, R. B. and Huang, C.-H. (1995) High performance liquid chromatography with light scattering detection and desorption chemical ionisation tandem mass spectrometry of milk fat triacylglycerols. *Lipids*, **30**, 85–90.

Stroud, M. R., Handa, K., Salyan, M. E. K. *et al.* (1996a) Monosialogangliosides of human myelogenous leukemia HL60 cells and normal human leukocytes. 1. Separation of E-selectin binding from non-binding gangliosides, and absence of sialosyl-Le^x having tetraosyl to octaosyl core. *Biochemistry*, **35**, 758–69.

Stroud, M. R., Handa, K., Salyan, M. E. K. *et al.* (1996b) Monosialogangliosides of human myelogenous leukemia HL60 cells and normal human leukocytes. 2. Characterization of E-selectin binding fractions, and structural requirements for physiological binding of E-selectin. *Biochemistry*, **35**, 770–8.

Sweeley, C. C. (1991) Sphingolipids, in *Biochemistry of Lipids, Lipoproteins and Membranes* (eds D. E. Vance and J. Vance), Elsevier, Amsterdam, pp. 327–61.

Takatsu, A. and Nishi, S. (1993) Determination of serum cholesterol by stable isotope dilution method using discharge-assisted thermospray liquid chromatography/mass spectrometry. *Biol. Mass Spectrom.*, **22**,,

Taki, T., Ishikawa, D., Handa, S. and Kasama, T. (1995) Direct mass spectrometric analysis of glycosphingolipid transferred to a polyvinylidene difluoride membrane by thin-layer chromatography blotting. *Anal. Biochem.*, **225**, 24–7.

Tanaka, T., Minamino, H., Unezaki, S. *et al.* (1993) Formation of platelet-activating factor-like phospholipids by Fe^{2+}/ascorbate/EDTA-induced lipid peroxidation. *Biochim. Biophys. Acta*, **1166**, 264–72.

Tokumura, A., Toujima, M., Yoshioka, Y. and Fukuzawa, K. (1996) Lipid peroxidation in low density lipoproteins from human plasma and egg yolk promotes accumulation of 1-acyl analogues of platelet-activating factor-like lipids. *Lipids*, **31**, 1251–8.

Valeur, A., Olsson, N. U., Kaufmann, P. *et al.* (1994) Quantification and comparison of some natural sphingomyelins by on-line HPLC/discharge assisted thermospray mass spectrometry. *Biol. Mass Spectrom.*, **23**, 313–19.

Van Rollins, M. and Knapp, H. R. (1995) Identification of arachidonate epoxides/diols by capillary chromatography–mass spectrometry. *J. Lipid Res.*, **36**, 952–66.

Voyksner, R. D. and Pack, T. (1991) Investigation of collisional–activation decomposition process and spectra in the transport region of an electrospray single quadrupole mass spectrometer. *Rapid Commun. Mass Spectrom.*, **5**, 263–8.

Voyksner, R. D., Linder, M. and Keever, J. (1994) Ultrasonic versus pneumatic nebulization electrospray LC/MS, in *Proceedings of the 42nd ASMS Conference on Mass Spectrometry and Allied Topics*, American Society for Mass Spectrometry, p. 998.

Wakeham, S. G. and Frew, N. M. (1982) Gas capillary gas chromatography–mass spectrometry of wax esters, steryl esters and triacylglycerols. *Lipids*, **17**, 831–43.

Wang, Y. and Cole, R. B. (1996) Acid and base hydrolysis of lipid A from *Enterobacter agglomerans*, as monitored by electrospray ionization mass spectrometry: pertinence to detoxification mechanisms. *J. Mass Spectrom.* **31**, 138–49.

Waugh, R. J. and Murphy, R. C. (1996) Mass spectrometric analysis of 4 regioisomers of F_2-isoprostanes formed by free-radical oxidation of arachidonic acid. *J. Am. Soc. Mass Spectrom.*, **7**, 490–9.

Wegener, R., Kobbe, B. and Stoffel, W. (1996) Quantification of gangliosides by microbore high performance liquid chromatography. *J. Lipid Res.*, **37**, 1823–9.

Weintraub, S. T., Lear, C. and Pinckard, R. N. (1993) Differential electron capture mass spectra response of pentafluorobenzoyl derivatives of platelet activating factor alkyl chain homologues. *Biol. Mass Spectrom.*, **22**, 559–64.

Wheelan, P., Murphy, R. C. and Simon, F. R. (1996) Gas chromatographic/mass spectrometric analysis of oxo and chain-shortened leukotriene B_4 metabolites. Leukotriene B_4 metabolism in Ito cells. *J. Mass Spectrom.*, **31**, 236–46.

Wheelan, P., Zirrolli, J. A. and Murphy, R. C. (1993) Low-energy fast atom bombardment tandem mass spectrometry of monohydroxy substituted unsaturated fatty acids. *Biol. Mass Spectrom.*, **22**, 465–73.

Wheelan, P., Zirrolli, J. A. and Murphy, R. C. (1996) Negative- ion electrospray tandem mass–spectrometric structural characterization of leukotriene B-4 (LTB-4) and LTB-4-derived metabolites. *J. Am. Soc. Mass Spectrom.*, **7**, 129–39.

Wissing, J. B. and Behrbohm, H. (1993) Diacylglycerol pyrophosphate, a novel phospholipid compound. *FEBS Letters*, **315**, 95–9.

Wolf, F. E., Dobson, R.L.M., Wehmeyer, K.R. et al. (1994) Simultaneous quantitation of peptide-leukotrienes and prostaglandin E_2 by HPLC/MS/MS: application to zymosan-induced mouse preparations, in *Proceedings of the 42nd ASMS Conference on Mass Spectrometry and Allied Topics*, American Society for Mass Spectrometry, p. 94.

Wolucka, B. A., Rozenberg, R. and de Hoffmann, E. (1996) Desorption chemical ionization tandem mass spectrometry of polyprenyl and dolichol phosphates. *J. Am. Soc. Mass Spectrom.*, **7**, 958–64.

Woodard, D. S., Mealey, B. L., Lear, C. S. et al. (1995) Molecular heterogeneity of PAF in normal human mixed saliva: quantitative mass spectral analysis after direct derivatization of PAF with pentafluorobenzoic anhydride. *Biochim. Biophys. Acta*, **1259**, 137–47.

Wu, Y., Li, L. Y.-T., Henion, J.D. and Krol, G.J. (1996) Determination of LTE_4 in human urine by liquid chromatography coupled with ionspray tandem mass spectrometry. *J. Mass Spectrom.*, **31**, 987–93.

Yamane, M., Abe, A. and Yamane, S. (1994) High performance liquid chromatography–thermospray mass spectrometry of epoxy polyunsaturated fatty acids and epoxyhydroxy polyunsaturated fatty acids from an incubation mixture of rat tissue homogenate. *J. Chromatogr. B*, **652**, 123–36.

Yang, Y.-L., Kuksis, A., Myher, J.J. and Pang, H. (1992) Surface components of chylomicrons from rats fed glyceryl and alkyl esters of fatty acids: minor components. *Lipids*, **27**, 1173–86.

Yang, L.-Y., Kuksis, A., Myher, J. J. and Steiner, G. (1995) Origin of the triacylglycerol moiety of plasma low density lipoproteins in the rat: structural studies. *J. Lipid Res.*, **35**, 125–36.

Yang, L.-Y., Kuksis, A., Myher, J. J. and Steiner, G. (1996) Contribution of de novo fatty acid synthesis to very low density lipoprotein triacylglycerols: evidence from mass isotopomer distribution analysis of fatty acids synthesised from $[^2H_6]$ ethanol. *J. Lipid Res.*, **37**, 262–74.

Zhang, J.-R., Cazers, A. R., Lutzke, B. S. and Hall, E. D. (1995) HPLC–chemiluminescence and thermospray LC/MS study of hydroperoxides generated from phosphatidylcholine. *Free Rad. Biol. Med.*, **18**, 1–10.

Zhang, J. Y., Nobes, B.J., Wang, J. and Blair, I.A. (1994) Characterization of hydroxyeicosatetraenoic acids and hydroxyeicosatetraenoic acid phosphatidylcholines by liquid secondary ion tandem mass spectrometry. *Biol. Mass Spectrom.*, **23**, 399–405.

Zoellner, P. and Lorbeer, E. (1995) Utility of nicotinoyl derivatives in structural studies of mono and diacylglycerols by gas chromatography/mass spectrometry. Part 2. *J. Mass Spectrom.*, **30**, 432–37.

Zoellner, P. and Schmid, R. (1996) Utility of nicotinoyl derivatives in structural studies of mono- and diacylglycerols by gas chromatography/mass spectrometry. *J. Mass Spectrom.*, **31**, 411–17.

Zoellner, P., Lorbeer, E. and Remberg, G. (1994) Utility of nicotinoyl derivatives in structural studies of mono- and diacylglycerols by gas chromatography/mass spectrometry. *Organic Mass Spectrom.*, **29**, 253–9.

7 Chromatography of food irradiation markers
J.-T. MÖRSEL

7.1 Introduction

Food irradiation is one of the most discussed modern technologies in today's food industry. The first attempts to apply ionizing radiation to food processing was made at the end of the 19th century. Direct application of radioactive isotopes, for example radon, in food was soon recognized to be hazardous to humans, so Schwartz (1921) obtained the first US patent for microbiological decontamination of food by use of X-rays. In the following years many new applications of food irradiation were recognized, and the practical application to some types of foods was tested. Three main fields are of greater interest today.

- prolongation of shelf life;
- impedance of seed germination;
- manipulation of technological properties.

Direct irradiation of foodstuffs is not the only area of commercial interest; treatment of packing materials and enzyme solutions with γ radiation or the application of irradiation to drugs and cosmetics are widely used. The polymerization of compound foils is also initiated with the help of irradiation, especially with use of electron beam sources.

Food irradiation is used today in over 40 countries with about 500 licensed applications. The legal situation is different from country to country, and generally accepted rules are not available. Therefore, in 1984 the UN Food and Agriculture Organisation/World Health Organization/International Atomic Energy Agency Joint Expert Committee on Food Safety (WHO, 1984) recommended general rules for the safe application of food irradiation:

- food (raw material, as well as final products) should only be irradiated up to a maximum dose of 10 kGy;
- maximum radiant energy should be 10 MeV for electron beams and 15 kGy for isotope sources;
- multiple irradiation should not be permitted.

In most countries these rules have been made law. It is also generally accepted that permission should be sought first before proceeding. In some places single products and in others groups of products need to be licensed.

In the interest of product safety and consumers' peace of mind methods for the determination of irradiated food have been developed. With the help of these methods it should be possible to identify irradiated food, to determine the applied dose and, perhaps in the future, to recognize multiple irradiation. There exist methods to identify irradiated foods for many applications but as yet there is no way to detect where multiple irradiation has occurred. In this chapter the aim is to demonstrate some recent results in the field of chromatographic food analysis.

7.2 Chemical changes in irradiated foods

Most food constituents undergo chemical changes during irradiation with X-rays and γ-rays. All kinds of chemical bonds in organic molecules are cleaved by the high quantum energy of ionizing radiation. Ionizing energies of typical chemical bonds in comparison with the quantum energy of the ^{60}Co isotope are given in Table 7.1. It can be seen that more then one bond can be cleaved by one quantum.

Four types of interaction between matter and radiation are under discussion – the photo effect, the Compton effect, the pair formation effect and the spurs model. In the following sections changes of the main food constituents (lipids, carbohydrates and proteins) will be summarized.

7.2.1 Lipids

Ionizing radiation causes two general types of change: initiation of autoxidation and scission of glycerides.

Radiochemical autoxidation leads to the same group of characteristic products well known from thermal autoxidation processes. Quantitative

Table 7.1 Ionizing energy of chemical bonds and the quantum energy of ^{60}Co isotopes

Type of bond	Binding energy (kJ mol^{-1})
C—C	350
C=C	620
C—H	415
C—O	355
C=O	715
O—H	465
^{60}Co isotope:	
E_1	3.108×10^6 (1.17 MeV)
E_2	3.524×10^6 (1.33 MeV)

Table 7.2 Isomer distribution of linoleic acid hydroperoxides after γ-irradiation

Hydroperoxide isomer	Radiochemical reaction (%)[a]	Thermal reaction (%)[b]
13-E,Z $C_{18:2}$—HPO	35.91	20.1
13-E,E $C_{18:2}$—HPO	15.06	28.9
9-E,Z $C_{18:2}$—HPO	37.12	21.3
9-E,E $C_{18:2}$—HPO	11.92	29.7
total 13-isomers	50.97	49.0
total 9-isomers	49.03	51.0
total E,Z-isomers	73.03	41.4
total E,E-isomers	26.98	58.6

[a] Source: Mörsel, 1996.
[b] Source: Chan and Levett, 1977

distribution of primary and secondary products may differ from those from thermal processes. My own experiments have shown that the hydroperoxide distribution differs slightly from that obtained under conventional autoxidation (Table 7.2). Even greater differences are observed in the ratio of Z (*cis*) and E (*trans*) isomers. The results indicate that the number of reactions following primary activation are fewer under irradiation-induced reactions compared with thermally-induced reactions. Thus the radiochemical reaction is the more selective reaction. The reaction rate of radiochemical oxidation is ten times higher than that of the thermal reaction at the same temperature.

Cleavage of glyceride bonds during irradiation has been investigated intensively by Le Tellier and Nawar (1972a, b, c; Le Tellier, 1974; Nawar,

Table 7.3 Formation of hydrocarbons during irradiation of lipids

$$\text{RO} \underset{e}{\overset{a \; b \| c \quad d}{\longrightarrow}} \text{O--C--CH}_2\text{--CH}_2\text{--CH}_2\text{--CH}_2\text{--R}$$
$$\text{RO---OR}$$

Reacting bond	Reaction	Characteristic product
a	+H.	fatty acid
	$-CO_2$; +H.	alkane C_{n-1}
b	+H.	aldehyde C_n
c	+H.	alkane C_{n-1}
	−H.	alkene C_{n-1}
d	+H.	alkane C_{n-2}
	−H.	alkane C_{n-2}

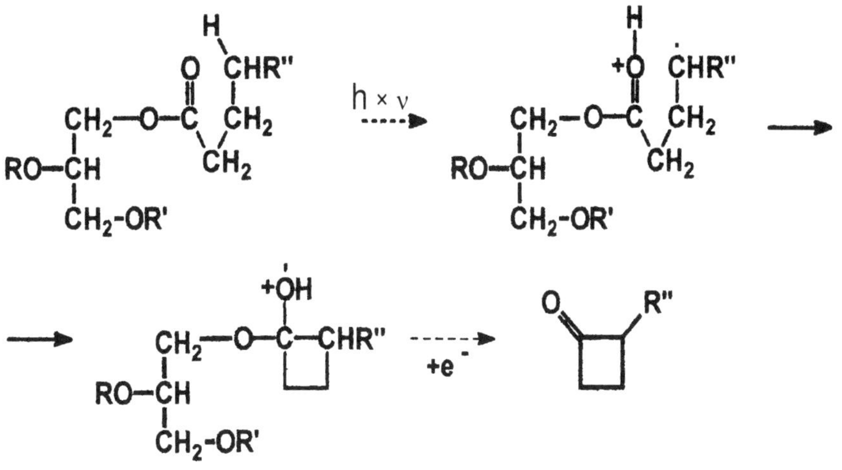

Scheme 7.1 Formation of cyclobutanes from fatty acids during irradiation.

1978). The main process is shown in Table 7.3. In this way hydrocarbons which act as characteristic marker substances are formed.

The formation of cyclobutanones has been reported by Nawar and Le Tellier (1972; Le Tellier and Nawar, 1972a) and Dubravcic (1968). The cyclization reaction (Scheme 7.1) produces characteristic cyclobutanones derived from the fatty acids of the lipid. In recent years, as well as the well-known 2-dodecylcyclobutanone, 2-tetradecylcyclobutanone, 2-tetradecenylcyclobutanone and 2-tetradecadienylcyclobutanone have been found and synthesized in irradiated lipids, so the cyclization products of all main fatty acids has been observed. All the products are widely used to identify irradiated foodstuffs and to estimate the applied dose.

7.2.2 Carbohydrates

The primary degradation reactions of carbohydrates during irradiation also involve radicals. Schertz et al. (1968, 1970) reported hydrogen abstraction to be the first reaction step (Scheme 7.2). Malondialdehyde is formed from the ketone under the influence of reducing agents. With oxygen hydroperoxides and ketosugars are formed (von Sonntag, 1988). Malondialdehyde has been proposed as a potential marker to detect irradiated food, but as it is also found in natural products it is no longer considered to be specific to the irradiation process.

The main products of glucose irradiation are 5-desoxigluconic acid, gluconic acid and arabinose. Polysaccharides are cleaved to form a great variety of products such as formic acid, acetaldehyde, formaldehyde and maltose, glyceraldehyde, malondialdehyde, aceton, furfural, dihydroxyacetone, methanol and hydrogen peroxide have been reported as by-

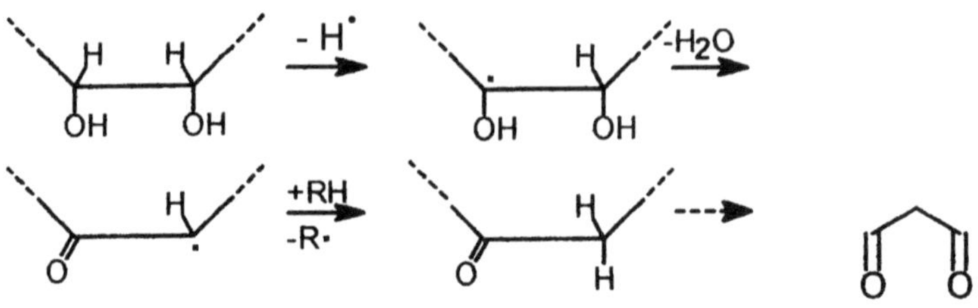

Scheme 7.2 Reaction of carbohydrates during γ–irradiation

products (Dauphin and Saint-Lebe, 1977). The irradiation of polymer carbohydrates has become of interest because of the better digestibility of cellulose-rich feed in animal production (Hübner, 1991).

7.2.3 Proteins

The basic reaction of amino acids under irradiation conditions is the formation of a radical intermediate on the α-carbon atom. From this, ketoacids, iminoacids and degradation products are formed (Scheme 7.3). Cysteine is decomposed to cystine, with hydrogen sulphide formation causing the 'off' flavour of irradiated protein-rich products.

Of special interest is the formation of o-tyrosine. Karam and Simic (1988a, b) reported the occurrence in irradiated phenylalanine, showing that the primary hydroxylation leads to an intermediate that can isomerize (Scheme 7.4). The determination of o- and m-tyrosine may be a way of identifying irradiated protein-rich food. This appears to be difficult, however, because of the dependence of reaction rate on oxygen partial pressure. Also, the occurrence of o-tyrosine traces in natural products have been reported (Pfeilsticker, 1991).

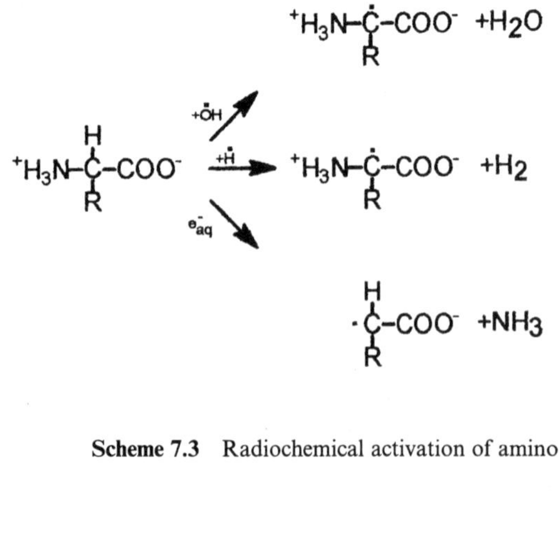

Scheme 7.3 Radiochemical activation of amino acids.

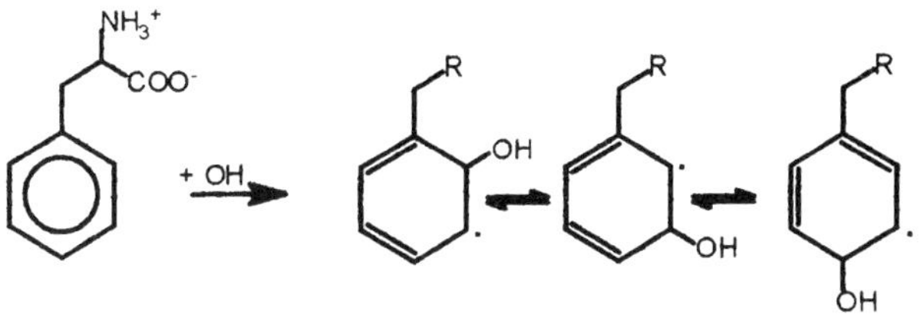

Scheme 7.4 Formation of o-tyrosine.

7.3 Determination of marker substances by gas chromatography

7.3.1 Determination of hydrocarbons

Hydrocarbons were identified as marker substances in the 1960s and early 1970s. Pioneer work was done by Kavalam and Nawar (1969) and by Champagne and Nawar (1969) who intensively investigated volatile components. Nawar (1986) reported the occurrence and quantitative distribution of hydrocarbons in irradiated poultry and meat. Some examples of concentrations of hydrocarbons are given in Table 7.4. A strong correlation between dose and hydrocarbon concentration can be observed. This effect can be used to estimate the applied dose.

Sample preparation. Determination of hydrocarbons can be done only after extraction and enrichment, and a number of different techniques have been developed for this purpose. The lipid can be extracted with organic solvents, for example hexane or pentane, from the matrix. Since

Table 7.4 Hydrocarbons in chicken after irradiation.

Hydrocarbon	Dose dependency[a]				
	0 kGy	0.5 kGy	1 kGy	2 kGy	3 kGy
14:1	7.9	27.8	50.0	92.1	148.8
14:2	0.0	5.7	15.6	28.8	63.6
15:0	14.1	24.6	52.3	75.2	139.1
15:1	0.0	3.2	9.2	18.0	35.3
16:1	2.6	9.0	14.4	23.0	41.1
16:2	0.0	31.9	69.3	140.5	263.0
16:3	0.0	6.6	17.2	31.6	100.7
17:0	15.3	15.5	20.4	26.7	46.0
17:1	0.2	21.5	46.0	97.8	168.2
17:2	0.0	8.9	19.9	36.8	78.2

[a] Amount of hydrocarbon per 100 g of sample (µg).
Source: Nawar et al. 1990

the analyte is very similar to the solvent, contamination of samples with hydrocarbon traces from the solvent must be avoided. Therefore solvents should be monitored before use. Pentane seems to be the more suitable solvent compared with hexane because of its lower boiling point. Samples with high fat contents can simply be prepared by melting the fat, followed by centrifugation. Since the triglyceride matrix is extracted along with the hydrocarbons before gas chromatography, triglycerides must be separated before analysis. In the first experiments (Nawar, Zhu and Yoo, 1990) cold-finger distillation of hydrocarbons was used to separate them from the triglycerides. First, the solvent is vaporized under normal pressure, for example in a rotary evaporator. The sample is then transferred to a cold-finger apparatus and distillation is carried out at temperatures below 80°C in high vacuum. Such sample preparation is difficult and time-consuming. More recently chromatographic methods have been investigated extensively to minimize analytical problems. Liquid chromatography using florisil, silica or aluminium oxide as the stationary phase have been tested. Chromatography with florisil gave the best results for separation of hydrocarbons from triglycerides. Therefore, after activation by heating, florisil has to be inactivated by means of about 3% water. This is essential not only for recovery but also to achieve the required retention behaviour. The development is commonly done with hexane or some other non-polar solvent. All chemicals used in the process have to be monitored regarding their purity, particularly they must not contain traces of hydrocarbon. Thus a blank reaction should be carried out. Sample preparation has also been performed with RP-18 solid-phase extraction cartridges. Thus glycerides and hydrocarbons were separated successfully. The great advantages of this are the low degree of solvent consumption and economy of time. Nevertheless, this method has not been standardized so far.

Gas chromatography. Chromatography of the cleaned samples can be carried out with the help of the different stationary phases. Owing to the properties of the analyte, apolar phases are preferred. Methyl-silicones with up to 10% phenyl groups and a 30 m column length are widely used. Linear gradients in the range of 50°C–100°C give optimal results. To determine the peaks and identify substances a flame ionization detector, mass spectrometry or mass selective detector (MSD), can be used. When applied in single ion monitoring (SIM) mode detection of substances is to a high sensitivity and selectivity. Characteristic ions (Table 7.4) for SIM are seen at *m/z* 196 (M^+) for 1-tetradecene (originating from palmitic acid), pentadecane 212 (M^+) (originating from palmitic acid) and 238 (M^+) for 1-heptadecene (originating from stearic acid). To verify the results non-irradiated control samples should be analysed. Sometimes it is also useful to irradiate samples again, to detect false positive findings and in order to estimate the applied dose.

A characteristic chromatogram from chicken meat irradiated with a 5 kGy dose at a dose rate of 0.9 kGy h^{-1} is shown in Fig. 7.1. The chromatogram indicates that as well as the characteristic hydrocarbons (1-tetradecene, pentadecane, 1,7-hexadecadiene, 1-heptadecene and 1,8-heptadecene) many non-characteristic volatile compounds have been isolated and detected, which demonstrates the necessity of blank experiments.

High-pressure liquid chromatography/gas chromatography coupling. As a result of the experimental problems, the coupling of high-pressure liquid chromatography (HPLC) with gas chromatography seemed to be a suitable solution. The application to hydrocarbon analysis has been possible since Grob *et al.* (1984) found a method for the direct transfer of HPLC eluent to GC. Biedermann, Grob and Meier (1989) published the first results indicating that a loss of analyte during evaporation is the main problem. Thus off-line coupling is an acceptable way of combining the advantages of HPLC with the accuracy of capillary GC while avoiding troubles with the equipment. Generally, highly non-polar solvents in reversed-phase HPLC must be chosen to separate hydrocarbons from triglycerides. This causes additional problems with residual non-polar components of the samples which will be eluted only very slowly. Therefore it is advantageous to include a back-flush step in the analysis design. Back-flushing can be done by changing the direction of the eluent by means of a switching valve (Fig. 7.2).

A second effect is the dependence of the column capacity on the polarity of the mobile phase. Small amounts of polar modifiers in mobile phase have been found to lower the column capacity (Table 7.5). On the whole, the solution selected will depend strongly on the general analytical problem and on the equipment available.

7.3.2 Determination of cyclobutanones

Sample preparation. To determine cyclobutanones in irradiated food it is necessary to isolate the lipids. This can be performed as described for hydrocarbons (Section 7.3.1). Because of the polar character of cyclobutanone the separation of cyclobutanones from triglycerides is more difficult. Cold-finger distillation can be used but chromatographic techniques seem to be more suitable. Column chromatography using florisil has been applied successfully in order to solve the problem. Therefore the sorbent has to be deactivated carefully with about 5% water. This method can be optimized by the operator. The chemicals must be tested to exclude contamination with interfering substances. Since cyclobutanones do not appear naturally they cannot contaminate the reagents.

Gas chromatography. GC can be carried out using the columns described for hydrocarbon analysis (Section 7.3.1). The temperature program can be

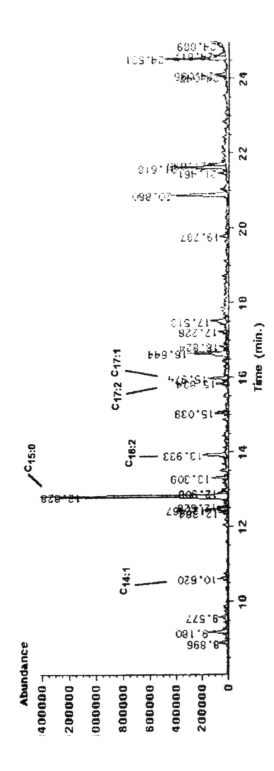

Figure 7.1 Gas chromatography–mass spectrometry chromatogram of hydrocarbons from chicken meat. Chromatograph HP 5890/5970 GC–MS system, DB-5 column, 12.5 m × 0.2 mm inner diameter, 0.33 μm film thickness; carrier gas = helium.

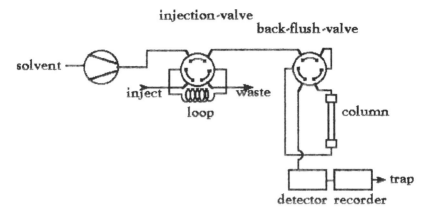

Figure 7.2 Back-flush design of analytical column, with switching valve.

Table 7.5 Dependence of relative column capacity on the polarity of the mobile phase (LiChrosphere^{TM} column, 250 mm × 4.6 mm, with methyl-tert-butylether (MTBE) modifier

Percentage MTBE in hexane	Column capacity (%)
0	100
1	20
2	10
4	5

changed if necessary. Identification of cyclobutanones should be made by use of GC–MS coupling or MSD. More complex chromatograms can be interpreted by working in the SIM mode of GC–MS. For quantitative and qualitative purposes, authentic reference materials are necessary. Today, 2-dodecylcyclobutanone (DCB), 2-tetradecylcyclobutanone (TCB) and 2-tetradecenylcyclobutanone (TECB) are commercially available. For quantitative determination, external calibration with reference samples and standard addition is useful. Positive samples can be re-tested by irradiation with γ-rays under verified conditions. Cyclobutanones in irradiated meat occur in the concentration range of ten to some hundreds of microgrammes per kilogramme. This indicates that cyclobutanones are typically trace components. Experimental results should therefore be estimated very carefully.

7.3.3 Determination of dose and method limits

For legal evaluation of food samples it is very important to know the applied dose. In general there is a correlation between dose and concentration of hydrocarbons or cyclobutanones. It is possible to estimate the dose by re-irradiation of the sample. The range of concentration is known for materials such as chicken meat. One must take into consideration that the concentration of marker substances can be reduced by evaporation as well as by reaction of the analyte during storage (including when frozen). Thus determination of concentrations can give only an estimation of the dose.

The method described can be used only in samples with a high content of fat. Total lipids should be present as over 10% in dry matter. The accuracy depends also on the fatty acid composition. Lipids with a characteristic main component will also show one main cyclobutanone. In the case of an equal distribution of different fatty acids the concentration of derived cyclobutanones will be low. In most cases DCB, TCB and TECB are useful indicator substances for identifying irradiated foods.

7.4 Determination of marker substances by high-pressure liquid chromatography

As described above, 2-alkylcyclobutanones are the only cyclic irradiation products produced under food irradiation conditions. Le Tellier and Nawar (1972a) reported the formation of 2-alkylcyclobutanones. It was shown that a series of homologous 2-alkylcyclobutanone is formed in simple triglycerides (fatty acids with 6–18 carbon atoms). The number of carbon atoms is identical to the number of carbon atoms in the precursor acid. In 1991 Boyd and Stevenson published a method for the determination of 2-dodecylcyclobutanone. It should be re-emphasized that 2-alkylcyclobutanones have never been detected in non-irradiated samples. Two general conclusions can be drawn from these results: irradiated foods can be detected, and these marker substances are the first that have not been detected in thermally treated foodstuffs.

Today there are two possible ways to analyse cyclobutanones – GC–MS after liquid chromatographic separation (Section 7.3) and HPLC of labelled derivatives. Because of the low concentration of cyclobutanones (below $0.5\,\mathrm{mg\,kg^{-1}}$), HPLC with fluorescent labelling seems to be a method that could solve the analytical problem without further enrichment of the analyte. Several fluorescent dyes have been tested. Dansylhydrazine and 7-diethylaminocoumarin-3-carbonylazide are highly sensitive dyes that have been used with good results. Dansylhydrazine can be reacted with 2-alkylcyclobutanones directly, but 7-diethylaminocoumarin-3-carbonylazide reacts only with alcohols. Therefore a reduction of the cyclic ketone (e.g. with sodium borohydride) is necessary (Scheme 7.5). After reduction the cyclobutanoles can be reacted with various derivatives. Labelling with 7-diethylaminocoumarin-3-carbonylazide resulted in a detection limit of about 5 ng. Given a concentration of $0.5\,\mathrm{mg\,kg^{-1}}$ of DCB in irradiated meat samples, about 10 mg will be necessary for the detection of irradiated samples. A characteristic chromatogram is shown in Fig. 7.3 in which 2-

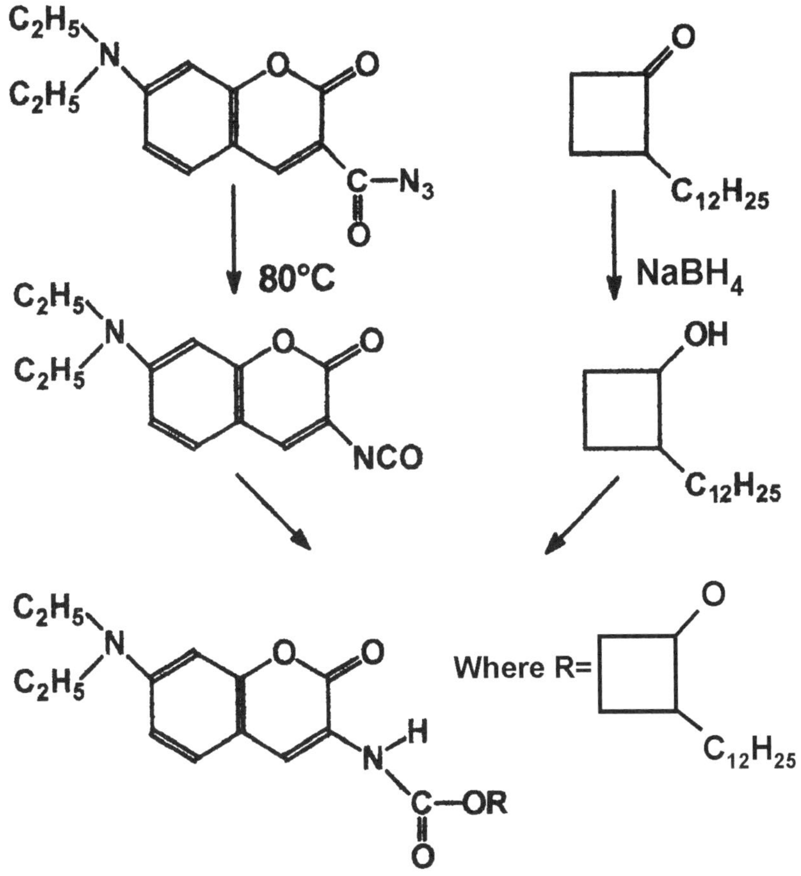

Scheme 7.5 The labelling of 2-dodecylcyclobutanone with 7-diethylaminocoumarin-3-carbonylazide.

dodecylcyclobutanone is clearly separated from the accompanying alcohols (reduction products of aldehydes and ketones).

Sometimes, with dansylhydrazine as the labelling reagent, interfering substances appeared in the chromatograms. Therefore screening of samples by thin-layer chromatography (TLC) is helpful. This method also allowed faster identification of irradiated samples. Reaction of fat and dansylhydrazine can be carried out directly on the TLC plates without further sample preparation. The TLC plate (e.g. RP 18 reversed-phase) is first spotted with the reagent solution and, after drying, samples applied to the reagent spots. With dansylhydrazine, reaction at 80°C for 30 min in an oven is sufficient. The samples are developed with medium polar aprotic solvent mixtures such as methanol/acetone (90:10). Figure 7.4 shows a series of dansylhydrazine-labelled samples. Without TLC separation DCB shows only as a shoulder on a big

262   LIPID ANALYSIS IN OILS AND FATS

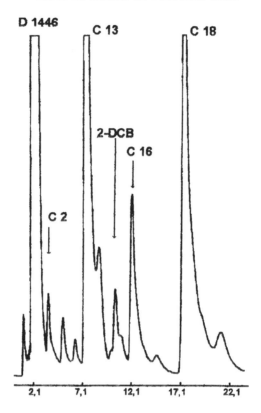

Figure 7.3 Separation of alcohols and of 2-dodecylcyclobutanol (labelled with 7-diethylaminocoumarin-3-carbonylazide). Chromatographic details: Nucleosil™ 100 packing; RP-18™ coating; 18.3 μm film thickness; 250 mm × 4.6 mm inner diameter; solvent system methanol/water (97.5:2.5 vol./vol.).

interfering peak [Fig. 7.4(a)]. However, after TLC pre-separation a clearly separated chromatogram is obtained [Fig. 7.4(b)]. Co-injection of reference materials established the right identification of the peak [Fig. 7.4(c)].

HPLC for hydrocarbon analysis is of no practical interest because of the detection problems: no sensitive light absorption, no electrochemical detectability. There have been some successful attempts with light-scattering detection but in total the detection limit is too insensitive for direct determination in irradiated samples. Thus, at present, only HPLC methods are available for the detection of 2-alkylcyclobutanones.

The determination of o-tyrosine (formed from proteins during irradiation) can be done with HPLC as well as with GC. There are no satisfactory methods because of the bad resolution in the chromatograms, interfering sample constituents and missing reference materials. This area of irradiation analysis will be of further interest in the future because of identification problems with low-fat-containing protein-rich foodstuffs.

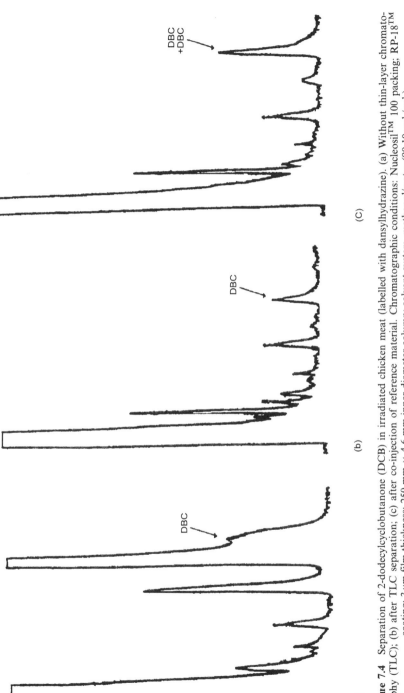

Figure 7.4 Separation of 2-dodecylcyclobutanone (DCB) in irradiated chicken meat (labelled with dansylhydrazine). (a) Without thin-layer chromatography (TLC); (b) after TLC separation; (c) after co-injection of reference material. Chromatographic conditions: Nucleosil™ 100 packing; RP-18™ coating; 3 μm film thickness; 250 mm × 4.6 mm inner diameter column; solvent system, methanol/water (90:10 vol./vol.).

7.5 Conclusions

GC and HPLC are commonly used methods that can be applied successfully to the analysis of irradiated food. GC needs expensive and time-consuming sample preparation; HPLC is sufficiently sensitive but only for analysis of derivatives. Fluorescent labelling can be used, and the higher sensitivity which results allows interference to be excluded; in addition, the sample sizes requested for analysis are much smaller. Today there is a great variety of labelling reagents commercially available, for example dansylhydrazine and 7-diethylaminocoumarin-3-carbonylazide have been used with good results. It can be assumed that further investigations in this field will uncover different and more sensitive analytical methods.

References

Boyd, N. D. and Stevenson, M. H. (1991) *J. Agric. Food Chem.*, **39**, 789–92.
Biedermann, M. K., Grob, K. and Meier, M. (1989) *J. High Res. Chrom.*, **12**, 591–8.
Champagne, J. R. and Nawar, W. W. (1969) *Food Sci.*, **34**, 335–9.
Chan, H. W. S. and Levett, G. (1977) *Lipids*, **12**, 99–104.
Dauphin, C. and Saint-Lebe, C. (1977) in *Radiation Chemistry of Carbohydrates*, 275, Elsevier, Amsterdam.
Dubravcic, M. F. and Nawar, W. W. (1968) *J. Am. Oil Chem. Soc.*, **45**, 656–60.
Grob, K. D., Fröhlich, B. and Schilling, H. P. (1984) *J. Chromatogr.*, **295**, 55–61.
Hübner, G. (1991) *SozEp-Hefte*, **2**, 43–66.
Karam, L. R. and Simic, M. G. (1988a) *ISH-Hefte*, **125**, 297–303.
Karam, L. R. and Simic, M. G. (1988b) *Analyt. Chem.*, **60**, 1117–19.
Kavalam, J. R. and Nawar, W. W. (1969) *J. Am. Oil Chem. Soc.*, **46**, 387–90.
Le Tellier, P. R. (1974) *Lipids*, **9**, 163–9.
Le Tellier, P. R. and Nawar, W. W. (1972a) *Agr. Food Chem.*, **20**, 129.
Le Tellier, P. R. and Nawar, W. W. (1972b) *J. Am Oil Chem. Soc.*, **49**, 259–63.
Le Tellier P. R. and Nawar, W. W. (1972c) *Lipids*, **7**, 75–6.
Mörsel, J.-T. (1996) Studies on changes in lipids during γ-irradiation of food. Technische Universität Berlin, Habitation, Berlin.
Nawar, W. W. (1978) *J. Agr. Food Chem.*, **26**, 21–5.
Nawar, W. W. (1986) *Food Review International*, **2**, 45–78.
Nawar, W. W. and Le Tellier, P. R. (1972) *J. Am. Oil Chem. Soc.*, **49**, 259–63.
Nawar, W. W., Zhu, Z. R. and Yoo, Y. J. (1990) in *Food Irradiation and the Chemist*, The Royal Society of Chemistry, London, pp. 13–24.
Pfeilsticker, K. (1991) *SozEp Hefte*, **7**, 178–80.
Schertz, H. (1970) *Radiat. Res.*, **43**, 12–23.
Schertz, H. (1968) *Natur*, **219**, 611–18.
Schwartz, R. J. (1921) *J. Agr. Res.*, **20**, 845.
von Sonntag, C. (1988) *ISH-Hefte*, **125**, 269–85.
WHO (1984) Report of the FAO/WHO/IAEA Joint Expert Committee on Food Safety. WHO, Genf.

8 Development of purity criteria for edible vegetable oils
J. B. ROSSELL

8.1 Introduction

At the time that this work started, in 1980, there were several large and important international problems concerning edible oil purity and authenticity, such as a problem of cottonseed adulteration with palm oil. It is claimed that a Singapore dealer had secured a contract with the Egyptian government to supply a large quantity of cottonseed oil in two lots. He obtained his cottonseed oil from Australia, but there was a slight shortfall in the quantity and he therefore bulked out the cottonseed oil with a small amount of palm olein, the liquid fraction from palm oil. This went unnoticed and, as a consequence, when the time came for a further delivery, he used a much larger quantity of palm olein. This time the deceit was detected. If he had been successful in this enterprise he would have netted an additional US$14 million in illegal profit. There were other problems of oil purity being faced or suspected by the trade in 1980. Palm oil was fractionated into hard and soft fractions in order to secure advantages of the Malaysian tax and duty structure, which favoured local industry. However, the Western world wished to purchase whole palm oil, and Singapore dealers therefore purchased the palm fractions from Malaysia and recombined them for export to Europe and North America. Unfortunately, the blending was seldom in the same proportions as when the fractions had first been generated and the quality of the 'palm oil' therefore varied considerably. This led to gross manufacturing difficulties for the food industries buying palm oil.

Other problems have related to the suspected presence of small amounts of soft oils such as soybean oil in groundnut oil and blends of soya and rapeseed oils, rape usually being cheaper but difficult to detect at levels of 5% or 10% in soya. One problem related to the purity of corn oil. Several English refineries had difficulty selling their corn oil as buyers were importing corn oil from mainland Europe at impossibly low prices and it was suspected that the corn oil might be impure, but nothing could be proved.

At that time, the main international basis for establishing oil authenticity was the Codex Alimentarius Specifications for Oils and Fats, which listed fatty acid ranges for the oils in question. These were generally considered to be too wide, permitting impure oils to pass as pure.

We therefore launched the authenticity project. We obtained authentic commercial samples of vegetable oilseeds, avoiding botanical curiosities, hand-picked specimens or experimental varieties. Whenever possible, we noted the geographical origin and history, including the year of harvest, of the oilseeds. The seeds were inspected visually to ensure that they were of the stated type. Any foreign seeds or other impurities were manually removed. Oil was then extracted in the laboratory from these purified oilseeds and analysed by appropriate criteria. The project was funded by the United Kingdom Ministry of Agriculture, Fisheries and Food (MAFF), the Federation of Oils, Fats and Seeds Associations Limited (FOSFA International), and the Leatherhead Food Research Association. Some supplementary funding was also provided by the Egyptian government and by CPC (UK) Ltd. Numerous companies and organizations gave valuable assistance in the provision of oilseed samples. The major topic of interest in Europe was rapeseed oil of the low erucic acid variety, as it was undergoing experimental development, with new varieties being introduced every year. As a consequence, we analysed 134 samples of rapeseed oil, far more than for any other oilseed type. In the case of palm oil, we could not easily obtain undamaged palm fruit from the production areas, but instead obtained palm oil directly from plantation and mill managers. In all cases, the geographical origin of the oilseed was noted, together with the year of harvest and any information about the agricultural strain or variety.

8.2 Materials and methods

As mentioned earlier, oilseed samples representative of commercial vegetable oil production were obtained from the main production areas, worldwide. Hand-picked specimens, botanical curiosities and experimental agricultural strains were avoided as far as possible. Table 8.1 lists the number of samples obtained for each oilseed type and the number of geographical harvest regions sampled in each case. The first item in the table, babassu kernel oil, was of minor importance, only five samples being analysed. However, it is a potential impurity in palm kernel and coconut oils and partly for this reason it was considered important to establish its chemical characteristics. In addition, the Codex Alimentarius Commission Fats and Oils Committee has issued a specification for babassu kernel oil and MAFF asked us to check the criteria published in this Codex specification.

As indicated above, oilseed samples were obtained in all cases except palm, for which oil samples were obtained directly from plantation or mill managers. The oilseed samples were cleaned manually to remove any foreign material, and oil was quantitatively extracted from the seeds with petroleum ether (boiling point 40°C–60°C) by a method technically similar to ISO 659. Samples of the oils were immediately analysed. When a delay was

Table 8.1 Samples analysed in the main authenticity project

Oil type	Number of samples	Number of geographic origins
Babassu kernel	5	2
Coconut	40	9
Cottonseed	56	16
Groundnut	71	16
Maize germ	36	10
Palm	55	11
Palm oil fractions	13	1
Palm kernel	86	20
Rapeseed		
low erucic acid	134	13
high erucic acid	12	6
Safflowerseed	31	9
Soybean	39	11
Sunflowerseed	49	14
Total	627	138

anticipated, oils were stored at subambient temperatures. Fatty acid methyl esters (FAME) were prepared from the oil by the method in ISO 5509 and in the early work analysed on a Perkin Elmer F17 or Sigma 2 instrument fitted with a 2 m × 2 mm inner diameter (i.d.) column packed with 6% Silar 5 CP^{TM} on 100/200 mesh Chromasorb W HP^{TM} using nitrogen carrier gas at 20 ml min^{-1} and an oven temperature of 200°C. Peak integration was via a Spectra-Physics SP400 chromatography data system. In more recent work, the methyl esters were analysed on a Carlo Erba model 4160 with an on-column injection system, fitted with a flexible fused silica WCOT 50 m × 0.32 mm i.d. column coated with a 0.2 µm layer of Silar 10C (Chrompack UK Ltd, London). Hydrogen at 0.25 ml min^{-1} was used as carrier gas. The column temperature was initially 140°C programmed to 200°C at 1.5°C min^{-1}, and the detector temperature was 250°C. Peak integration was via a Perkin Elmer Nelson V4 Turbochrom unit.

Figure 8.1 shows a chromatographic trace of a standard rapeseed oil analysed on the Carlo Erba instrument. The flame ionization detector (FID) signal was recorded in the memory of the Nelson Turbochrom for subsequent printout. Reported fatty acid compositions are given as a percentage of the total peak area. The apparatus was regularly calibrated with standard methyl ester mixtures, as shown in Fig. 8.1.

Compositions of the fatty acids at the glycerol 2-position were determined by the IUPAC method (1987). Harder fats were incubated at 42°C for 5 min prior to lipase addition. This removed the need for hexane addition in most cases. We found that hexane addition, to dissolve high-melting fats, gave results of lower reproducibility.

Carbon number triglyceride compositions were determined by direct injection of 12% (m V^{-1}) solutions of the sample in chloroform onto a Pye

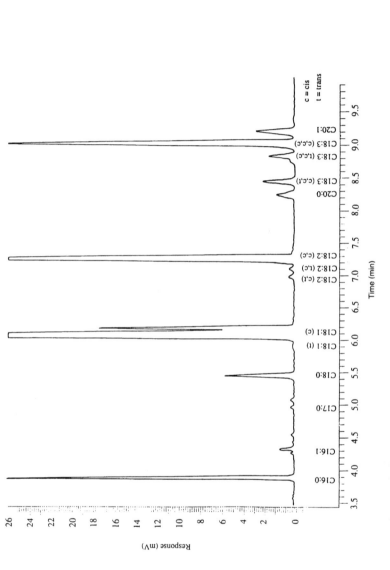

Figure 8.1 Chromatogram of rapeseed oil standard analysed on a Carlo Elba model 4160 instrument (details of the set-up are given in text, Section 8.2). Composition (%m m^{-1}): C16:0, 5.34; C16:1, 0.29; C17:0, 0.12; C18:0, 1.45; C18:1 (*cis*), 58.70; C18:2 (*trans*), 0.25; C18:2 (*cis*), 21.24; C18:3 (*trans*), 1.57; C18:3 (*cis*), 8.72; C20:0, 0.55; C20:1, 1.18.

Unicam GCD gas chromatograph fitted with a 0.6 m × 3 mm i.d. glass column packed with 3% OVI on 100/200 mesh Gas Chrom Q. Nitrogen carrier gas was used at a flow rate of 55 ml min^{-1}. The column temperature was programmed at 300°C for 4 min, thereafter rising to 355°C at 4°C min^{-1}.

Tocol analysis was essentially by the high-pressure liquid chromatography (HPLC) and fluorescence detection method of Thompson and Hatina (1979) using a Spectra-Physics SP 8100 HPLC instrument coupled to a Perkin Elmer LS3 fluorescence detector with an excitation wavelength of 290 nm and emission of 330 nm. A 250 mm × 49 mm column packed with Partisil 5^{TM} (5 µm), fitted with a 50 mm × 50 mm guard column, separated the tocols when eluted with a solvent comprising dry heptane/damp heptane/isopropanol solvent (49.55%:49.55%:0.9%), at a flow rate of 1 ml min^{-1}. Peak integration was by Spectra-Physics SP 4100. Calibration was with pure samples of α-tocopherol, α-tocotrienol and γ-tocopherol, supplied by Roche Products. Values for other tocols were obtained by reference to the literature, as explained previously (Rossell, King and Downes, 1983).

Sterol analysis was also as described previously (Rossell, King and Downes, 1983) except that derivatization was by means of trimethylchlorosilane (Rossell, King and Downes, 1985).

The bulk stable carbon isotope ratio (SCIR) values were obtained by combusting the material to generate CO_2, purifying the CO_2 by gas–liquid chromatography (GLC) and analysing the mass distribution in the CO_2 in a sensitive isotope mass spectrograph. The technique has been widely described elsewhere.* Although the technique is applied in only a limited number of laboratories, it is sufficiently routine to be no more expensive than fatty acid analysis.

The technique of SCIR analysis has not previously been applied to the analysis of oils and fats, and it will therefore be strange to those familiar with the more traditional methods of lipid analysis. It is therefore appropriate to explain the background in a little more depth.

Carbon has several isotopes, the most abundant of which has an atomic mass of 12 Daltons and is designated ^{12}C. The isotope ^{13}C is a stable, naturally occurring isotope, much less abundant than ^{12}C, constituting about 1.1% of the carbon in atmospheric CO_2 (Winkler and Schmidt, 1980). Photosynthetic fixation of CO_2 by plants takes place via three different routes, which depend on the nature of the plant (Brause and Raterman, 1982; Brause et al., 1984; Bricout and Koziet, 1987; Pollard, 1993; van der Merwe, 1992; Winkler and Schmidt, 1980). The higher plants fix CO_2 via the carbon cycle to form a three-carbon compound, the C_3 pathway. In some

---

* See Braunsdorf et al., 1993; Brause and Raterman, 1982; Brause et al., 1984; Bricout and Koziet, 1987; Craig, 1957; Guarino, 1982; Lee and Wrolstad, 1988; McGaw, Milne and Duncan, 1988; Rossell, 1994; Schmid, Grundman and Fogy, 1981; Soter, 1980; van der Merwe, 1992; Winkler and Schmidt, 1980.

plants the initial 'Hatch–Slack' step is via a dicarboxylic acid, a four-carbon compound or C_4 pathway. A third and unusual fixation path is via crassulacean acid, the CAM pathway. Although chemically the end products of these pathways may be identical, there is isotopic bias, and the isotopic composition of the final product shows distinct variation between these different types of plant.

Virtually all terrestrial plants photosynthesize using the C_3 pathway, but a few plants, namely tropical grasses, have evolved the C_4 pathway as an adaptation to a hot dry environment (Pollard, 1993). Maize is one such tropical grass, and its different SCIR ratio has been used to plot the archaeological spread of maize agriculture in prehistoric American cultures (van der Merwe, 1992).

The ratio of ^{13}C to ^{12}C is expressed as ‰ (per mil, or parts per thousand) with respect to an international standard PDB (Pee Dee Belemnite) according to the formula

$$\delta(^{13}C) = \frac{(^{13}C/^{12}C)_{sample} - (^{13}C/^{12}C)_{standard}}{(^{13}C/^{12}C)_{standard}} \times 10^3. \qquad (8.1)$$

In some ways, it is unfortunate that the standard accepted for this work (PDB) has a ^{13}C content higher than that of most other analytes, with the result that the majority of values calculated according to the above formula are negative; in fact all SCIR values of vegetable oils are negative. This can cause some confusion when comparing different values; for instance, what does 'bigger' mean when it relates to a negative number?

Isotope ratios of individual fatty acids were obtained by the gas chromatography combustion/isotope ratio mass spectrometry technique described by Woodbury *et al.* (1995).

8.3 Results and discussion

8.3.1 *Fatty acid analyses*

Fatty acid compositions of palm oil are shown in Table 8.2. The samples included 21 from Malaysia, with further samples from the Ivory Coast, Sumatra, Papua New Guinea, Solomon Islands, New Britain and Nigeria. The fatty acid composition does not vary much by geographical harvest region – information that we passed onto the Codex Alimentarius Fats and Oils Committee which was able to update and revise its published fatty acid composition specifications for palm oil.

Unfortunately, none of these fatty acid composition results enabled us to determine with any confidence which blends of palm oil had been reconstituted by mixing or recombining the fractions produced in Malaysia. Fortunately, a plot of the iodine value (IV) of the oil against the slip melting

Table 8.2 Distribution of fatty acid compositions (wt%) for whole palm oil and fractions

Origin (no. of samples)	C12	C14	C16	C16:1	C18	C18:1	C18:2	C18:3	C20
Malaysia (21)	ND–0.1	0.9–1.1	43.1–45.3	0.1–0.3	4.0–4.8	38.4–40.8	9.4–11.1	0.1–0.4	0.1–0.4
Mainland Malaysia (11)	ND–0.1	0.9–1.0	43.1–45.0	0.1–0.3	4.0–4.8	39.2–40.8	9.4–10.8	0.1–0.4	0.1–0.4
Sabah (8)	ND–0.1	0.9–1.1	43.2–45.3	0.1–0.3	4.3–4.7	38.4–40.1	10.1–11.1	0.2–0.4	0.1–0.4
Sarawak (2)	ND[a]	0.9–1.0	43.5–45.0	0.2–0.3	4.5–4.8	38.5–39.4	10.4–10.5	0.2–0.3	0.2–0.4
Ivory Coast (8)	0.1–0.2	0.8–1.0	43.4–45.2	0.1–0.2	4.9–5.5	37.1–39.9	9.6–10.9	0.2–0.4	0.2–0.4
Sumatra (4)	ND–0.2	1.1–1.3	44.6–46.3	0.1–0.2	4.1–4.6	36.7–38.6	10.1–11.5	0.2–0.3	0.2–0.4
Papua New Guinea (3)	ND–0.1	0.9–1.0	43.4–45.4	Tr–0.1	4.4–4.6	37.5–39.4	11.0–11.1	0.3[a]	0.2–0.4
Solomon Islands (4)	ND–0.1	1.0–1.1	44.4–44.7	Tr–0.1	4.4–4.8	37.6–38.9	10.3–11.0	0.3–0.4	0.4[a]
New Britain (4)	0.1[a]	1.0–1.3	43.5–44.1	Tr–0.2	4.6–5.0	37.4–38.3	11.8–11.9	0.2–0.3	0.4[a]
Nigeria (1)	0.2	1.0	45.4	0.1	4.6	37.7	10.6	0.2	0.3
Overall range (45)	ND–0.2	0.8–1.3	43.1–46.3	Tr–0.3	4.0–5.5	36.7–40.8	9.4–11.9	0.1–0.4	0.1–0.4
Mean	0.1	1.0	44.3	0.15	4.6	38.7	10.5	0.3	0.3
Stearins (8) (Mainland Malaysia)	0.1–0.2	1.0–1.3	46.5–68.9	Tr–0.2	4.4–5.5	19.9–38.4	4.1–9.3	0.1–0.2	0.1–0.3
Oleins (5) (Mainland Malaysia)	0.1–0.2	0.9–1.0	39.5–40.8	Tr–0.2	3.9–4.4	42.7–43.9	10.6–11.4	ND–0.4	0.1–0.3

[a] All samples had the same value within experimental error.
ND = not detected; Tr = trace (unquantified level of less than 0.05%).

point is a very sensitive technique, as shown in Fig. 8.2, and is relatively simple to perform and can be carried out in less technically advanced laboratories situated in some of the production areas (Rossell, King and Downes, 1983, 1985).

The IV is readily determined by the standard titration technique known to many oil chemists, or it can be calculated from the fatty acid composition, and the slip melting point is measured by observing a small plug of fat in a glass capillary held vertically under water in a water bath. When about 96% of the fat is melted, the hydrostatic pressure forces the plug of fat to rise in the capillary tube. The movement is quite sudden, easily observed and takes place at reproducible temperatures. The values of IV and slip melting point are plotted, as in Fig. 8.2, forming distinct clusters for palm oil and palm olein and two lines (A and B) in the case of palm stearin. Any points not conforming to the established clusters are judged to have been contaminated or adulterated and, if need be, subjected to further investigation. Two lines are shown for palm stearin, depending on the importance attributed to the stearin with IVs of below 36 units. These are seldom seen commercially and may be discounted in assessing the probability of palm oil adulteration with commercial palm stearin.

Cocoa butter substitutes are prepared by separating out a middle melting fraction from palm oil and blending the resulting product with other exotic tropical fats. These fats are called palm mid-fractions. The slip melting points and iodine values of a few such fats are also plotted in Fig. 8.2 for added interest. It is unlikely that palm oil will ever be adulterated or contaminated with a palm mid-fraction as these fractions command a high price and are traded separately from the conventional bulk oils.

One of the difficulties with the Codex Alimentarius specifications in the 1980s was with groundnut (arachis) oil, as the Codex data (1983) allowed samples with up to 1% linolenic (C18:3) and 2% erucic (C22:1) acids to pass as pure oil. Some 71 samples of groundnut oil from 16 different geographic locations were studied and found to have a maximum level of linolenic acid of 0.1%, the maximum level of erucic acid being 0.3% (Table 8.3). The higher value for linoleic acid in earlier Codex data (1983) was probably due to overlap of C18:3 with C20:0 in earlier packed-column analyses. The probable reason for the high value for erucic acid shown for groundnut oil in earlier Codex tables (1983) is that an oxidized impurity can co-elute with the fatty acid methyl esters of erucic acid. Figure 8.3 illustrates the chromatogram of the FAME of a groundnut oil slightly contaminated with soybean oil (about 3%). This illustrates that such contamination is easily detected from the linolenic acid component in the analysis. Furthermore, it is just apparent that the peak corresponding to erucic acid at 34.7 min is a doublet. With the Nelson Turbochrom the peaks can be enhanced, as shown with the C22:1 peak in Fig. 8.3. The impurity can be removed by column or thin-layer chromatographic purification of the original triglyceride oil or of the derived

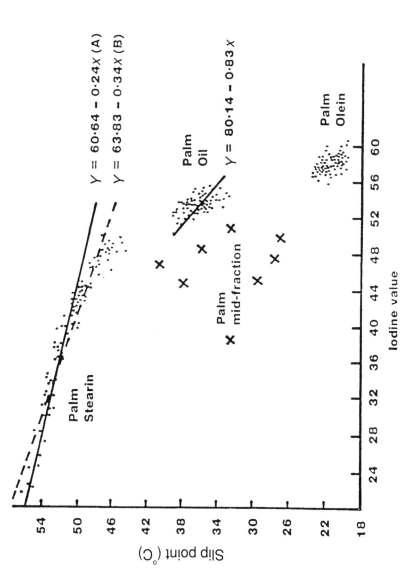

Figure 8.2 Slip point and iodine value.

Table 8.3 Fatty acid composition as determined by gas–liquid chromatography. Revised Codex (1993) ranges

Fatty acid	Arachis oil[a]	Babassu oil	Coconut oil	Cottonseed oil	Grapeseed oil	Maize oil	Mustardseed oil	Palm oil	Palm kernel oil
C6:0	ND	ND	0.0–0.6	ND	ND	ND	0.0–0.5[b]	ND	0.0–0.8
C8:0	ND	2.6–7.3	4.6–9.4	ND	ND	ND		ND	2.4–6.2
C10:0	ND	1.2–7.6	5.5–7.8	ND	ND	ND		ND	2.6–5.0
C12:0	0.0–0.1	40.0–55.0	45.1–50.3	0.0–0.2	0.0–0.5	0.0–0.3	0.0–1.0	0.0–0.4	41.0–55.0
C14:0	0.0–0.1	11.0–27.0	16.8–20.6	0.6–1.0	0.0–0.3	0.0–0.3	0.5–4.5	0.5–2.0	14.0–18.0
C16:0	8.3–14.0	5.2–11.0	7.7–10.2	21.4–26.4	5.5–11	8.6–16.5	0.0–0.5	41.0–47.5	6.5–10.0
C16:1	0.0–0.2	ND	ND	0.0–1.2	0.0–1.2	0.0–0.4	ND	0.0–0.6	ND
C17:0	ND	ND	ND	ND	ND	ND	ND	ND	ND
C17:1	ND	ND	ND	ND	ND	ND	ND	ND	ND
C18:0	1.9–4.4	1.8–7.4	2.3–3.5	2.1–3.3	3.0–6.0	1.0–3.3	0.5–2.0	3.5–6.0	1.3–3.0
C18:1	36.4–67.1	9.0–20.0	5.4–8.1	14.7–21.7	12–28	20.0–42.2	8.0–23	36.0–44.0	12.0–19.0
C18:2	14.0–43.0	1.4–6.6	1.0–2.1	46.7–58.2	58–78	39.4–62.5	10–24	6.5–12.0	1.0–3.5
C18:3	0.0–0.1	ND	0.0–0.2	0.0–0.4	0.0–1.0	0.5–1.5	6.0–18	0.0–0.5	
C20:0	1.1–1.7	ND	0.0–0.2	0.2–0.5	0.0–1.0	0.3–0.6	0.0–1.5	0.0–1.0	
C20:1	0.7–1.7	ND	0.0–0.2	0.0–0.1	ND	0.2–0.4	5.0–13	ND	
C20:2	ND	ND	ND	0.0–0.1	ND	0.0–0.1	0.0–0.1	ND	
C22:0	2.1–4.4	ND	ND	0.0–0.6	0.0–0.3	0.0–0.5	0.2–2.5	ND	0.0–0.1[c]
C22:1	0.0–0.3	ND	ND	0.0–0.3	ND	0.0–0.1	22–50	ND	
C22:2	ND	ND	ND	0.0–0.1	ND	ND	0.0–0.1	ND	
C24:0	1.1–2.2	ND	ND	0.0–0.1	0.0–0.1	0.0–0.4	0.0–0.5	ND	
C24:1	0.0–0.3	ND	ND	ND	ND	ND	0.5–2.5	ND	

Fatty acid	Palm olein	Palm stearin	Rapeseed oil	Rapeseed oil (low erucic acid)	Safflowerseed oil	Sesameseed oil	Soybean oil	Sunflowerseed oil
C6:0	ND	ND	} 0.1[b]	ND	ND	ND	ND	ND
C8:0	ND	ND		ND	ND	ND	ND	ND
C19:0	ND	ND		ND	ND	ND	ND	ND
C12:0	0.1–0.5	0.1–0.4	0.2	0.0–0.2	0.0–0.2	0.0–0.1	0.0–0.1	0.0–0.1
C14:0	0.9–1.4	1.1–1.8	1.5–6.0	3.3–6.0	5.3–8.0	7.9–10.2	0.0–0.2	0.0–0.2
C16:0	38.2–42.9	48.4–73.8	0.0–3.0	0.1–0.6	0.0–0.2	0.1–0.2	8.0–13.3	5.6–7.6
C16:1	0.1–0.3	0.05–0.2	ND	0.0–0.3	ND	0.0–0.2	0.0–0.2	0.0–0.3
C17:0	ND	ND	ND	ND	ND	0.0–0.1	ND	ND
C17:1	ND	ND	0.5–3.1	1.1–2.5	1.9–2.9	4.8–6.1	2.4–5.4	2.7–6.5
C18:0	3.7–4.8	3.9–5.6	8–60	52.0–66.9	8.4–21.3	35.9–42.3	17.7–26.1	14.0–39.4
C18:1	39.8–43.9	15.6–36.0	11–23	16.1–24.8	67.8–83.2	41.5–47.9	49.8–57.1	48.3–74.0
C18:2	10.4–13.4	3.2–9.8	5–13	6.4–14.1	0.0–0.1	0.3–0.4	5.5–9.5	0.0–0.2
C18:3	0.1–0.6	0.1–0.6	0.0–3.0	0.2–0.8	0.2–0.4	0.3–0.6	0.1–0.6	0.2–0.4
C20:0	0.2–0.6	0.3–0.6	3–15	0.1–3.4	0.1–0.3	0.0–0.3	0.0–0.3	0.0–0.2
C20:1	ND	ND	0.0–1.0	0.0–0.1	ND	ND	0.0–0.1	ND
C20:2	ND	ND	0.0–2.0	0.0–0.5	0.2–0.8	0.0–0.3	0.3–0.7	0.5–1.3
C22:0	ND	ND	5–60[c]	0.0–2.0	0.0–1.8	ND	0.0–0.3	0.0–0.2
C22:1	ND	ND	0.0–2.0	0.0–0.1	ND	ND	ND	0.0–0.3
C22:2	ND	ND	0.0–2.0	0.0–0.2	0.0–0.2	0.0–0.3	0.0–0.4	0.2–0.3
C24:0	ND	ND	0.0–3.0	0.0–0.4	0.0–0.2	ND	ND	ND
C24:1	ND	ND						

[a] Also known as groundnut or peanut oil
[b] Range for total of acids C6:0 to C12:0.
[c] Range for total of acids C18:3 to C24:1.
ND = not detected.

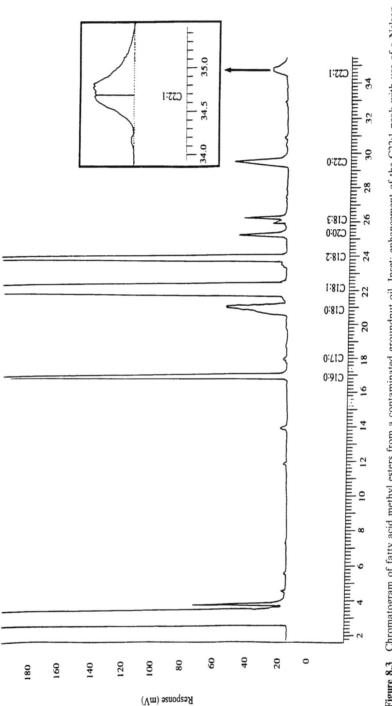

Figure 8.3 Chromatogram of fatty acid methyl esters from a contaminated groundnut oil. Inset: enhancement of the C22:1 peak with use of a Nelson Turbochrom to show the doublet. Composition (%m m^{-1}): C16:0, 12.0; C18:0, 4.0; C18:1, 55.0; C18:2, 21.3; C18:3, 0.3; C20:0, 1.5; C20:1, 1.0; C22:0, 2.6; C22:1, 0.1; C24:0, 1.3. Experimental details are given in text, Section 8.2.

fatty acid methyl esters, as described elsewhere (Rossell, King and Downes, 1983). We never clarified the exact nature of the impurity, but presumed that it was an oxidized component as it was encountered more frequently in oxidized oils. When these and other aspects were taken into account, a complete revision of the Codex fatty acid composition ranges was accomplished (Table 8.3; Codex, 1993).

8.3.2 Other traditional analyses

As many will appreciate, the detection of small amounts of rapeseed oil in soybean oil can be easily accomplished by determination of sterol composition. Table 8.4 shows sterol compositions of nine oils, including rapeseed and soybean oil; rapeseed oil contains quite significant quantities of brassicasterol, whereas the level in soybean oil is almost zero. Table 8.4 also shows that rapeseed oil contains a small amount of cholesterol. In some countries it may be claimed that vegetable oils are free from cholesterol. Although this may be permitted under the local law, it is not, of course, scientifically correct.

The concentrations of fatty acids at the triglyceride 2-position were also determined and compared with the concentrations of the same acid in the oil overall. This enabled derivation of an enrichment factor (Rossell, King and Downes, 1983, 1985). The enrichment factor is constant for an oil irrespective of the geographical origin. It is therefore possible to construct a straight-line plot of the concentration of an acid at the 2-position versus its concentration in the oil overall (Fig. 8.4). This gives a straight line of different gradient for each oil. Determination of the enrichment factor can be valuable in substantiating evidence in cases where a suspected blend falls into a grey area with regard to some of the other tests.

Tocopherol concentrations in the different oils are shown in Table 8.5. These differ from oil to oil and can therefore be used as purity criteria. They must be used with caution, however, as tocopherols act as antioxidants and levels can fall as an oil ages as a result of oxidation. Levels cannot rise, however. Thus groundnut oil contains only a low level ($3-22\,\text{mg}\,\text{kg}^{-1}$) of δ-tocopherol, whereas soybean oil contains very high concentrations ($154-932\,\text{mg}\,\text{kg}^{-1}$). Any contamination of groundnut oil with soybean oil is therefore likely to increase the amount of δ-tocopherol in the oil. The ratio of α/δ-tocopherol has been shown to be useful in detecting low levels of sunflowerseed oil in groundnut oil (Rossell, King and Downes, 1983). Cottonseed oil contains no tocotrienols, but palm olein, a common adulterant, contains quite high concentrations.

Another problem, not yet mentioned, in the introduction of authenticity difficulties is the comingling of coconut oil and palm kernel oil. These both have high levels of lauric acid but can be distinguished by the difference in their relative contents of the short-chain acids with 6, 8 and 10 carbon atoms. Unfortunately, the methyl esters of these acids are quite volatile

Table 8.4 Ranges and mean values of sterol compositions in vegetable oils determined on an OV-17 column in the gas–liquid chromatography stage of analysis.

Sterol[a]	Palm kernel oil	Coconut oil	Cottonseed oil	Soybean oil	Maize oil	Groundnut oil	Palm oil	Sunflowerseed oil	Rapeseed oil
Cholesterol									
range (%)	1.0–3.7	0.6–3.0	0.7–2.3	0.6–1.4	0.2–0.6	0.6–3.8	2.7–4.9	0.2–1.3	0.4–1.3
mean (%)	1.7	1.7	1.0	0.9	0.4	1.5	3.5	0.5	0.7
Brassicasterol									
range (%)	ND–0.3	ND–0.9	0.1–0.9	ND–0.3	ND–0.2	ND–0.2	ND	ND–0.2	5.0–13.0
mean (%)	0.1	0.5	0.3	0.1	0.05	0.0	ND	<1	9.6
Campesterol									
range (%)	8.4–12.7	7.5–10.2	7.2–8.4	15.8–24.2	18.6–24.1	12.3–19.8	20.6–24.2	7.4–12.9	18.2–38.6
mean (%)	10.0	8.7	7.9	19.5	21.3	17.0	23.0	9.5	34.1
Stigmasterol									
range (%)	12.3–16.1	11.4–13.7	1.2–1.8	15.9–19.1	4.3–7.7	5.4–13.3	11.4–11.8	8.6–10.8	ND–0.7
mean (%)	13.7	12.5	1.4	17.5	5.5	8.7	11.7	9.4	Tr
Sitosterol									
range (%)	62.6–70.4	42.0–52.7	80.8–85.1	51.7–57.6	54.8–66.6	48.0–64.7	56.7–58.4	56.2–62.8	45.1–57.9
mean (%)	67.0	46.7	83.2	54.3	63.4	58.5	57.7	59.9	49.9
Δ^5-Avenasterol									
range (%)	4.0–9.0	20.4–35.7	1.9–3.8	1.9–3.7	4.2–8.2	8.3–18.8	2.1–2.7	1.9–6.9	ND–6.6
mean (%)	6.2	26.6	2.6	2.4	5.7	12.3	2.5	3.4	4.5
Δ^7-Stigmasterol									
range (%)	ND–2.1	NS–3.0	0.7–1.4	1.4–5.2	1.0–4.2	ND–5.2	0.4–0.8	7.0–13.4	ND–1.3
mean (%)	0.6	2.4	1.0	2.9	1.8	2.1	0.7	10.4	0.5
Δ^7-Avenasterol									
range (%)	ND–1.4	0.6–3.0	1.4–3.3	1.0–4.6	0.7–2.7	ND–5.5	0.4–2.5	3.1–6.5	ND–0.8
mean (%)	0.1	1.1	0.7	0.6	0.5	1.2	1.0	4.8	0.4
Others									
range (%)	ND–2.7	ND–3.6	ND–1.5	ND–1.8	ND–2.4	ND–1.4		ND–5.3	ND–4.2
mean (%)	0.7	1.1	0.7	0.6	0.5			2.0	0.5
Total									
range (mg kg^{-1})	792–1187	470–1110	2690–5915	1837–4089	7931–22137	901–2854	389–481	2437–4545	4824–11 276
mean (mg kg^{-1})	1025	807	4490	3199	13 776	1575	446	3387	7516

[a] % of total sterol fraction.
ND = not detected; NS = not separated; Tr = trace.

and are easily lost during analysis. This not only makes the analytical results somewhat less reliable, but also can throw doubt on the database with which the results are compared. The technique of measuring the triglyceride composition by high-temperature GLC was therefore developed (Rossell, King and Downes, 1985) and extended to give a plot of so-called K values (Rossell, King and Downes, 1983, 1985) for coconut and palm kernel oils, which showed a very clear distinction, as shown in Fig. 8.5.

Triglyceride carbon numbers are determined by GLC, as described in the methods section. The results for the triglycerides with carbon numbers

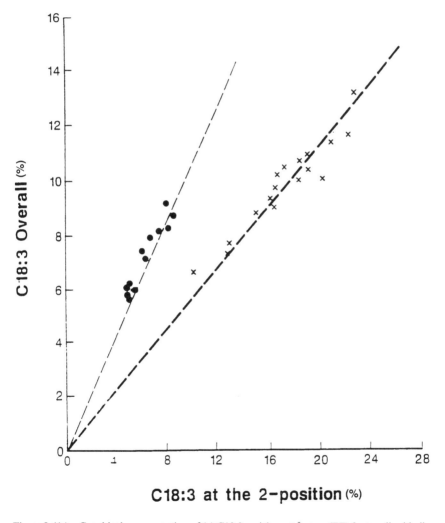

Figure 8.4(a) Graphical representation of (a) C18:3 enrichment factors (EF) for two liquid oils. ($\times$ = low erucic acid rapeseed oil, EF $\approx$ 1.75; $\bullet$ = soybean oil, EF $\approx$ 0.9).

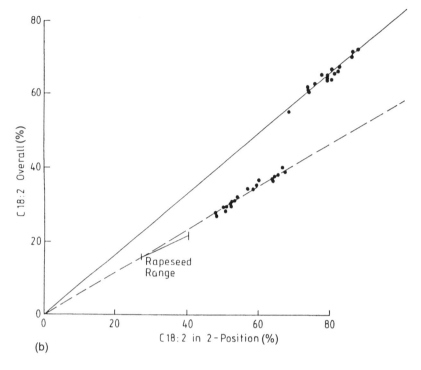

Figure 8.4(b) C18:2 enrichment factors for three liquid oils (——— = sunflower oil, EF = 1.2; - - - - = groundnut oil, EF = 1.7).

ranging from C34 to C40 are then normalized to equal 100, and the resulting values are called K values (Rossell, King and Downes, 1983, 1985). The sum of the K values for triglycerides with carbon numbers 34 and 40 (i.e. $K_{34} + K_{40}$) are then plotted against $(K_{36} + K_{38})$, to give the graph in Fig. 8.5. This shows a clear distinction between coconut and palm kernel oil. The difference is, of course, related to the higher content of fatty acids with 10 or fewer carbons in coconut oil in comparison with palm kernel oil, but the approach avoids the experimental danger of losing, by volatilization, the methyl esters of these short-chain fatty acids during analytical manipulation (Rossell, King and Downes, 1983, 1985). Samples with points appearing intermediate between the two clusters in Fig. 8.5 are immediately suspected of being blends and may be subjected to more stringent tests.

All in all, about 30 detailed reports have been issued to the Leatherhead Food Research Association members on these various analytical criteria, information that our members have found extremely useful. The information was also presented as a whole to the Codex Alimentarius Committee, which used it to update its fatty acid composition ranges in two stages. The latest revision, not yet published in fully approved form by the Codex

Table 8.5 Ranges and means (mg kg^{-1}) of tocopherol (T) and tocotrienol (T$_3$) levels in vegetable oils

Tocol[a]	Relative retention times		Palm kernel oil[a, b]	Coconut oil[a]	Cottonseed oil	Soybean oil	Maize oil	Ground-nut oil	Palm oil	Sunflowerseed oil	High erucic acid rapeseed oil	Low erucic acid rapeseed oil
αT	range	1.0	ND–44	ND–17	136–543	9–352	23–573	49–304	4–185	403–855	39–305	100–320
	mean				338	99.5	282	178	89	670	NA	202
βT	range	1.5	ND–248	ND–11	ND–29	ND–36	ND–356	0–41	–	9–45	24–158	16–140
	mean				16.9	7.7	54	8.8		27	NA	65
γT	range	1.8	ND–257	ND–14	158–594	409–2397	268–2468	99–389	6–36	ND–34	230–500	287–753
	mean				429	1021	1034	213	18	11	NA	490
δT	range	2.7	–	ND–2	ND–17	154–932	23–75	3–22	–	ND–7	5–14	4–22
	mean				3.3	421	54	7.6		0.6		9
αT$_3$	range	1.1	ND–Tr	ND–5	–	–	ND–239	–	4–336	–	–	–
	mean						49		128			
βT$_3$	range	1.7	–	–	–	–	ND–52	–	–	–	–	–
	mean						8					
γT$_3$	range	1.95	ND–60	ND–1	–	–	ND–450	–	42–710	–	–	–
	mean						161		323			
δT$_3$	range	3.2	–	–	–	–	ND–20	–	Tr–148	–	–	–
	mean						6		72			
Total	range		ND–257	Tr–31	410–1169	575–3320	331–3402	176–696	98–1327	447–900	312–928	424–1054
	mean				788	1549	1647	407.4	630	709	NA	766

[a] Means are negligible
[b] High values may be the result of migration of some palm oil into the palm kernels before separation.
ND = not detectable; Tr = trace, NA = not available.

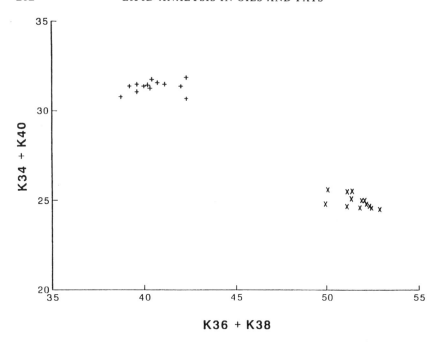

Figure 8.5 Plot of K34 + K40 against K36 + K38 for palm kernel oil (×) and coconut oil (+). K34, K36, K38 and K40 (K_{34}, K_{36}, K_{38} and K_{40} in text) = the K values for triglycerides with carbon numbers 34, 36, 38 and 40, respectively. Source: Rossell, J. B., King, B. and Downes, M. J., Composition of oil, *J. Am. Oil Chem. Soc.*, **62**, 221.

Alimentarius Commission, and at present due for further consideration and possible amendment, by the Fats and Oils Committee, is shown in Table 8.3.

8.3.3 *Problems of maize oil analysis*

In many respects, the fatty acids of groundnut and sunflowerseed oils lie on either side of that for maize oil. This means that blends of groundnut and sunflowerseed oil can have fatty acid compositions very close to that of maize oil and the mixture of groundnut and sunflower oils is therefore very difficult to detect if used as an adulterant for maize oil. The purity of maize oil has in fact been very difficult to establish for a variety of reasons. As just explained, the fatty acid composition data are difficult to interpret, whilst sterol data are equally difficult because maize oil has far more sterols than the other vegetable oils. In any blend, the maize oil sterols therefore swamp those of any adulterant, with the result that the final mixture still corresponds to that of maize. Other methods of establishing maize oil purity were therefore explored, leading to the technique of stable carbon isotope ratio measurement.

8.3.4 Stable carbon isotope ratio analysis

The principles of SCIR were discussed in Section 8.2. C_3 plants have $\delta(^{13}C)$ values [equation (8.1)] falling in the range $-24‰$ to $-30‰$; C_4 plants have $\delta(^{13}C)$ values falling in the range $-9‰$ to $-14‰$; and CAM plants have $\delta(^{13}C)$ values in the range $-12‰$ to $30‰$. Maize, as already discussed in Section 8.4, is a C_4 plant and is thus distinguished from all other commercial vegetable oil sources, which are C_3 plants. This isotopic ratio is maintained throughout the metabolic chain and can be seen in animals that eat the different plants.

The technique of SCIR analysis has been widely reported in the analysis of foods suspected of adulteration. Thus Guarino (1982) applied it to the analysis of vanillin; Brause et al. (1984), Bricout and Koziet (1987) and Braunsdorf et al. (1993) to the authentication of orange juice; and Brause and Raterman (1982) and Lee and Wrolstad (1988) to the analysis of apple juice. Gaffney et al. (1979) discussed the possible application of SCIR analysis to the differentiation of maize-fed animal protein from soya protein in soya–meat mixtures; as explained previously, maize is a C_4 plant whereas soya is a C_3 plant. In reviews of the subject, Gaffney et al. (1979), Winkler and Schmidt (1980) and Schmid, Grundman and Fogy (1981) also draw attention to the possibilities of SCIR analysis of sugar, honey, syrup, ethyl alcohol and alcoholic beverages, vinegar, vegetable fibres such as cotton, jute and coconut fibre, and edible oils and fats.

The application to the authentication of maize oil is thus strongly indicated, but, until the Leatherhead Food Research Association project was undertaken it had not been sufficiently studied to provide reliable ranges of the natural variation in maize and non-maize oil sources for the technique to be used reliably.

This chapter therefore presents the results of research to establish the fatty acid compositions (FACs) and $\delta(^{13}C)$ stable carbon isotope ratios (SCIRs) of 130 samples comprising a variety of edible oils. An attempt was made to link the SCIR with the FAC in such a way as to reduce the limit of detection (increase the sensitivity) of an adulterant oil in maize oil.

Table 8.6 shows the carbon isotope ratio results for a range of vegetable oils. The 73 non-maize oils studied all had isotope ratios between -25.38 and -32.39. They were therefore clearly differentiated from those of the 42 maize oil samples studied, which had ratios from -13.71 to -16.36. Figure 8.6 shows a histogram of maize samples plotted alongside those for non-maize samples, and the differentiation can again be clearly seen. A curious feature was that, when we looked at the histogram for the maize samples in greater depth, there were two peaks. At first it was difficult to explain this but then we recalled an earlier observation in which maize had been distinguished with respect to the hemisphere of growth in terms of its FAC and IV. We therefore replotted the carbon isotope ratio measurements in terms of

Table 8.6 Ranges and means of the stable isotope ratio according to sample type

Sample type (no. of samples)	Stable isotope ratio	
	Range	Mean
Cottonseed (8)	−27.4 to −28.28	−27.78
Groundnut (7)	−26.48 to −28.69	−27.87
Palm olein (6)	−29.51 to −29.84	−29.65
Palm kernel (10)	−27.49 to −30.27	−29.47
Palm oil (6)	−29.25 to −29.91	−29.64
Rapeseed (7)	−27.47 to −29.4	−28.56
Safflower (6)	−27.87 to −30.17	−28.94
Sesame (7)[a]	−25.38 to −29.28	−27.93
Soybean (6)	−29.67 to −30.55	−30.09
Sunflower (5)	−27.94 to −29.76	−28.95
Others (5)[b]	−28.90 to −32.39	−30.79
All vegetable oils excluding maize (73)	−25.38 to −32.39	−28.99
Animal fats (5)	−27.56 to −32.01	−30.28
Fish oils (4)	−25.37 to −27.95	−26.66
Maize (42)	−13.71 to −16.36	−14.95

[a] Two Sudanese sesame-seed oils (41459 and 46132) had fewer negative results than other samples, and were therefore checked several times.
[b] Others = barley, oat and wheatgerm oils, rice bran oil and virgin olive oil.

the hemisphere of growth and found that the stable carbon isotope ratio is also dependent on the hemisphere of growth, as shown in Fig. 8.7. There is therefore a more sensitive method of identifying maize oil purity, as one can use a narrower range for the oil related to its hemisphere of growth. If the hemisphere of growth of the maize oil is not known, the IV of the maize

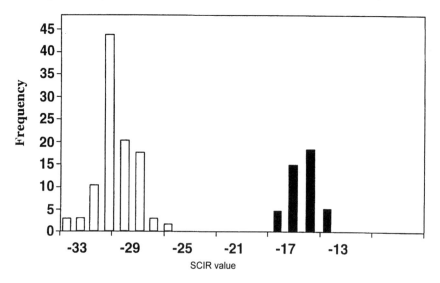

Figure 8.6 Frequency distribution (histograms) of stable carbon isotope ratio (SCIR) values for non-maize oils (open histograms) and maize oil (shaded histograms).

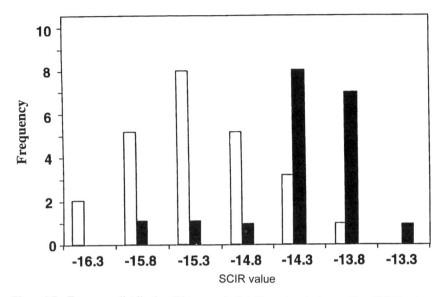

Figure 8.7 Frequency distribution (histograms) of stable carbon isotope ration (SCIR) values for maize oil grown in the northern hemisphere (open histograms) and southern hemisphere (filled histograms).

Table 8.7 Detection limits of adulterant vegetable oils in maize oil by use of the quantity $R = \frac{1}{5}[IV/\delta(^{13}C)]^{2.5}$, where IV is the iodine value and $\delta(^{13}C)$ is the stable isotope carbon ratio [calculated from equation (8.1) in text] and of the fatty acid composition (FAC)

Adulterant oil	Detection limit by use of R (%)	Detection limit by FAC (%)
Palm olein	4	9.8
Palm kernel	4	0.6
Palm oil	4	8.6
Cottonseed	7	21.9
Groundnut	6	12.5
Safflower	8	43.2
Sunflower	8	40.2
Rapeseed	7	33.3
Sesame	7	41.1
Soybean	7	9.6

oil can be determined and by this means its hemisphere of growth clarified; the narrower range appropriate can then be used. An alternative is an arithmetic combination of IV and SCIR, several techniques having been employed. The most sensitive appeared to be the ratio shown in Table 8.7. Using this technique, we can identify between 4% and 7% of impurity in maize oil, a considerable improvement on the limits of detection by fatty acid composition (Table 8.8).

The question then arises of why maize oil should have a different IV, different FAC and different SCIR depending on its hemisphere of growth.

Table 8.8 Limits of detection of adulteration in maize oil by analysis of fatty acid composition (FAC), of stable carbon isotope ratio (SCIR), by the ratio, $R = \frac{1}{5}[IV/\delta(^{13}C)]^{2.5}$ [where IV is the iodine value and $\delta(^{13}C)$ is the SCIR calculated by means of equation (8.1) in text] and by gas chromatography/combustion/SCIR (GC/C/SCIR)

Oil	FAC (%)	SCIR (%)	R (%)	GC/C/SCIR (%)
Palm olein	9.8	10	4	
Palm oil	8.6	10	4	
Cotton	21.9	10	7	
Groundnut	12.5	10	6	4–5
Safflower	43.2	10	8	
Sunflower	40.2	10	8	
Rape	33.3	10	7	5
Sesame	41.1	10	7	

Comparing these various factors one may deduce that linoleic acid might have a more negative SCIR than oleic acid, but why this should be is still a mystery. We are at present involved in a collaborative exercise (Woodbury et al., 1995) with Dr Richard Evershed and his group at the University of Bristol, where there is a sensitive GC–MS apparatus for measuring the isotope ratio of individual peaks in a GC chromatogram.

The prediction that linoleic acid has a more negative SCIR than oleic acid has been tested and found not to be the case (Woodbury et al., 1995). Instead, all the fatty acids show a bias, depending on the hemisphere of growth. Furthermore, it was found that C16:0 is more depleted in ^{13}C than either C18:1 or C18:2 (Woodbury et al., 1995). Work is at present in hand to see if this applies to all other oils and fats as well as to the few studied so far. One important observation from this most recent work is that the plots showing variation of isotope ratio for an individual fatty acid as the blend ratio is varied are curved rather than linear. This is because the isotope ratio of a fatty acid in a blend is influenced by the proportion of fatty acid in each of the components and the isotope ratio of the fatty acid in each of the oils. The equation for the calculation of the isotope ratio of an individual fatty acid in a mixture of oils, $\delta(^{13}C)_{calc}$ is as follows:

$$\delta(^{13}C)_{calc} = \delta(^{13}C)_A \left(\frac{P_A p_A}{P_A p_A + P_B p_B}\right) + \delta(^{13}C)_B \left(\frac{P_B p_B}{P_A p_A + P_B p_B}\right) \quad (8.2)$$

where

$\delta(^{13}C)_A$ = SCIR of fatty acid from adulterant oil (A);
$\delta(^{13}C)_B$ = SCIR of fatty acid from maize oil (B);
P_A = proportion of oil mixture that is the adulterant oil;
P_B = proportion of oil mixture that is maize oil;
p_A = amount of fatty acid in the adulterant oil as a percentage of total fatty acids;
p_B = amount of fatty acid in maize oil as a percentage of total fatty acid.

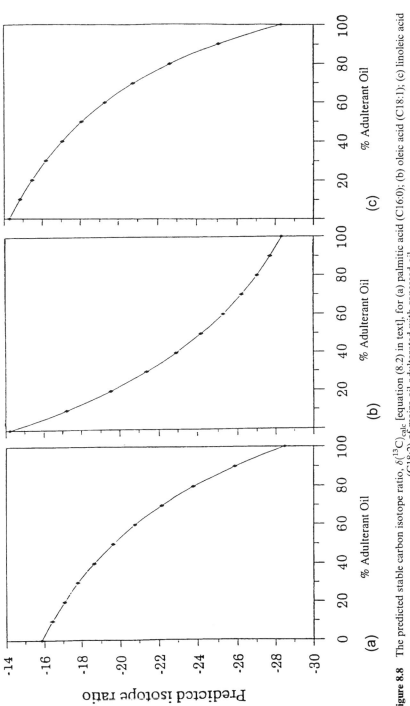

Figure 8.8 The predicted stable carbon isotope ratio, $\delta(^{13}C)_{calc}$ [equation (8.2) in text], for (a) palmitic acid (C16:0); (b) oleic acid (C18:1); (c) linoleic acid (C18:2) of maize oil adulterated with rapeseed oil.

Corresponding curves (Woodbury et al., 1995) relating to the adulteration of maize oil with rapeseed oil are shown in Fig. 8.8. Thus, when small amounts of rapeseed oil are added to maize oil, the influence on the isotope ratio of the palmitic acid in the blend is initially quite small [Fig. 8.8(a)] as rapeseed oil contains less palmitic acid than does maize oil. The opposite is true with oleic acid [Fig. 8.8(b)] since rapeseed oil has more oleic acid than does maize oil. Linoleic acid has a similar mixing curve [Fig. 8.8(c)] to that of palmitic acid. The use of oleic acid to detect adulteration of maize oil with rapeseed oil is thus the most sensitive, as this is the steepest of the curves in the region of interest (Woodbury et al., 1995). Other cases of adulteration may merit consideration of different mixing curves to get maximum sensitivity, depending on the fatty acid compositions of the components.

The collaborative work with Bristol University is continuing, and it is planned in the future to study the isotope ratios of the component sterols and tocopherols, and whether there is any difference in the isotope ratios of an acid located at the triglyceride 2-position in comparison with the same acid at the 1- and 3-positions. It may also be possible to develop views on whether there is any isotopic bias on the desaturation and/or chain-elongation mechanisms involved in the biosynthesis of palmitic, oleic and linoleic acids.

Acknowledgements

The author wishes to express thanks for experimental contributions made by Isobel Bently (a sandwich student), Iris Edgar and Ann Probert, who extracted the oils from the seeds, to M. J. Downes and M. A. Jordan, who made contributions in the early stages of the work, and to Richard Evershed of Bristol University for suggesting the joint project. Thanks are also due to the many industrial companies and other outside bodies who donated oilseed samples for the work.

References

Braunsdorf, R., Hener, U., Przibilla, G. et al. (1993) Analytische und Technologische Einflusse auf das $^{13}C/^{12}C$ Isotopenverhaltnis von Orangen oil-Komponenten. *Lebensmittelunters. und Forsch.*, **197**, 24–8.

Brause, A. R. and Raterman, J. M. (1982) Verification of authenticity of apple juice. *J. Assoc. off Analyt. Chem.*, **65** (4), 846.

Brause, A. R., Raterman, J. M., Petrus, D. R. and Doner, L.W. (1984) Verification of authenticity of orange juice *J. Assoc. Off. Analyt. Chem.*, **67** (3), 5359.

Bricout, J. and Koziet, J. (1987) Control of authenticity of orange juice by isotopic analysis *J. Agric. Food Chem.*, **35**, 758.

Codex (1983) *Codex Standard for Edible Fats and Oils – Supplement 1 to Codex Alimentarius Volume XI*. Codex Stan 21-1981.

Codex (1993) Codex Alimentarius Alinorm 95-17 – Report on the Fourteenth Session of the Codex Committee on Fats and Oils held London, September 1993.

Craig, H. (1957) *Geochim. cosmochim Acta*, **12**, 133.

Gaffney, J., Irsa, A., Friedman, L. and Emken, E. (1979). $^{13}C/^{12}C$ analysis of vegetable oils, starches, proteins, and soy–meat mixtures. *J. Agric. Food Chem.*, **27**, 475–8.

Guarino, P. A. (1982) Isolation of vanilla extract for stable carbon isotope analysis: interlaboratory study. *J. Assoc. off. Analyt. Chem.*, **65**, 835.

IUPAC (1987) *IUPAC Standard Methods for the Analysis of Oils, Fats and Derivatives*, 7th Edn, Blackwell Scientific Publications, Oxford and London.

Lee, H. S. and Wrolstad, R. E. (1988) Stable isotopic carbon composition of apples and their subfractions – juice, seeds, sugars and non-volatile acids. *J. Assoc. Off. Analyt. Chem.*, **71**, 795.

McGaw, B. A., Milne, E. and Duncan, G. J. (1988) A rapid method for the preparation of combustion samples for stable carbon isotope analysis by isotope ratio mass spectrometry. *Biomedical and Environmental Mass Spectrometry*, **16**, 269–73.

Pollard, A. M. (1993) Tales told by dry bones. *Chemy Ind.*, 359–62.

Rossell, J. B. (1994) Stable carbon isotope ratios in establishing maize oil purity. *Fat Sci. Technol.*, **96**, 304.

Rossell, J. B., King, B. and Downes, M. J. (1983) Detection of adulteration. *J. Am. Oil Chem. Soc.* **60**, 333.

Rossell, J. B., King, B. and Downes, M. J. (1985) Composition of oil. *J. Am. Oil Chem. Soc.* **62**, 221.

Schmid, E. R., Grundman, H. and Fogy, I. (1981) Determination of isotope ratios and their application to food analysis. *Ernahrung*, **10**, 459.

Sofer, Z. (1980) Preparation of carbon dioxide for stable isotope analysis of petroleum fractions. *Analyt. Chem.*, **52**, 1389–91.

Thompson, J. N. and Hatina, G. (1979) *J. Liquid Chromatogr.*, **2**, 327.

van der Merwe, N. J. (1992) New developments in archeological science, in *Proceedings of the British Academy*, **77**, 247–64.

Winkler, F. J. and Schmidt, H. L. (1980) Possible application of carbon-13 isotopic massspectrometry in food analysis. *Z. Lebensmittelunters. und Forsch.*, **171**, 85–94.

Woodbury, S. E., Evershed, R. P., Rossell, J. B. *et al.* (1995) Detection of vegetable oil adulteration using gas chromatography combustion/isotope ratio mass spectrometry. *Analytical Chemistry*, **67**, 2685.

9 Analysis of intact polar lipids by high-pressure liquid chromatography–mass spectrometry/tandem mass spectrometry with use of thermospray or atmospheric pressure ionization

A. Å. KARLSSON

9.1 Introduction

The separation of intact polar lipids by liquid chromatography (LC) and the subsequent detection by mass spectrometry (MS) has today become straightforward. LC–MS is no longer a sophisticated technique only in the hands of specialists. Today, it is a routinely used, although advanced, analytical technique. The fields of application are expanding and today the use of LC–MS with electrospray (ES) ionization grows at the expense of other ionization techniques, at least where analysis of intact polar lipids is concerned.

The ease with which LC–MS is used today was for many years only in people's dreams. Coupling a liquid chromatograph to a mass spectrometer involved several problems that needed to be solved. All these have now been solved in principle and one may say that the users of LC–MS today are restricted only by their imagination.

The content of this chapter focuses on the analysis of intact polar lipids by high-pressure liquid chromatography (HPLC) with flow or loop injection – and mass spectrometry (MS) or tandem mass spectrometry (MS–MS) using thermospray (TS), discharge-assisted TS [or 'plasmaspray' (PSP)], electrospray (ES) and atmospheric pressure chemical ionization (APCI). It was intended to include only those papers describing the analysis of intact polar lipids by liquid chromatography on-line with MS. However, many papers describe flow or loop injection with MS and/or the analysis of derivatives of polar lipids. These papers, describing excellent applications of MS and/or MS–MS of polar lipids, are included since this chapter would not have been complete without them.

The focus lies on the applications of LC–MS methods in the analyses of intact polar lipids rather than on technical aspects of liquid chromatography or mass spectrometry. There are several reviews and books on the theory of different mass spectrometric techniques (Constantin and Schnell, 1991; Desiderio, 1992; McLafferty, 1973; Niessen and van der Greef, 1992; Yergey *et al.*, 1990) and of MS of polar lipids (Adams and Ann, 1993; Gelpí, 1995; Jensen and Gross, 1988; Murphy, 1993).

Finally, basic knowledge of polar lipid structure, liquid chromatography and mass spectrometry is assumed.

9.2 Theory

9.2.1 Polar lipids

Phospholipids are major constituents of membranes in plants, animals and micro-organisms. In addition to their structural role, some phospholipids participate in various ways in biological processes.

Polyphosphoinositides are important in cellular signalling regulatory systems. Phospholipids also serve as a source of arachidonic acid and other polyunsaturated fatty acids that can be metabolized to biologically active eicosanoids. Sphingomyelins also participate in signal transduction events and are important constituents in nervous tissue and blood. Ceramides can act both as precursors in the biosynthesis and as intermediates in the degradation of sphingolipids. Many glycolipids are major components of plant cell membranes. They also participate in carbohydrate–carbohydrate interactions across opposing membrane surfaces. Neutral glycolipids can serve as cellular differentiation markers and as cell surface antigens. Bacterial glycolipids are also involved in pathogenic processes contributing to human diseases. The commercial use of polar lipids is increasing in fields such as biomembranes, skin-care formulations and drug delivery. Accordingly, sensitive methods for the determination of polar lipids are of great importance and interest in many scientific fields.

The nomenclature of polar lipids is often confusing. In this chapter polar lipids with glycerol as the molecular backbone are referred to as phospholipids, and those derived from sphingosine – sphingolipids, and all lipids linked to any type of carbohydrate moiety – as glycolipids. Thus, in this chapter the term 'polar lipids' refers to phospholipids, sphingolipids and glycolipids.

Phospholipids. Phospholipids are fats in which glycerol is esterified with two fatty acids and a phosphoric acid. Such monophosphate esters are called phosphatidic acids (PAs). Most naturally occurring phospholipids contain one saturated and one unsaturated fatty acid. Free PAs are rare in nature. If found one may suspect them to be a degradation product from the hydrolysis of the more common phospholipids. The most common phospholipids are: phosphatidylcholine (PC), phosphatidylethanolamine (PE), phosphatidylglycerol (PG), phosphatidylinositol (PI) and phosphatidylserine (PS) (Fig. 9.1).

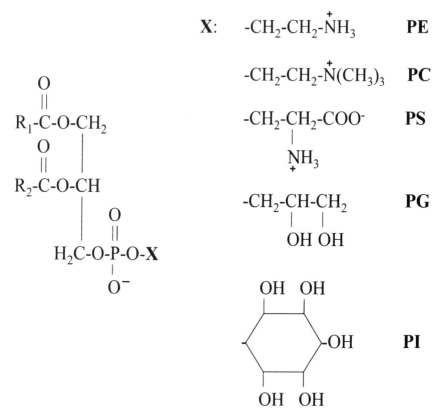

Figure 9.1 Structure of the most common phospholipids. PA = phosphatidic acid; PE = phosphatidylethanolamine; PC = phosphatidylcholine; PS = phosphatidylserine; PG = phosphatidylglycerol; PI = phosphatidylinositol.

Sphingomyelin. Sphingomyelin is a common member of the sphingolipid family. It is derived from sphingosine to which a fatty acid is linked by an amide bond. In addition, the primary hydroxyl group of sphingosine is esterified to phosphoryl choline (Fig. 9.2).

Glycolipids. These contain a carbohydrate moiety and may have either glycerol or sphingosine as the molecular backbone (Fig. 9.3).

9.2.2 Liquid chromatography

Normal-phase chromatography is usually employed when separating a lipid extract into polar lipid classes. Columns packed with silica, diol-modified silica, polyvinylalcohol (PVA)–silica or other materials have been successful in separating all kinds of polar lipids (Arnoldsson and Kaufman, 1994;

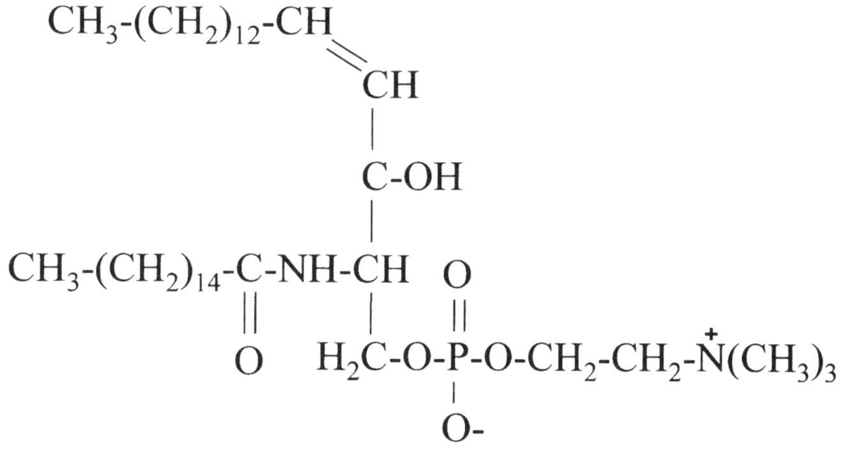

ANALYSIS OF INTACT POLAR LIPIDS 293

$$CH_3\text{-}(CH_2)_{12}\text{-}CH=CH$$
$$|$$
$$C\text{-}OH$$
$$|$$
$$CH_3\text{-}(CH_2)_{14}\text{-}\underset{\underset{O}{\|}}{C}\text{-}NH\text{-}CH$$
$$|$$
$$H_2C\text{-}O\text{-}\underset{\underset{O^-}{|}}{\overset{\overset{O}{\|}}{P}}\text{-}O\text{-}CH_2\text{-}CH_2\text{-}\overset{+}{N}(CH_3)_3$$

Figure 9.2 Structure of *d*-18:1, 16:0-sphingomyelin.

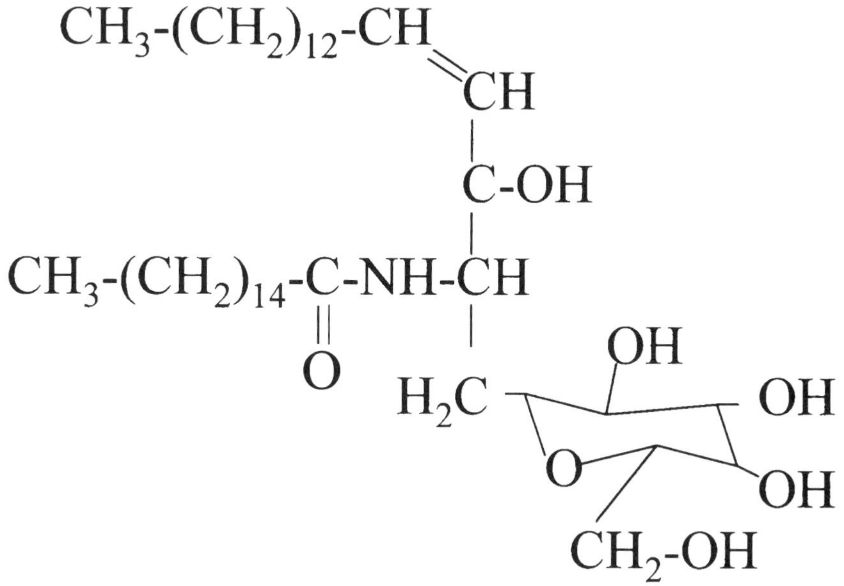

Figure 9.3 Structure of a galactosyl ceramide.

Karlsson *et al.*, 1996; Valeur, Michélsen and Odham, 1996). Reversed-phase chromatography (using C18 columns) is employed when separation into molecular species of one class is desired. It can also be used for class separation even though separation of some molecular species is also obtained. Usually, a binary gradient of two solvent mixtures are used, ranging in analysis time from a few minutes to an hour.

9.2.3 Mass spectrometry

Thermospray and plasmaspray. Thermospray (TS) requires that the sample contain a volatile electrolyte such as ammonium acetate. It also requires heat in order to evaporate the solvent entering the mass spectrometer. The concentration of the electrolyte is approximately 0.01–0.1 M. The LC effluent is introduced into the mass spectrometer through a heated stainless-steel capillary tube. At the entrance to the heated evacuated chamber of the MS a supersonic jet of fine droplets is formed. When the solvent evaporates the analyte molecules are surrounded by a sheath of electrolyte, generating high local electrical fields. This results in a fairly soft ionization of the analyte. The ions are admitted into the analyser of the mass spectrometer through a sampling cone. Often a repeller electrode is placed in the ionization chamber opposite the sampling cone to improve ion extraction efficiency. One disadvantage with TS is that no hydrophobic solvent can be used.

A special version of thermospray is the discharge-assisted thermospray or 'plasmaspray' (PSP). A plasma discharge electrode is then placed in the ionization chamber and the probe is maintained at a potential of −600 V relative to the discharge electrode. The ionization method resembles that of chemical ionization and no volatile electrolyte is needed. The ionization is 'harder' compared with true thermospray but a higher total ion current is often obtained. Since no electrolyte is needed an extended range of solvents can be used, making it possible to use highly hydrophobic solvents. No molecular weight derived ions are usually obtained using a stainless-steel capillary tube. By using a laser-drilled sapphire tip, $[M + 1]^+$ ions are obtained in PSP (Valeur, 1992). The term 'plasmaspray' is used instead of discharge-assisted thermospray throughout the rest of this chapter.

Electrospray. In contrast to TS and PSP, electrospray (ES) is a low-temperature, atmospheric-pressure ionization (API) process. The effluent from the LC enters the atmospheric pressure region of the MS where a strong electrostatic field is created between the LC probe tip and a counter electrode. This produces a mist of fine charged droplets, which can be facilitated by a nebulizer gas (nitrogen). The solvents of the droplets evaporate and the charge of each droplet increases rapidly and charged sample ions are formed. The ions are then transferred into the analyser. Electrospray ionization is a much softer process compared with TS and PSP, making it easier to obtain molecular ions of intact polar lipids. Modern API sources accommodate flow rates up to 2000 µl min^{-1}.

Atmospheric pressure chemical ionization. Atmospheric pressure chemical ionization (APCI) is based on chemical ionization by ion-molecule or electron capture reactions that are carried out in an ion source operating at atmospheric pressure. Today, a heated nebulizer interface is used in combination with an API source. The heated solvent and analyte mixture is introduced in the source, where the chemical reaction takes place. The chemical ionization reactions are usually initiated by means of electrons produced by a corona discharge. With the development of the modern API sources there has been a renewed interest in the use of APCI.

9.2.4 Liquid chromatography–mass spectrometry

Many papers today still describe MS analysis of intact polar lipids with use of flow or loop injection without the subsequent separation by liquid chromatography. It is of course possible to monitor specific ions and perform MS or MS–MS analyses on a biological sample containing polar lipids. However, since molecular species of different phospholipids may overlap in molecular weight, flow or loop injection is a risky business that should be avoided when looking for classes and/or molecular species of polar lipids! By choosing a separation method that clearly separates polar lipid classes the tedious work of interpreting spectra is greatly facilitated and molecular species determination can then be accomplished by mass spectrometry. Concentration can then be focused directly on molecular species identification and/or comparison of the molecular species composition of different classes. Thus, if a detailed determination of molecular species composition is desired, a second separation into molecular species is not necessary when using HPLC.

In addition, co-eluting matrix compounds are often present in a sample from biological material. This will most likely suppress the total ion current formed in the ionization chamber, thus lowering the sensitivity of the method of analysis (Buhrman, Price and Rudewicz, 1996; Chan, 1996; Knebel, Sharp and Madigan, 1995). By using chromatography, ion suppression is reduced. Furthermore, unexpected compounds, metabolites and/or artefacts may be easier to identify with HPLC–MS. With flow or loop injection, these compounds may be overlooked with loss of (important) data as a result.

Many mass spectrometers today can work in Windows or Macintosh environments. The development of computer hardware and software over the years has made data handling and processing fast and easy. It is now common practice to produce high-quality copies of spectra collected during an analysis run to be copied and pasted into a document while still performing the analysis. This was an impossible task less than ten years ago when one had to wait for the end of analysis before processing any data obtained from that analysis.

9.3 Applications

9.3.1 Thermospray and plasmaspray

The papers cited in this section describe the analysis of intact polar lipids and different derivatives thereof. The analysis of derivatives is included to present a more complete picture of compounds that can be analysed by HPLC–TS (or PSP)–MS (or MS–MS). Several papers describe a chromatographic separation of only a few phospholipids, applicable to a specific area of interest, for example the determination of aminophospholipids (Hullin, Kim and Salem, 1989), molecular species determination and the determination of positional isomers of fatty acyl groups in the sn-1 and sn-2 positions (Kuypers, Bütikofer and Shackleton, 1991), identification of molecular species already quantified by HPLC–ultra violet (UV) detection (Bütikofer et al., 1990) and the determination of methyl ester derivatives of glycolipids from virulent and avirulent strains of bacteria (Ioneda et al., 1993). Others describe the separation of many, or all major, phospholipids, including sphingomyelin, and apply the method(s) to specific problems. Kim and Salem (1986, 1987) have described several methods of analysis of phospholipids. Kim, Yergey and Salem (1987) analysed phospholipids, including hydrogenated mono- and digalactosyldiglycerides (MGDG and DGDG, respectively) from plants. Furthermore, the molecular species composition of phospholipids (PS, PI, PE and PC) in rat brain have been determined by Ma and Kim (1995). Odham et al. (1988) have worked with HPLC–PSP–MS in the determination of phospholipids, including polyphosphoinositides [phosphatidylinositol mono- and bisphosphate (PIP and PIP-2, respectively)]. Valeur and co-workers (Valeur, Michélsen and Odham, 1993; Valeur et al., 1994) continued working with HPLC–PSP–MS, determining phospholipids from *Pseudomonas fluorescens*, and quantifying and characterizing of sphingomyelins from bovine brain, bovine milk and chicken egg yolk. The limit of detection (LOD) reported is usually in the low nanogramme range for TS–MS in single-ion monitoring (SIM) mode. LODs in the high picogramme range using PSP–MS (SIM) have been reported (Kim and Salem, 1986; Kim, Yergey and Salem, 1987; Valeur, Michélsen and Odham, 1993).

Papers describing the use of high-pressure liquid chromatography with thermospray and (tandem) mass spectrometry. Kim and Salem (1986, 1987) wrote two of the earliest papers describing the use of filament-on TS (similar to chemical ionization) in the analysis of intact polar lipids. Reversed-phase chromatography was used for molecular species separation of several classes of lipids. For PE, PC, PI, PS, SM and platelet activating factor molecular species separation was achieved within 20, 32, 4, 8, 16 and 3 min, respectively. Molecular ions were detected for all lipids except PI, and the diglyceride ions (for sphingomyelin the ceramide ions) were the base peaks in the spectra for all lipids (Fig. 9.4), monoglyceride ions (for sphingomyelin the fatty acid amide ions) as well as ions derived from the polar head-groups were also detected.

Hydrogenated samples of MGDG and DGDG were analysed by Kim, Yergey and Salem (1987). The classes were separated within 3 min, and

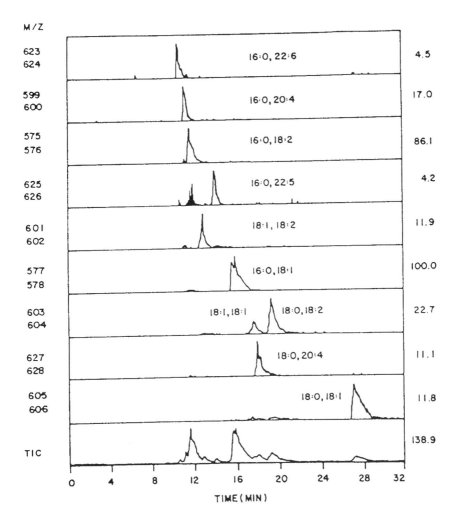

Figure 9.4 High-pressure liquid chromatography separation of 50 μg of a natural phosphatidylcholine mixture from egg yolk. The reconstructed ion chromatograms of diglyceride ions were selected from data acquired by full mass scanning from 120 amu to 820 amu. The relative intensity is shown based on the peak height. Column: 3 μm Ultrasphere-ODS (4.6 mm × 7.5 cm). Mobile phase: McOH/hexane/0.1 M NH_4OAc (71.5.7), 1 ml min^{-1}. Reprinted with permission from Kim, H. Y. and Salem, N. Jr, Phospholipid molecular species analysis by thermospray liquid chromatography/mass spectrometry, *Anal. Chem.*, **58** (1), 9–14, 1986.

mono- and diglyceride ions were detected as well as the sodium adduct molecular ions (Fig. 9.5).

Odham *et al.* (1988) compared plasmaspray with thermospray by using ammonium acetate as a buffer, where PSP required no addition of buffer prior to ionization. The total ion current obtained in PSP was significantly higher than previously reported for TS (filament-on). No or very few molecular ions were observed as a result of the harder ionization in PSP compared with TS. Diacylglycerol- and monoacylglycerol-derived fragments of phospholipids were obtained. Cation-exchange HPLC separated PI and PE in a phospholipid extract from bacterial cells (*Pseudomonas fluorescens*). Furthermore, PC, PE, PA, PS, cardiolipin (CL) and polyphosphoinositides (PIP and PIP-2) were also studied.

Trinitrobenzene derivatives of PE and PS from human red blood cells and rat brain were analysed by Hullin, Kim and Salem (1989). For quantification purposes, reversed-phase HPLC with UV detection was used. HPLC–TS (filament-on) was used for confirmation of molecular species separation.

Reversed-phase HPLC–TSP–MS was used by Bütikofer *et al.* (1990) for the determination of glycerobenzoate derivatives of diradylglycerols of phospholipids. Neither intact phospholipids nor classes of polar lipids were determined. The PI moiety of the glycosylphosphatidylinositol (GPI)

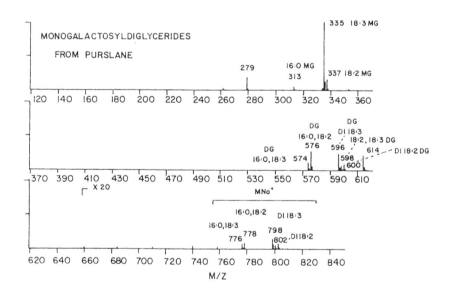

Figure 9.5 Positive ion spectrum of a purslane extract containing monogalactosyldiglycerides. Reprinted from *J. Chromatogr.*, **394** (1), Kim, H. Y., Yergey, J. A. and Salem, N. J., Determination of eicosanoids, phospholipids and related compounds by thermospray liquid chromatography–mass spectrometry, pp. 155–70, 1987, with kind permission of Elsevier Science – NL, Sara Burgerhartstraat 25, 1055 KV Amsterdam, The Netherlands.

anchor of *Torpedo marmorata* (flounder) was compared with that of the membrane, and the other major phospholipid classes of *Torpedo marmorata* electrocytes were also studied. Quantification by HPLC with UV detection was followed by identification of individual molecular species by TS–MS. Bütikofer *et al.* found that the composition of the molecular species varied significantly between the PI moiety of the GPI anchor and the PI from electrocyte membranes. They also discussed the biophysical properties of glycerophospholipids in membranes in terms of their dependence on the type of linkage of the side chains to the glycerol backbone, on the fatty acyl chain structure and on the positional distribution of the side chains. They stressed that detailed structural knowledge is essential for assessment of the behaviour of polar lipids in the membrane, and that only part of this information is obtained by determination of the total fatty acyl composition. [Koshy and Boggs (1996) discussed the formation of the compacted myelin membrane in the central nervous system myelin.]

HPLC–TS–MS was used by Kuypers, Bütikofer and Shackleton (1991) for the determination of glycerobenzoate derivatives of diradylglycerols of phospholipids; they did not determine intact phospholipids or classes of polar lipids. Quantification was by HPLC with UV detection, with subsequent identification of individual molecular species by TS. They describe the possibility of quantification of molecular species as well as positional isomers of individual molecular species by HPLC–TS–MS. They also discussed the possibility of determining positional isomers, that is the sn-1 or sn-2 position of a fatty acyl group. The method was applied to lipids from human erythrocytes.

Valeur, Michélsen and Odham (1993) showed how diacylglycerol- and monoacylglycerol-derived fragments of phospholipids were obtained. PG and PE were found in a phospholipid extract from bacterial cells (*Pseudomonas fluorescens*). They achieved class separation of the phospholipids but did not apparently separate molecular species (Fig. 9.6). Collision-induced dissociation (CID)–MS–MS analyses of PI from soybean were performed which revealed specific fatty acid compositions of the selected diacylglycerol-derived fragment.

Valeur *et al.* (1994) elucidated the principal cleavage pathways of sphingomyelin in positive ion PSP. Sphingomyelin from bovine milk, bovine brain and chicken egg yolk were analysed by reversed-phase HPLC–PSP–MS. Spectra of individual molecular species showed prominent ceramide-derived fragment ions as well as ions of long-chain base (LCB) and carboxylic acid origin of the ceramide unit.

Phospholipid molecular species from rat brain were determined by Ma and Kim (1995). Several separations of phospholipids are described in which not only class separation but also some molecular species separation were obtained. For example, PI, PE and PE plasmalogen (PE.e) overlapped, making straightforward identification somewhat difficult. Ma and Kim

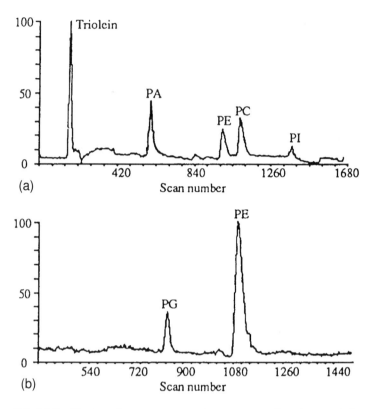

Figure 9.6 (a) Positive ion plasmaspray mass chromatogram of a mixture of glycerolipids consisting of: triolein, 1,2-dipalmitoyl-*sn*-glycero-3-phosphate (PA), 1,2-dioleoyl-*sn*-glycero-3-phosphoethanolamine (PE), (1,2-distearoyl-*sn*-glycero-3-phosphocholine (PC) and phosphatydylinositol (PI) from soybean. (b) Positive ion plasmaspray mass chromatogram of a total Bligh and Dyer lipid extract from *Pseudomonas fluorescens* cells showing two peaks containing phosphatidylglycerol (PG) and PE species. Reprinted with permission from Valeur, A., Michélsen, P. and Odham, G., Online straight-phase liquid chromatography/plasmaspray tandem mass spectrometry of glycerolipids, *Lipids*, **28** (3), 255–9, 1993.

report significant differences in the molecular species composition of each phospholipid class in rat whole-brain lipid.

Papers describing flow or loop injection with thermospray and (tandem) mass spectrometry. The diethylether soluble lipids from two strains of *Nocardia asteroides* (mycolic-acid-containing bacteria) were analysed in terms of composition and toxicity by Ioneda *et al.* (1993). Glycolipids from GUH-2 were highly toxic to mice. TSP–MS was used to determine the molecular weight of the methyl ester derivatives of the mycolic acid moiety (C_{50}–C_{56}) of glycolipids. Several chromatographic separation techniques were used (TLC, paper chromatography and column chromatography) prior to analyses by TSP–MS.

9.3.2 Electrospray – atmospheric pressure chemical ionization

The development of so-called high-flow, pneumatically-assisted, combined APCI–ES devices that allow flow rates up to $1000\,\mu l\,min^{-1}$ have made API techniques the fastest growing group of mass spectrometric ionization techniques today (Desiderio, 1992). ES is now replacing the widely used TS technique. Solvent mixtures for both reversed-phase and normal-phase HPLC are now easily used with ES–MS. Hydrophobic solvents, highly viscous solvents and solvent mixtures containing bases and/or acids as mobile phase 'modifiers' or volatile buffers do not present any real problem today (although there is still the limitation associated with using involatile mobile phase buffers). In our laboratory we have used a solvent mixture composed of hexane, 1-propanol, butanol, tetrahydrofuran, iso-octane, water and ammonium acetate for the analysis of phospholipid and glycolipid extracts. Spectra of all compounds previously detected by HPLC with evaporative light-scattering detection were collected. A higher source temperature than routinely used (160°C instead of 120°C) was needed for proper ionization of this highly viscous solvent mixture. The increase in temperature was the only necessary parameter change compared with our routinely used solvent system of hexane, 1-propanol, water, formic acid and triethylamine.

The authors of the papers cited below describe different applications of both ES and APCI mass spectrometry. The LODs reported in these papers differ and are also hard to compare because of different ways of acquiring mass spectral data. Some report LODs as the smallest total amount detected, for example picogramme or femtomol, whereas others express LODs as the concentration injected ($\mu g\,\mu l^{-1}$). Furthermore, the solvent flow rates differ substantially, from $1\,\mu l\,min^{-1}$ to $1000\,\mu l\,min^{-1}$. Finally, both the mass span over which data are collected and the total time of acquiring the data differ considerably. Typically, at a solvent flow rate of $400\,\mu l\,min^{-1}$ and using gradient LC separation (20 µl loop) with mass spectrometric detection in SIM mode, a LOD of 5–10 $fmol\,\mu l^{-1}$ is achieved. Use of very low solvent flow rates ($1\,\mu l\,min^{-1}$) and the acquisition of data over, say, 3 min may result in lower LODs. This approach is very useful when the amount of sample is limited.

The papers cited below describe methods of analysis with or without specific applications. Smith, Snyder and Harden (1995) and Sweetman *et al.* (1996) describe the possibility of bacterial profiling using ES. Kim, Wang and Ma (1994), Karlsson *et al.* (1996), Kerwin, Tuininga and Ericsson (1994) and Han and Gross (1995) describe general methods of analyses of phospholipids. Han *et al.* (1996) elucidate the storage depot of arachidonic acid in human platelets, Han and Gross (1994) analysed human erythrocyte plasma membrane phospholipids and Kerwin *et al.* (1995) analysed phospholipids from fungal mycelium. Sphingomyelin structure is discussed by

Kwon et al. (1996), Kerwin, Tuininga and Ericsson (1994), Karlsson et al. (1997) and Han and Gross (1995). Lipooligosaccharides are analysed by Gibson et al. (1996) and Cole (1996). Polyphosphoinositides are determined by Michélsen, Jergil and Odham (1995), whereas Ii et al. (1993), Reinhold et al. (1994) and Haigh et al. (1996) analysed, respectively, galactosylceramides, glycolipids and diacylglyceryl-N,N,N-trimethylhomoserine. Koshy and Boggs (1996) contribute an interesting discussion on membrane structure and stability from results obtained when analysing the complexation of galactosylceramide and cerebroside sulphate.

Papers describing the use of high-pressure liquid chromatography with atmospheric pressure ionization and (tandem) mass spectrometry. Kim, Wang and Ma (1994) used HPLC to separate the phospholipids in the order PS, PI, PA, PE and PC containing a given fatty acyl composition (Fig. 9.7). Some chromatographic overlapping of classes occurred, especially for PS and PI. In general, PC species were detected with the greatest sensitivity, followed by PE. In positive ion mode under the conditions used the sensitivity of PS detection was approximately 20 times less than that of PC detection. Kim, Wang and Ma report a sensitivity in the positive ion SIM ($[M + H]^+$) mode in the 0.5–1 pmol range for PC – 20–50 times more sensitive than LC–TS–MS which generates diglyceride fragments as the major ions. In-source fragmentation was induced by increasing the sampling cone voltage (Fig. 9.8). When increasing the voltage, the intensity of the diacylglyceride fragment ions increased while at the same time the molecular ion intensity decreased (Karlsson et al., 1996; discussed next).

Karlsson et al. (1996) used normal-phase HPLC with ES–MS to detect PG, PE, PC, PS and PI within 22 min but there was some overlapping of PS and PI. Molecular ions, diacylglyceride ions and fatty acid ions of all phospholipids were obtained in a single run (Fig. 9.9). Molecular ions and fatty acid fragment ions were obtained in negative ion mode at low (60 V) and high (110 V) cone voltages, respectively. Diacylglyceride ions were obtained in positive ion mode at a high cone voltage (110 V). Thus, at a high cone voltage, in-source fragmentation is obtained obviating the need for MS–MS in most cases unless the specific fatty acid composition of a certain molecule is desired. The spectra from each of the modes described above can be combined as in Fig. 9.10. The method was applied to a human gastric juice sample. PI molecular species from the sample were subjected to daughter ion MS–MS for the determination of species fatty acid combinations. Also, parent ion MS–MS of the in-source induced 241 Da e^{-1} (mass:charge ratio) (PI-specific) fragment was performed, confirming the identity of the PI molecular ions previously obtained in MS mode.

Both reversed-phase HPLC–ES–MS and flow injection ES–MS were used by Han et al. (1996) for the investigation of the alterations in individual molecular species of human platelet phospholipids during thrombin stimula-

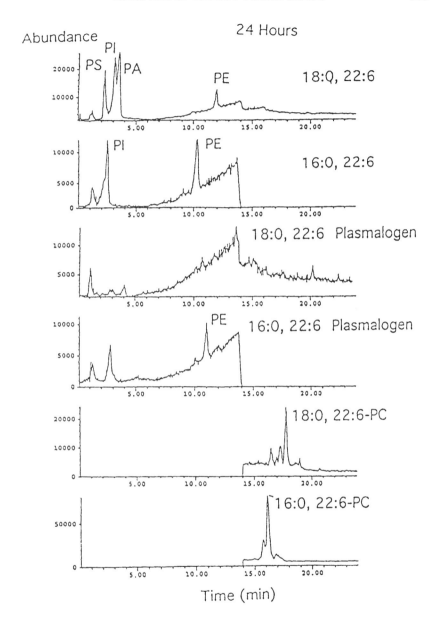

Figure 9.7 Ion chromatograms of 22:6n3 containing phospholipids obtained after incubation of C-6 glioma cells with 100 μM 22:6n3 for 24 h. Reprinted with permission from Kim, H. Y., Wang, T. C. L. and Ma, Y. C., Liquid chromatography/mass spectrometry of phospholipids using electrospray ionization, *Anal. Chem.*, **66** (22), 3977–82, 1994.

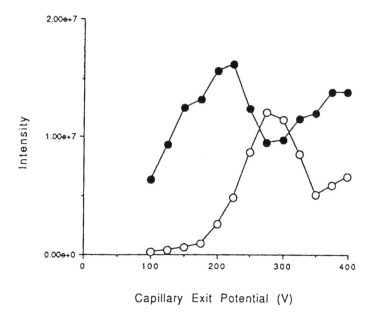

Figure 9.8 Effect of capillary exit voltage on diglyceride (DG) ion production. Approximately 125 pmol of 18:0, 22:6-PE was injected by the flow injection technique. (Compare with Figure 9.9.) –O–, DG, m/z 651, –●–, [M+H]$^+$, m/z 792. Reprinted with permission from Kim, H. Y., Wang, T. C. L. and Ma, Y. C., Liquid chromatography/mass spectrometry of phospholipids using electrospray ionization, *Anal. Chem.*, **66** (22), 3977–82, 1994.

tion. The plasmenyl–PE molecular species were found to be the largest endogenous storage depot of arachidonic acid in resting human platelets as well as the major source of arachidonic acid mobilized after thrombin stimulation of human platelets. Han *et al.* presented detailed tables of alterations in PC, PE, PS and PI molecular species during thrombin stimulation of human platelets. For example, 30 different molecular species of PE, both plasmenyl-PE and diacyl-PE, were quantified from 10^9 human platelets. This paper shows the excellent capability of ES–MS to determine the molecular species composition of a given phospholipid.

Kwon *et al.* (1996) used normal-phase HPLC to separate and quantify sphingomyelin (SPH). Flow injection ES–MS was then used to determine the SPH molecular species composition of isolated rat pancreatic islets. Kwon *et al.* concluded that sphingomyelin hydrolysis is not involved in the signalling pathway whereby cytokine interleukin-1 induces the overproduction of nitric oxide by pancreatic islets. They reported four molecular species of SPH: SPH (16:0/d-18:1), (18:0/d-18:1), (22:0/d-18:1) and (24:0/d-18:1). They assumed that sphingosine (18:1) is the LCB; however, studies of SPH from bovine brain (Jungalwala, Evans and McCluer, 1984), bovine milk (Morrison, 1969) and human plasma (Sweeley, 1963), show that there exist several LCB species, saturated as well as unsaturated, with one or two hydroxyl

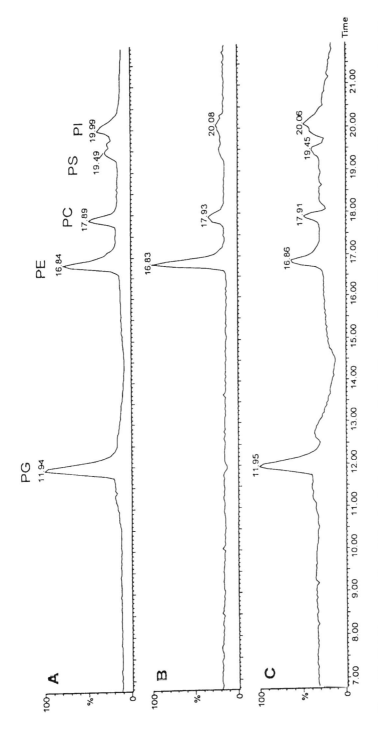

Figure 9.9 Total ion current from: (A) molecular ion region (negative ion ES, cone voltage 110 V); (B) diacylglyceride related fragment ion region (positive ion ES, cone voltage 60 V); (C) fatty acid fragment ion region (negative ion ES, cone voltage 110 V). (Compare with Fig. 9.8.) Reprinted with permission from Karlsson, A. Å., Michélsen, P., Lasson, Å. and Odham, G., Normal-phase liquid chromatography class separation and species determination of phospholipids utilizing electrospray mass spectrometry/tandem mass spectrometry, *Rapid Commun. Mass Spectrom.* **10**(7), 775–80, Copyright John Wiley & Sons Limited, 1996.

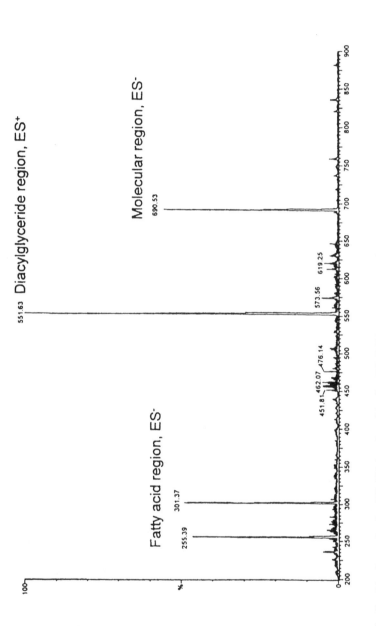

Figure 9.10 Combined negative ion ES (ES⁻) and positive ion ES (ES⁺) mass spectra of PE-di 16:0, obtained from Fig. 9.9. The peak at m/z 301.37 represents a fatty acid adduct ion. Reprinted from Karlsson, A. Å., Michélsen, P., Larsen, Å. and Odham, G., Normal-phase liquid chromatography class separation and species determination of phospholipids utilizing electrospray mass spectrometry/tandem mass spectrometry, *Rapid Commun. Mass Spectrom.*, **10**(7), 775–80. Copyright John Wiley & Sons Limited, 1996.

groups. It may be assumed that there are more LCBs in SPH from rat pancreatic islets than the 18:1-LCB reported here. Similar comments are made by Kerwin, Tuininga and Ericsson (1994), Han and Gross (1995) and Karlsson et al. (1997; discussed next).

Karlsson et al. (1997) characterized SPH from bovine milk by HPLC–ES–MS (positive ion mode), obtaining at least 36 different molecular ions. By using HPLC–APCI–MS (positive ion mode), they obtained molecular as well as ceramide fragment ions, although with lower total ion current than in ES–MS. HPLC–APCI–MS–MS (positive ion mode) monitoring of the ceramide units made it possible to generate fragment ions corresponding to the LCB and the N-acylated fatty acid (FA) of the ceramide unit (Fig. 9.11). Thus, specific LCB–FA compositions of SPH molecular species were obtained. Of the 11 molecular ions investigated at least 26 different LCB–FA combinations were detected. It was not possible to generate the LCB or FA ions by means of ES, high-cone-voltage CID–MS, or MS–MS. [More on SPH may be found in Kwon et al. (1996), discussed above, and in Kerwin, Tuininga and Ericsson (1994) and Han and Gross (1995), both discussed below.]

Papers describing flow or loop injection with atmospheric pressure ionization and (tandem) mass spectroscopy. In a short communication on ES–MS of the N-palmitoylgalactosylceramide Ii et al. (1993) focus on adduct ion formation with use of methanol/chloroform in negative ion mode.

A structural determination and quantitative analysis of individual phospholipid molecular species from human erythrocyte plasma membrane phospholipids, mainly PE and PC, were made by Han and Gross (1994). Injection was made directly from chloroform extracts of biological samples. More than 50 human erythrocyte plasma membrane phospholipid constituents were identified from the equivalent of less than 1 µl of whole blood (Fig. 9.12).

A thorough treatment of fragmentation patterns for different phospholipids with use of both positive and negative ion mode ES is given by Kerwin, Tuininga and Ericsson (1994) . Tables on molecular species composition for PC, PE, PI and PS, including SPH from bovine brain, are given. Kerwin, Tuininga and Ericsson report sensitivities in the low picogramme range: 'Over 20 parent ions were obtained using 0.1 pg of PE, but minor species were lost'. However, the accuracy of the molecular species composition of SPH may be questioned since no peaks characteristic of the N-acyl and sphingenine/sphinganine were generated. There may be more molecular species of SPH than reported since the sphingoid base may also vary. The inability of ES to generate these fragments is reported by Han and Gross (1995), discussed below and by Karlsson et al. (1997) and Kwon et al. (1996), discussed in the previous subsection.

Reinhold et al. (1994) applied ES–MS to methylated glycosphingolipid samples from bovine brain. Several molecular ions as well as structural

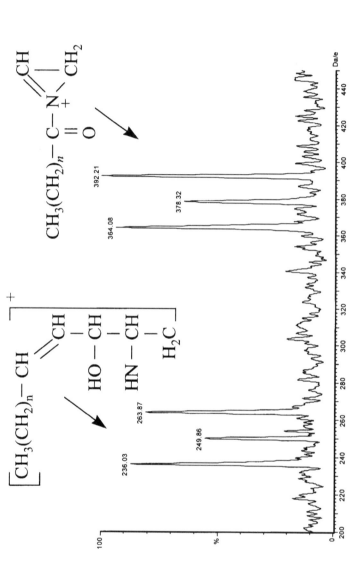

Figure 9.11 Daughter ions (obtained in positive ion mode atmospheric pressure chemical ionization–tandem mass spectrometry) of a ceramide fragment ion ($m/z = 604$) from sphingomyelin. Ions of m/z 236, 250 and 264 represent long-chain bases with 16, 17 and 18 carbon atoms, respectively. Ions of m/z 364, 378 and 392 represent fatty acids with 22, 23 and 24 carbon atoms, respectively. Reproduced from Karlsson, A. Å., Michelsen, P., Westerdahl, G. and Odham, G., Determination of molecular species of sphingomyelin in bovine milk using HPLC–MS–MS with electrospray ionization and atmospheric pressure chemical ionisation (unpublished manuscript).

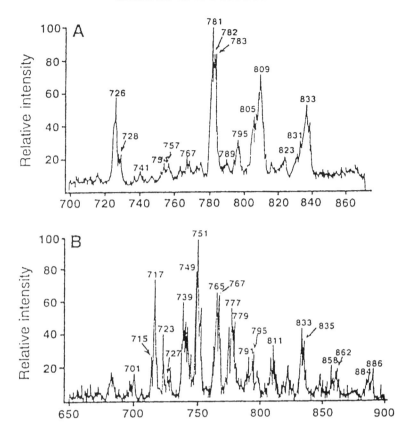

Figure 9.12 Direct electrospray ionization–mass spectrometry analysis of human erythrocyte plasma membrane phospholipids: (A) A positive-ion electrospray ionization (ESI) mass spectrum of erythrocyte plasma membrane phospholipid extract showing 14 molecular species of glycerophospholipids and 4 molecular species of sphingomyelin; (B) A negative-ion ESI mass spectrum of the same extract of plasma membrane phospholipids showing more than 25 molecular species of ethanolamine glycerophospholipids and 8 molecular species of serine and inositol glycerophospholipids. Reprinted with permission from Han, X. and Gross, R. W., Electrospray ionization mass spectroscopic analysis of human erythrocyte plasma membrane phospholipids, *Proc. Natl. Acad. Sci. USA*, **91**(22), 10635–9. Copyright (1994) National Academy of Sciences, USA.

details of major components were obtained. Low-energy CID provided higher product ion abundance than did high-energy CID. Furthermore, much of the ion current in high-energy CID appeared as small mass losses from the parent ion.

Han and Gross (1995) described methods of analysis for phospholipids and showed the excellent potential of ES–MS and ES–MS–MS for phospholipid analysis, giving fragmentation patterns for different phospholipids, particularly for PE, PC, SPH and CL with use of positive and negative ionization ES. Several proposed CID pathways of the phospholipids after electrospray ionization are noted. Also, CL is accounted for, which is not often done. Among other things, Han and Gross conclude that there is insufficient accessible collision energy in the quadropole system to dissociate the ceramide unit of the SPH molecule. Further descriptions of the mass spectrometry of SPHs can be found in Kerwin, Tuininga and Ericsson (1994), Kwon et al. (1996) and Karlsson et al. (1997), all discussed above.

ES–MS and ES–MS–MS were used by Kerwin et al. (1995) to examine molecular species of PC, PE and PI from fungal mycelium, and nuclei grown with and without isoprenoids which induce or do not induce reproduction.

Michélsen, Jergil and Odham (1995) described ES–MS and ES–MS–MS (negative ion mode) of polyphosphoinositides, PIP and PIP-2, showing singly and doubly charged deprotonated molecular ions. ES–MS–MS of the molecular ions showed fatty acid fragment ions.

Smith, Snyder and Harden (1995) illustrated the profiles of phospholipids (PA, PE, plasmenyl-PE, PC, PG, PS and PI) in extracts of four different bacterial species. They identified both class and molecular weight of the two fatty acyl moieties. They reported different profiles for each bacterium and suggested the possibility of detection and identification of bacteria by ES–MS.

Lipid A from lipooligosaccharides (LOSs) was analysed by means of ES–MS in both positive and negative ion mode by Cole (1996). Singly and doubly deprotonated molecules of the various lipid A forms were obtained, producing a profile of a crude lipid A extract.

ES–MS and ES–MS–MS (negative ion mode) were used by Gibson et al. (1996) for the structural determination of LOSs. Both O-deacylated and intact LOS were analysed. Mass spectra from O-deacylated LOS were more readily interpretable than mass spectra obtained from intact LOS. The paper is focused on the heterogeneity of the oligosaccharide moiety rather than on the fatty acid composition of the lipid A region.

ES–MS and ES–MS–MS (positive and negative mode) were used by Haigh et al. (1996) in conjunction with NMR to determine the identity of diacylglyceryl-N,N,N-trimethylhomoserine (DGTS) in the marine alga *Chlorella minutissima*. DGTS is a widely distributed lipid in ferns, mosses, liverworts, green algae, fungi and protozoans.

Koshy and Boggs (1996) use ES–MS for the investigation of the complexation–oligomerization (Fig. 9.13) of galactosyl ceramide (GalCer) and cerebroside I^3 sulphate (CBS) in the presence of calcium (Ca^{2+}). By altering the declustering potential Koshy and Boggs were able to show the stability of several oligomers, both homotypic and heterotypic. The heterotypic dimer had greater stability than any other complex. Since GalCer and CBS are present in high concentrations in myelin, Koshy and Boggs concluded that the Ca^{2+}-mediated carbohydrate–carbohydrate interaction, which can bridge opposing bilayers, may be involved in adhesion of the extracellular

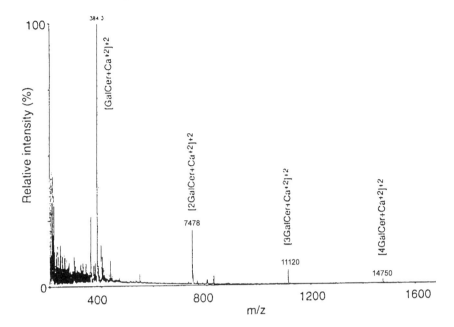

Figure 9.13 Positive ion spray mass spectrum of galactosyl ceramide (Galcer) (50 nmol ml^{-1}) with Ca^{2+} (500 nmol ml^{-1}) in methanol. The declustering potential was 80 V. Reproduced with permission from Koshy, K. M. and Boggs, J. M., Investigation of the calcium-mediated association between the carbohydrate head groups of galactosylceramide and galactosylceramide I^3 sulphate by electrospray ionization mass spectrometry, *J. Biol. Chem.*, **271**(7), 3496–9, 1996.

surfaces of the myelin sheath. Thus ES–MS is used for the understanding of membrane structure and stability [Bütikofer *et al.* (1990; Section 9.3.1) have also discussed biophysical properties of glycerophospholipids in membranes.]

Flow injection ES–MS of a lipid extract of *Escherichia coli* is described in a microcorrespondence article by Sweetman *et al.* (1996). In-source fragmentation was used to fragment molecular ions into their constituent groups. This enabled the identification of six fatty acid chains and two ions corresponding to PG and PE head-groups.

9.4 Practical experiences of analysis of polar lipids by means of liquid chromatography with (tandem) mass spectrometry

In this section I will describe some aspects of my experiences with LC–MS analysis of intact polar lipids over the years. My co-workers and I have found the following knowledge vital for performing the analysis described in this chapter, but have rarely seen these points in print.

9.4.1 Polyetheretherketone compared with stainless-steel tubing

Owing to the highly 'sticky' character of a polar lipid, it readily adheres to the tubing walls of the HPLC system. We have replaced all stainless-steel tubing (including the injection loop) with polyetheretherketone (PEEK) tubing. We have found it absolutely necessary to use only PEEK tubing, especially when performing analyses near the LOD or limit of quantification (LOQ). If stainless-steel tubing phospholipids will stick to the tubing walls resulting in high LODs and LOQs and 'memory effects', that is the appearance of ions not present in the sample analysed. If one runs a gradient after changing to PEEK one usually finds that most of the lipids sticking to the tubing walls will elute. If one runs one or two gradients with blank injections, one will obtain a true blank for the lipid in question. Furthermore, the rinsing cup of the auto-injector should be changed or washed regularly, since the polar lipids may stick to the walls of the rinsing cup as well.

Example. We recently analysed a Folch extract containing polyphosphoinositides but we obtained no ions at the expected retention times. We found that we had forgotten to replace part of the steel tubing with PEEK. Replacement of this remaining part resulted in sharp peaks of polyphosphoinositides. We estimated that, in this case, the decrease in sensitivity, arising from the stainless-steel tubing was at least one order of magnitude. Although most of the tubing was of PEEK material, the presence of only a short length of stainless steel was enough to adsorb most of the polar lipids present in the sample. This shows the importance of only using PEEK tubing when analysing polar lipids, especially since the compounds are often present in minute amounts.

9.4.2 Column packing

In our laboratory we make our own columns for normal-phase chromatography (diol-modified silica). We have noticed that by packing the columns under a high pressure, that is, 950–1000 bar, instead of at the recommended 500–600 bar, we obtained far better separation of the phospholipid classes. Accordingly, this way of packing columns may be a way to achieve better separation of intact polar lipids, if choice of column packing, solvent mixture gradient and sample clean-up have already been optimized.

9.4.3 Solvent quality

Solvents should always be of HPLC-grade, if possible, and freshly mixed. The solvents may of course be contaminated from time to time, giving high background levels, unwanted peaks and/or adduct ions. Different brands of the same solvent may also differ in purity from time to time.

9.4.4 Adduct ions

The number of adduct ions observed in an ES mass spectrum seems to increase when more unusual solvent mixtures are being used in combination with ES. Apart from the usual protonated, sodium and potassium adduct ions observed in positive ion mode, and the deprotonated and chlorine adduct ions in the negative ion mode, others may well be observed. When interpreting mass spectra, the presence of adduct ions may be considered either a challenge or a nuisance, depending on the mood of the analyst!

9.4.5 Wasting

A sample usually contains many unwanted substances, for example non-polar lipids such as triacylglycerols. It is not unusual to find that the compound(s) of interest forms only a few per cent of the total sample. By flushing out ('wasting') the unwanted compounds instead of letting them enter the mass spectrometer the time needed for cleaning the ion source is reduced substantially, as are ion current suppression effects. Wasting is performed by installing a second back-up column parallel to the analytical column. The flow is doubled, since two column lines must be provided with solvent flow. A six-port switch valve is used for automated switching of the flows, either to waste or into the mass spectrometer. With this set-up it is possible to switch between wasting and analysis several times during one run. This may be the only way to carry out automated analysis overnight (or longer) if the samples contain many compounds of no interest to the analyst. Without wasting, the ion source may have to be cleaned after every other analysis.

9.5 Summary

The papers listed in this chapter have I hope given a good picture of the use of HPLC–MS and flow or loop injection–MS in the analysis of intact polar lipids today. Clearly, the complex diversity of polar lipid molecular species is now more evident than ever, thanks to mass spectrometry and particularly to the modern high-flow combined APCI–ES devices. The fatty acids (ester- or ether-linked) of the glycerol backbone of glycerophospholipids, and the long-chain bases and fatty acids of sphingolipids, can now be determined both quantitatively and qualitatively. Positional isomers, that is, the *sn*-1 or *sn*-2 position of a fatty acyl group, can be determined as well as the molecular species composition of a specific polar lipid. Furthermore, the structure of polar head-groups, for example the sugar moiety of glycolipids, can be determined. All this can make it easier to understand cellular signalling in determining the low levels of the polar lipids involved in these processes.

Furthermore, membrane dynamics can be better understood when knowing more about membrane phospholipid composition.

Finally, on the market today there are ion trap mass spectrometers for on-line HPLC coupling which have been reported to be able to perform several MS analyses one after another, that is MS–MS–MS...MS ('MS^n'). Mass spectrometers consisting of a quadropole (Q) coupled to a time-of-flight (TOF) analyser may be interesting for the analysis of intact polar lipids. A Q-TOF mass spectrometer is reported to be more sensitive in full scan mode than is a double quadropole mass spectrometer. No or very little work has been done on intact polar lipids with the above-mentioned mass spectrometers.

The mass spectrometric techniques described in this chapter have greatly increased the knowledge of polar lipid structure and function. Still, there are I hope some things left to be discovered with the techniques of today and those of tomorrow.

References

Adams, J. and Ann, Q. (1993) Structure determination of sphingolipids by mass spectrometry. *Mass Spectrometry Reviews*, **12**, 51–85.

Arnoldsson, K. C. and Kaufmann, P. (1994) Lipid class analysis by normal phase high performance liquid chromatography: development and optimization using multivariate methods. *Chromatographia*, **38** (5/6), 317–24.

Buhrman, D.L., Price, P.I. and Rudewicz, P.J. (1996) Quantitation of SR 27417 in human plasma using electrospray liquid chromatography–tandem mass spectrometry: a study of ion suppression. *J. Am. Soc. Mass Spectrom.*, **7**, 1099–105.

Bütikofer, P., Kuypers, F. A., Shackleton, C. et al. (1990) Molecular species analysis of the glycosylphosphatidylinositol anchor of *Torpedo marmorata* acetylcholinesterase. *J. Biol. Chem.*, **265**(31), 18983–7.

Chan, K. (1996) Biological matrix effects. Proceedings of the 44th ASMS Conference on Mass Spectrometry and Allied Topics, Portland, OR, 12–16 May.

Cole, R. B. (1996) Electrospray ionisation mass spectrometry for structural characterisation of the lipid A component in bacterial endotoxins. *ACS Symp. Ser.*, **619** (Biochemical and Biotechnological Applications of Electrospray Ionisation Mass Spectrometry), 185–206.

Constantin, E. and Schnell, A. (1991) *Mass spectrometry*, Ellis Horwood series in Analytical Chemistry (eds J. F. Tyson and M. Masson) Ellis Horwood, New York.

Desiderio, D. M. (ed.) (1992) *Mass Spectrometry – Clinical and Biomedical Applications*, vols 1 and 2, Plenum Press, New York and London.

Gelpí, E. (1995) Biomedical and biochemical applications of liquid chromatography–mass spectrometry. *J. Chromatogr. A*, **703**, 59–80.

Gibson, B. W., Phillips, N. J., Melaugh, W. and Engstrom, J. J. (1996) Determining structures and functions of surface glycolipids in pathogenic *Haemophilus* bacteria by electrospray-ionisation mass spectrometry. *ACS Symp. Ser.*, **619** (Biochemical and Biotechnological Applications of Electrospray Ionisation Mass Spectrometry), 166–84.

Haigh, W. G., Yoder, T. F., Ericson, L. et al. (1996) The characterisation and cyclic production of a highly unsaturated homoserine lipid in *Chlorella minutissima*. *Biochim. Biophys. Acta*, **1299**(2), 183–90.

Han, X. and Gross, R. W. (1994) Electrospray ionisation mass spectroscopic analysis of human erythrocyte plasma membrane phospholipids. *Proc. Natl. Acad. Sci. USA*, **91**(22), 10635–9.

Han, X. and Gross, R. W. (1995) Structural determination of picomole amounts of phospholipids via electrospray ionisation tandem mass spectrometry. *J. Am. Soc. Mass Spectrom.*, **6**(12), 1202–10.
Han, X., Gubitosi-Klug, R. A., Collins, B. J. and Gross, R. W. (1996) Alterations in individual molecular species of human platelet phospholipids during thrombin stimulation: electrospray ionisation mass spectrometry-facilitated identification of the boundary conditions for the magnitude and selectivity of thrombin-induced platelet phospholipid hydrolysis. *Biochemistry*, **35**(18), 5822–32.
Hullin, F., Kim, H. Y. and Salem, N. Jr (1989) Analysis of aminophospholipid molecular species by high performance liquid chromatography. *J. Lipid Res.*, **30**(12), 1963–75.
Ii, T., Ohashi, Y., Matsuzaki, Y., Ogawa, T. and Nagai, Y. (1993) Electrospray mass spectrometry of pentacosasaccharides of blood group I-activity and related compounds. *Org. Mass Spectrom.*, **28**(11), 1340–4.
Ioneda, T., Beaman, B. L., Viscaya, L. and Almeida, E. T. (1993) Composition and toxicity of diethyl ether soluble lipids from *Nocardia asteroides* GUH-2 and *Nocardia asteroides* 10905. *Chem. Phys. Lipids*, **65**(3), 171–8.
Jensen, N.J. and Gross, M.L. (1988) A comparison of mass spectrometry methods for structural determination and analysis of phospholipids. *Mass Spectrometry Reviews*, **7**, 41–69.
Jungalwala, F.B., Evans, J.E. and McCluer, R.H. (1984) Compositional and molecular species analysis of phospholipids by high performance liquid chromatography coupled with chemical ionisation mass spectrometry. *J. Lipid Res.*, **25**, 738–49.
Karlsson, A.Å., Michélsen, P., Larsen, Å. and Odham, G. (1996) Normal-phase liquid chromatography class separation and species determination of phospholipids utilizing electrospray mass spectrometry/tandem mass spectrometry. *Rapid Commun. Mass Spectrom.*, **10**(7), 775–80.
Karlsson, A.Å., Michélsen, P., Westerdahl, G. and Odham, G. (1997) Determination of molecular species of sphingomyelin in bovine milk using HPLC–MS–MS with electrospray ionisation and atmospheric pressure chemical ionisation, unpublished manuscript.
Kerwin, J.L., Tuininga, A.R. and Ericsson, L.H. (1994) Identification of molecular species of glycerophospholipids and sphingomyelin using electrospray mass spectrometry. *J. Lipid Res.*, **35**(6), 1102–14.
Kerwin, J.L., Tuininga, A.R., Wiens, A.M. et al. (1995) Isoprenoid-mediated changes in the glycerophospholipid molecular species of the sterol auxotrophic fungus *Lagenidium giganteum*. *Microbiology* (Reading, UK), **141**(2), 399–410.
Kim, H.Y. and Salem, N. Jr (1986) Phospholipid molecular species analysis by thermospray liquid chromatography/mass spectrometry. *Anal. Chem.*, **58**(1), 9–14.
Kim, H.Y. and Salem, N. Jr (1987) Application of thermospray high-performance liquid chromatography/mass spectrometry for the determination of phospholipids and related compounds. *Anal. Chem.*, **59**(5), 722–6.
Kim, H.Y., Wang, T.C.L. and Ma, Y.C. (1994) Liquid chromatography/mass spectrometry of phospholipids using electrospray ionisation. *Anal. Chem.*, **66**(22), 3977–82.
Kim, H.Y., Yergey, J.A. and Salem, N. Jr (1987) Determination of eicosanoids, phospholipids and related compounds by thermospray liquid chromatography–mass spectrometry. *J. Chromatogr.*, **394**(1), 155–70.
Knebel, N.G., Sharp, S.R. and Madigan, M.J., (1995) Quantification of the anti-HIV drug Saquinavir by high-speed on-line high-performance liquid chromatography/tandem mass spectrometry. *J. Mass Spectrom.*, **30**, 1153.
Koshy, K.M. and Boggs, J.M. (1996) Investigation of the calcium-mediated association between the carbohydrate head groups of galactosylceramide and galactosylceramide I^3 sulfate by electrospray ionisation mass spectrometry. *J. Biol. Chem.*, **271**(7), 3496–9.
Kuypers, F.A., Bütikofer, P. and Shackleton, C.H.L. (1991) Application of liquid chromatography–thermospray mass spectrometry in the analysis of glycerophospholipid molecular species. *J. Chromatogr.*, **562**(1–2), 191–206.
Kwon, G., Bohrer, A., Han, X. et al. (1996) Characterisation of the sphingomyelin content of isolated pancreatic islets. Evaluation of the role of sphingomyelin hydrolysis in the action of interleukin-1 to induce islet overproduction of nitric oxide. *Biochim. Biophys. Acta*, **1300**(1), 63–72.

Ma, Y.C. and Kim, H.Y. (1995) Development of the online high-performance liquid chromatography/thermospray mass spectrometry method for the analysis of phospholipid molecular species in rat brain. *Anal. Biochem.*, **226**(2), 293–301.

McLafferty, F.W. (1973) *Interpretation of Mass Spectra*, 2nd edn, W.A. Benjamin, Reading, MA.

Michélsen, P., Jergil, B. and Odham, G. (1995) Quantification of polyphosphoinositides using selected ion monitoring electrospray mass spectrometry. *Rapid Commun. Mass Spectrom.* **9**, 1109–14.

Morrison, W.R. (1969) Polar lipids in bovine milk I. Long-chain bases in sphingomyelin. *Biochim. Biophys. Acta*, **176**, 537–46.

Murphy, R.C. (1993) Mass spectrometry of lipids, in *Handbook of Lipid Research 7*, Plenum Press, New York and London.

Niessen, W.M.A. and van der Greef, J. (1992) *Liquid Chromatography–Mass Spectrometry: Principles and Applications*, Marcel Dekker, New York.

Odham, G., Valeur, A., Michélsen, P. *et al.* (1988) Highly sensitive determination and characterisation of intact cellular ester-linked phospholipids using liquid chromatography–plasma spray mass spectrometry. *J. Chromatogr.*, **434**(1), 31–41.

Reinhold, B.B., Chan, S.Y., Chan, S. and Reinhold, V.N. (1994) Profiling glycosphingolipid structural detail: periodate oxidation, electrospray, collision-induced dissociation and tandem mass spectrometry. *Org. Mass Spectrom.*, **29**(12), 736–46.

Smith, P.B.W., Snyder, A.P. and Harden, C.S. (1995) Characterisation of bacterial phospholipids by electrospray ionisation tandem mass spectrometry. *Anal. Chem.*, **67**(11), 1824–30.

Sweeley, C.C. (1963) Purification and partial characterisation of sphingomyelin from human plasma. *J. Lipid Res.*, **4**(4), 402–6.

Sweetman, G. Trinei, M., Modha, J. *et al.* (1996) Electrospray ionisation mass spectrometric analysis of phospholipids of *Escherichia coli*. *Mol. Microbiol.*, **20**(1), 233–4.

Valeur, A. (1992) Utilisation of chromatography and mass spectrometry for the estimation of microbial dynamics. Dissertation, Lund University, Sweden.

Valeur, A., Michélsen, P. and Odham, G. (1993) Online straight-phase liquid chromatography/plasmaspray tandem mass spectrometry of glycerolipids. *Lipids*, **28**(3), 255–9.

Valeur, A., Olsson, N.U., Kaufmann, P. *et al.* (1994) Quantification and comparison of some natural sphingomyelins by online high-performance liquid chromatography/discharge-assisted thermospray mass spectrometry. *Biol. Mass Spectrom.*, **23**(6), 313–19.

Yergey, A.L., Edmonds, C.G., Lewis, I.A.S. and Vestal, M.L. (1990) *Liquid Chromatography/Mass Spectrometry – Techniques and Applications*, Plenum Press, New York and London.

10 The exploitation of chemometric methods in the analysis of spectroscopic data: application to olive oils

A. JONES, A. D. SHAW, G. J. SALTER, G. BIANCHI and D. B. KELL

10.1 Introduction

Multivariate analysis is the term used to describe the analysis of data where numerous observations or variables are obtained for each object studied (Afifi and Clark, 1996). The identity or value of any one object sample will be reflected in some or all of the variables measured to a greater or lesser degree. With spectroscopic data it is not generally possible to identify or quantify an object from one variable (Mark, 1991). This may, however, be achieved by disentangling the complicated interrelationships between a number of the variables by means of multivariate statistical methods (Martens and Næs, 1989).

Classification problems are those where the aim is to identify objects, for example the region of origin of an olive oil. Quantification problems are those where the aim is to predict the magnitude of a quantity, for example the amount of an adulterant in an olive oil. Multivariate methods may be applied equally to either.

In recent years, more powerful computers and widely available statistical software have led to a tremendous increase in the use of multivariate data analysis. With the power of modern computers, most efforts have focused on analysing the whole spectrum of data (Brereton and Elbergali, 1994). This approach relies on statistical software to produce optimum results from the data fed into it; much of these data could contain little information, and therefore be of no value to the model.

Recent research has shown that judicious variable selection can improve statistical predictions of models (Baroni *et al.*, 1992; Brereton, 1995; Brereton and Elbergali, 1994; Broadhurst *et al.*, 1997; Brown, 1993; Cruciani and Watson, 1994; Defalguerolles and Jmel, 1993; Hazen, Arnold and Small, 1994; Heikka, Minkkinen and Taavitsainen, 1994; Kubinyi, 1994a, b, 1996; Lindgren *et al.*, 1995; Norinder, 1996; Shaw *et al.*, 1996, 1997; Sreerama and Woody, 1994); statistical theory, in particular the parsimony principle

(Flury and Riedwyl, 1988; Seasholtz and Kowalski, 1993), supports these results.

Olive oil is an ideal candidate for multivariate analysis. For economic reasons, the labelling of olive oils is frequently falsified (Collins, 1993; Firestone and Reina, 1987; Firestone, Carson and Reina, 1988; Firestone *et al.*, 1985; Li-Chan, 1994; Simpkins and Harrison, 1995b; Zamora, Navarro and Hidalgo, 1994), so there is a need for easy and cheap methods for identification. This chapter concentrates on the application of multivariate methods to nuclear magnetic resonance (NMR) and pyrolysis mass spectrometry (PyMS) data. It provides a brief introduction to principal components analysis (PCA), principal components regression (PCR), partial least squares regression (PLS) and the use of artificial neural networks (ANNs), then moves on to variable selection and its application to olive oil data.

10.2 Olive oil

10.2.1 *Economics*

The value of olive oil produced annually is around US$2.5 billion (Kiritsakis, 1991), other olive products amounting to around US$300 million. A total of 9.4 million tonnes of olive fruit are produced per annum from 805 million olive trees worldwide, occupying some 24 million acres of land. Some 98% of these trees are in the Mediterranean area. Of the 60 million tonnes of seed oil consumed worldwide every year, 2 million tonnes are olive oil (Anon., 1994).

Almost 25% of the farming income in the Mediterranean basin as a whole comes from olive products, Spain and Italy being far and away the largest producers, with Greece (with around half the production of Spain and Italy) coming third. In 1987, Italy and Spain contributed about 65% of world olive production (Salunkhe *et al.*, 1991).

Olive production often follows a two-year cycle, a good crop one year being followed by a poor or medium crop the next year. This is probably the biggest problem facing the olive industry (Kiritsakis, 1991).

10.2.2 *Chemistry*

The olive fruit is a drupe, that is to say it contains a stone, pulp and an outer skin (like a plum). The chemical composition of the fruit (Bianchi, Giansarte and Lazzari, 1996) is approximately as given in Table 10.1. Table olives generally have a lower oil content (around 10%–14%) than olives used for oil production.

The main fatty acids contained in olive oil, which are attached (esterified) to the glycerol backbone in one of the three locations α, β or α' are shown in Table 10.2.

Table 10.1 Chemical composition of the olive fruit

Component	Percentage
Water	48
Oil	21
Mono- and disaccharides	3
Polysaccharides	27
Waxes, triterpenes and phenols	1
Other minor components	Trace
Total	100

Table 10.2 The main fatty acids contained in olive oil

Acid	Abbreviation	Percentage range	Structure
Saturated			
Palmitic acid	C16:0	7.5–20	$CH_3—(CH_2)_{14}—COOH$
Stearic acid	C18:0	0.5–5	$CH_3—(CH_2)_{16}—COOH$
Lignoceric acid	C24:0	1.0 (max.)	$CH_3—(CH_2)_{22}—COOH$
Monounsaturated			
Palmitoleic acid	C16:1Δ9	0.3–3.5	$CH_3—(CH_2)_5 HC=CH—(CH_2)_7—COOH$
Oleic acid	C18:1Δ9	56–83	$CH_3—(CH_2)_7—HC=CH—(CH_2)_7—COOH$
Eicosenoic acid	C20:1Δ11	trace	$CH_3—(CH_2)_7—HC=CH—(CH_2)_9—COOH$
Polyunsaturated			
Linoleic acid	C18:2Δ9, 12	3.5–20	$CH_3—(CH_2)_4—HC=CH—CH_2—HC$ $=CH—(CH_2)_7—COOH$
Linolenic acid	C18:3Δ9, 12, 15	0.0–1.5	$CH_3—CH_2—(HC=CH—CH_2)_3$ $—(CH_2)_6—COOH$
Arachidonic acid	C20:4Δ5, 8, 11, 14	0.8 (max.)	$CH_3—(CH_2)_4—(HC=CH—CH_2)_4$ $—(CH_2)_2—COOH$

Other saponifiable constituents include phosphatides. Minor constituents, together called the 'unsaponifiable fraction', include hydrocarbons, terpenes, fatty alcohols, wax, phenols and amino acids. In addition, there will be a small amount of free fatty acids (FFAs), the amount being dependent on the grade of oil.

Virgin olive oil is the oil extracted by purely mechanical means from sound, ripe fruits of the olive tree (*Olea europaea* L.). Extra virgin olive oil is absolutely perfect in flavour and odour, and has a maximum free fatty acid content in terms of oleic acid of 1 g per 100 g (EC, 1991; Goodacre, Kell and Bianchi, 1993; Kiritsakis, 1991).

Compared with other edible oils, olive oil contains a low percentage of saturated fatty acids (that is, fatty acids with no double bonds in the carbon chain) at around 16% (mainly palmitic, 16:0). It contains a high percentage of monounsaturated fatty acids, around 70% (mainly oleic, 18:1) (MAFF, 1995; Mottram, 1979) and around 15% polyunsaturated fatty acids (mainly linoleic, 18:2).

Virgin olive oil generally conserves for a longer time than do most other vegetable oils (the maximum duration of optimal usage is often around 18 months). It is suggested that this is a result of the combined effect of a high monounsaturate content and some of the minor constituents of the oil, which act as antioxidants. Perrin (1992) identifies phenolic compounds as the main antioxidants; Kiritsakis and Dugan (1985) additionally mention carotene, whilst noting that chlorophyll has the opposite effect.

Garcia et al. (1996) note the effects of storage temperature of the fruits before oil extraction on the quality of olive oil. Prolonged storage at the wrong temperature (which, at least in Spain, is typically in great heaps in the open air) can, they point out, increase the amount of free oleic acid; this affects the grade of the oil. This is a problem in Spain, as there are insufficient mills to process all the olives at the peak of the harvesting season. The problem may be overcome by storage in cool buildings (Kiritsakis, 1991).

Other factors affecting the chemistry of the oil are the extraction method used (Aparicio, Navarro and Ferreiro, 1991; Rade et al., 1995; Ranalli and Martinelli, 1994), storage conditions (Garcia et al., 1996; Kiritsakis, 1984; Rade et al., 1995), orography (e.g. distance from sea, altitude) (Aparicio, Ferreiro and Alonso, 1994; Armanino, Leardi and Lanteri, 1989) and the time of harvest (Boschelle et al., 1994; Haumann, 1996; Tsimidou and Karakostas, 1993).

There are various accepted classifications of olive oil (Kiritsakis, 1991). The European Union rules governing olive oil classification are very comprehensive, covering in 83 pages not only the oil characteristics but also the methods of analysis to be used, right down to the selection of tasters (EC, 1991).

10.2.3 Health aspects

Much has been made in recent years of the so-called 'Mediterranean diet' (Gussow, 1995; Trichopoulou et al., 1995b; Tsimidou, 1995), of which olive oil is a basic ingredient. Olive oil has a fine aroma and a pleasant taste, which is generally agreed to be at its best in extra virgin olive oils, and is considered to have many nutritional and health benefits (Kiritsakis, 1991). It is almost the only vegetable oil to be consumed as it is, that is without raffination (excluding the little-consumed nut and sesame oils) (Perrin, 1992).

There are many varied claims and suggested reasons as to the health benefits. There is very strong evidence that olive oil consumption reduces the risk of death due to circulatory system diseases (Fraser, 1994; Kafatos and Comas, 1991). Visioli and Galli (Galli, Petroni and Visioli, 1994; Visioli and Galli, 1994, 1995) suggest that this is at least partially because of the presence of natural antioxidants (including the bitter-tasting glycosidic compound oleuropein) and micronutrients preventing low-density lipoprotein (LDL) cholesterol from oxidizing [oxidized LDL particles are particularly atherogenic (Fraser, 1994)] and so retarding the formation of atherosclerotic lesions (MAFF, 1995). They say that these antioxidants and micronutrients are present in large amounts in extra virgin olive oil (lesser amounts are found in other grades of olive oil). They imply that these properties may be more important than the high monounsaturated/saturated fatty acid ratio, and even suggest (Visioli, Vinceri and Galli, 1995) that the wastewater used for washing the olive paste during oil production could be utilized as it too is high in antioxidants.

A diet relatively high in monounsaturated fatty acids (MUFAs) does in any case reduce the levels of the undesirable LDLs in the body (Bosaeus et al., 1992; MAFF, 1995; Shepherd and Packard, 1992); indeed the reduction is as great as that of a low-fat, high-carbohydrate diet (Kafatos and Comas, 1991). There is also evidence that a high MUFA diet increases the beneficial high-density lipoprotein (HDL) cholesterol, in contrast to polyunsaturated fats, which decrease both LDL and HDL levels (Kafatos and Comas, 1991), although this finding is not universally accepted (Haumann, 1996).

It has also been suggested that increased olive oil consumption helps prevent the onset of rheumatoid arthritis, and reduces its severity (Linos et al., 1991).

The importance of monounsaturated fats has been recognized only in recent years; not so long ago they were completely overlooked with regard to blood cholesterol levels in favour of polyunsaturated fats (Mottram, 1979, pp. 52–3).

Martin-Moreno et al. (1994) also note that olive oils contain a 'generous amount of antioxidants' and speculate that 'diets high in monounsaturated fats presumably yield tissue structures that are less susceptible to antioxidative damage than would be the case in high polyunsaturated diets' (p. 778). They identify an inverse correlation between breast cancer and olive oil intake, as do Trichopoulou and co-workers (Trichopoulou, 1995; Trichopoulou et al., 1995a, b), who also claim that margarine consumption increases this risk. Trichopoulou et al. (1995c) suggest that olive oil consumption is one of the factors in the traditional Greek diet that is a cause of the longevity of those elderly people in a study group who followed that diet. Greece has the highest consumption of olive oil in the world, at 20.8 kg per person per year (Kiritsakis and Markakis, 1991), or 57 g a day (at 500 calories, around 20%–25% of the energy requirement of an adult).

Murphy (1995, p. 302) notes that 'over the next decade and beyond, we face the prospect of being able to engineer most major oil crops to produce the fatty acid composition of our choice'. So, if monounsaturates in olive oils were the only reason for the health benefits claimed, this selling point might not have much of a future.

10.2.4 Analysis

As a consequence of the benefits mentioned, and because of the amount of labour and land needed to produce a given amount of oil, olive oil commands a much higher price than do most other edible oils. This in turn means that there is a great temptation to adulterate the oil with a cheaper oil, such as olive pomace oil, corn oil, sunflower oil, or even lard or castor oil (Firestone and Reina, 1987; Firestone, Carson and Reina, 1988; Firestone *et al.*, 1985; Zamora, Navarro and Hidalgo, 1994). Rapeseed oil, which can have a similar quantity of oleic acid (Shahidi, 1990), and high oleic sunflower oil, are also popular choices as adulterants. In addition, it is claimed that many oils purported to be extra virgin had been processed in order to reduce the acidity level and so gain this classification. Firestone *et al.* (1985) reported on a US survey in which 4 out of 5 virgin olive oils were correctly labelled, compared with only 3 out of 20 olive oils. In 1988 they followed up the 1985 report (Firestone, Carson and Reina, 1988), noting some improvement. This time, although only 17 out of 31 virgin olive oils were correctly labelled, so were 15 out of 26 olive oils; over 40% were incorrectly labelled. A British Broadcasting Corporation (BBC) radio investigation into olive oils in May 1994 suggested similar figures for the British market.

The necessity to be able to detect adulterations in oils in general was highlighted in May 1981, when 20 000 people became ill and 350 died in Spain after consuming oils containing 'refined' aniline denatured rapeseed oil (Aldridge, 1992).

One problem faced today is that, as detection methods become more sophisticated so too do the methods of the adulterators. Some methods, which rely on the detection of compounds which do not appear in the genuine product, are useless if the adulterator knows the technique and therefore removes the offending compound (Aparicio, Alonso and Morales, 1996).

Grob *et al.* (1994a) report that extra virgin oils can be distinguished by the presence of a substantial quantity of volatile components (i.e. they have not been deodorized). If none of these volatiles is present, the oil has been treated. 'Pure' oils, being a blend, are more difficult to distinguish. Grob *et al.* (1994b) were able to detect adulteration of olive oils down to 10% (even lower for most oils) by using LC–GC–FID (liquid chromatography–gas chromatography–flame ionization detection) by direct analysis of these minor components. They do note, however, that strong raffination made adulteration difficult to detect.

Historically, and indeed to the present day, panels of trained assessors have been used to differentiate olive oils (Aparicio and Morales, 1995; Aparicio, Gutierrez and Morales, 1992; Lyon and Watson, 1994; Morales *et al.*, 1995). Chemometric methods applied to the results of such panels do indeed give good results, but it is slow and restricted to the sensory characteristics. Problems arising from different panels describing the same sensory attribute with different terms (which is quite understandable when panels are from different cultures and use different languages) may be overcome by means of sensory wheels, a technique which explores the relationships between sensory attributes (Aparicio and Morales, 1995; Boskou, 1996). Also, in the present scientific climate, where the culprits' methods are becoming ever more sophisticated, it is not possible to use taste panels to detect reliably the adulteration or misclassification of origin. Such a method cannot, according to Peri and Rastelli (1994), be used as a legal tool for evaluating quality or origin, although it still is (EC, 1991).

Goodacre, Kell and Bianchi (1993) were successful in detecting adulteration of extra virgin olive oil by using ANNs and PyMS. Extensions to this work are described below in the section on pyrolysis mass spectrometry.

The Institute of Food Research (IFR, 1994) used Fourier transform infrared (FTIR) spectroscopy and NMR (not stated, but presumably ^{13}C) for the identification of oils. Both methods successfully differentiated olive oils of 'differing botanical origin' and could also discriminate extra virgin and other grades of olive oil. Fatty acid composition was found to be the main factor for discriminating the origin, and 'other trace analytes' were used to distinguish the grade of oil. Other workers have also successfully applied FTIR to the analysis of olive oils and other edible oils (Ismail et al., 1993; Lai, Kemsley and Wilson, 1994, 1995; van de Voort, 1994; van de Voort, Ismail and Sedman, 1995; van de Voort et al., 1994a, b).

Despite the IFR results, Zamora, Navarro and Hidalgo (1994) were also able to distinguish between different grades of oil by means of ^{13}C NMR alone, which relies largely on the fatty acid composition. Forina and Tiscornia (1982) agree that fatty acid content is important in the geographical classification of olive oils.

Vlahov (1996), at the Istituto Sperimentale per la Elaiotecnica, has used ^{13}C NMR to determine the quantities of diacylglycerols (DGs), both 1,2-DG and 1,3-DG, in olive varieties Grossa di Cassano, Nebbio, Coratina, Leccino, Dritta and Caroleo. It was found that the later-ripening olive varieties, Coratina and Nebbio, had significantly lower DG totals than had the other varieties. The ratio of 1,2-DG and 1,3-DG also varied significantly between varieties. She concludes that the results may represent preliminary new parameters for future analysis. No monoacylglyceride signals were found in the samples analysed. Other preliminary work by Valhov and Angelo (1996) has suggested that the position of fatty acids within the triglycerides may be important for discrimination (Forina and Tiscornia, 1982; Zupan and Gasteiger, 1993).

Very recently, high-field ^{1}H NMR has been used with success by one group for regional and variety discrimination of virgin olive oil (Segre et al., 1996) and oil quality (Sacchi, 1996). They suggest that proton NMR at very high field (they were using 600 MHz) can be a more powerful technique than ^{13}C NMR for quality control of virgin olive oils. Their results for region and variety are promising but suggest that further work is yet required. Their variety results were obtained with oils from Umbria only, rather than oils from various regions.

Gigliotti, Daghetta and Sidoli (1994) used high-pressure liquid chromatography (HPLC) for geographical characterization of olive oils. Using ratios of OOO, SOO, LLL and OOL (O = oleic, S = stearic and L = linoleic acid) and ECN (equivalent carbon number) they were able to distinguish Moroccan, Tunisian, Greek, Spanish and Sicilian oils. Their results also show that many bought oils do not fit into any of their categories, suggesting the possibility of adulteration. Perrin (1992) notes that Tunisian olive oil has a much lower monounsaturate/polyunsaturate ratio than do oils of other countries (less than 3:1 compared with 4:1–10:1 generally).

Boschelle et al. (1994) applied chemometric methods to chemophysical data to identify the cultivar of olives from the Gulf of Trieste area. They were successful in distinguishing the local variety Bianchera from those newly introduced into the region.

Tsimidou and Karakostas (1993) used data on the percentage of five fatty acids (palmitic, palmitoleic, stearic, oleic and linoleic – data on minor acids were not available) to classify Greek olive oils by region. Their results show that year of harvest is more influential in PCA than is variety or region of origin. In contrast to the results shown later in this chapter (Section 10.4), they found more difficulty in distinguishing variety than region.

Zupan and Gasteiger (1993) used Kohonen ANNs to discriminate Italian olive oils by region, suggesting that this is much better than PCA for mapping onto a two-dimensional plane. The network inputs were the fatty acid content of 572 olive oils from nine different areas of Italy (north and south Apulia, Calabria, Sicily, inner Sardinia, coastal Sardinia, east and west Liguria and Umbria). The analysis used the percentage of eight fatty acids (palmitic, palmitoleic, stearic, oleic, linoleic, arachidic, linolenic and eicosenoic) in the oil, determined previously by unstated methods (presumably GC). Although not able to identify any of the regions with 100% accuracy, the Kohonen nets were able to predict correctly up to 302 oils from a test set of 322 (although 51 of these 302 were only correctly assigned by using a K nearest-neighbour decision for empty space hits). Forina and Tiscornia (1982), using the same data set, were able to predict oils with 94.5% accuracy by using a K nearest-neighbour decision projected onto the hyperspace of the training set variables. The test and training sets were randomly selected and were repeated ten times.

Garcia and López (1993) and Aparicio, Alonso and Morales (1994b) were able to distinguish between Italian, Spanish and Portuguese olive oils by using an expert system, SEXIA (Aparicio and Alonso, 1994). For the Italian regions, Garcia and López were able to distinguish 99% of their Sardinian samples, 91.3% of the northern Italian samples, but only 77.4% of the southern Italian samples. They note that there is evidently greater disparity between oils from the south. These figures would appear to be in line with our results shown later in this chapter (Section 10.4.12), where Toscana region oils were much easier to predict than southern oils.

Guinda, Lanzón and Albi (1996) were able to distinguish five varieties of Spanish olive oils by examining the hydrocarbon fraction (excluding squalene) of 50 oils from six provinces of Spain. They did not use chemometric techniques, but a simple decision tree (e.g. *if Percentage First Fraction $C25 \geq 28$ then variety = Empeltre else...*).

Sato (1994) showed that near infra-red spectroscopy can be used with PCA to discriminate many vegetable oils from each other, including olive oil. Schwaiger and Vojir (1994) had similar success with GC analysis and PCA, the first two principal components separating olive oil well from the other oils.

Although the Raman effect was discovered in 1928 by Sir Chandrasekhara Venkata Raman, it has not until recently been applied to food adulteration problems (Baeten *et al.*, 1996; Li-Chan, 1994; Ozaki *et al.*, 1992; Sadeghi-Jorabchi *et al.*, 1990, 1991). Baeten *et al.* (1996) used FT-Raman which, they claim, produces fluorescence-free spectra, using a 1.064 μm laser. They were able to detect adulteration with soybean, corn and olive pomace with 100% accuracy down to 1% adulterant. In fact 780 nm excitation in a confocal instrument (Williams, 1994; Williams *et al.*, 1994) produces excellent dispersive Raman spectra from olive oils in a wholly non-invasive fashion (N. Kaderbhai and the authors, unpublished observations). Baeten *et al.* (1996) comment that at present liquid and gas chromatography is the most accurate technique to determine adulteration, and it is this method that is the European Union adulteration standard (EC, 1991), but that FT-Raman has the potential for detecting adulterants beyond the limits of liquid and gas chromatography.

Not surprisingly, climate has been shown to affect the chemical composition of olive oil (Aparicio, Ferreiro and Alonso, 1994). Variations in chemical composition between regions are presumably largely explained by this factor (although there are probably other factors).

Simpkins and Harrison (1995a) summarize many of the methods used for detection of authenticity in olive oils and many other food products. They note that most new applications they reviewed relied on advanced statistical procedures for data analysis.

Li-Chan (1994) reviews current developments in the detection of adulteration of olive oil, pointing out that multivariate spectroscopic methods of analysis derive their power from the simultaneous use of multiple variables in the spectrum. The effect of interference in some variables can then be reduced by the calibration method used (PCR, PLS, etc.). This type of approach has previously been used for crude petrochemical oil, as described by Kvalhcim *et al.* (1985) and Brekke *et al.* (1990), with very promising results.

Aparicio, Alonso and Morales (1996) suggest that the future lies in spectroscopic (probably, they say, FT-Raman) and chromatographic techniques, coupled with mathematical algorithms.

10.3 Data acquisition methods

10.3.1 Nuclear magnetic resonance

According to Yoder and Schaeffer (1987), NMR spectroscopy is probably the most powerful tool available to the chemist for the probing of the structure of molecules. It can rapidly produce data from which structural formulas and even some three-dimensional aspects of the structure of molecules can be deduced. Its use is in the detection of nuclear-spin reorientation in an applied magnetic field (Campbell and Dwek, 1984). Indeed Harris (1986) claims that it is arguably the single most important tool for obtaining detailed information on chemical systems at the molecular level. It is certainly recognized as a valuable technique for analysing food products (Belton, 1995).

Nuclei of certain isotopes (e.g. ^{1}H, ^{13}C, ^{19}F) possess intrinsic angular momentum or spin (they are said to be spin-active and have the ability to resonate). A nucleus which is spinning also possesses an associated magnetic moment μ. When placed in a strong magnetic field, these nuclei can absorb electromagnetic radiation in the radio frequency range. An NMR spectrometer picks up and displays the precise frequency at which resonance takes place (Williams, 1986).

Carbon forms the backbone of all organic molecules, and Carbon-13 (^{13}C) is the only magnetic carbon isotope (Wehrli, Marchand and Wehrli, 1988). From the point of view of the organic chemist, it is fortunate that such an isotope exists, forming some 1.1% by weight of naturally occurring carbon (Stryer, 1981).

In the field of olive oils, NMR has been applied before (Anon., 1994; Bianchi et al., 1993, 1994a; Gussoni et al., 1993; IFR, 1994; Zamora, Navarro and Hidalgo, 1994), demonstrating the potential of this technique for analysing olive oils. Brekke et al. (1990) and Kvalheim et al. (1985) also show that a combination of NMR and PCA can be used to distinguish between different North Sea crude oils.

Theory. The magnitude of the spin (angular momentum) of the nucleus is $\hbar[I(I+1)]^{1/2}$, where I is the nuclear spin quantum number and $\hbar$ is the reduced Planck's constant $h/2\pi$. I may have only integral or half-integral values $(0, \frac{1}{2}, 1, 1\frac{1}{2}), \ldots, 6$, in units of $h/2\pi$ (Friebolin, 1993), the value being dependent upon the isotope. For ^{13}C and ^{1}H, $I = \frac{1}{2}$. When $I = 0$, the nucleus has no angular momentum. Transitions between nuclear spin energy levels give rise to the resonance phenomenon of NMR.

In a nucleus containing an even number of both protons and neutrons, $I = 0$. This includes the common atoms ^{12}C and ^{16}O. This leads to considerable simplification of the spectra of organic molecules. The reason for this is that nucleons with opposite spin can pair (though neutrons can pair only with neutrons, and protons with protons), just as electrons pair. If the numbers of neutrons and/or protons is odd, then the spin is non-zero, though the actual value depends upon orbital-type internucleon interactions (Akitt, 1983).

Because it is difficult to know to sufficient accuracy the value of the magnetic field applied (Harris, 1986), a standard of known resonance frequency is usually used. The most convenient standard for ^{13}C and proton NMR is tetramethylsilane (TMS) because it contains four equivalent carbon atoms, and the resultant strong signal means that only a small amount (1%–5%) need be added; it gives rise to sharp signals, is chemically inert and is soluble in most organic materials (Kemp, 1986). The TMS peak is taken as 0 in the δ scale and increases in a downfield direction.

The resonance frequency v_0 of an isolated nucleus is:

$$v_0 = \frac{\gamma B_0}{2\pi} \quad (10.1)$$

where B_0 is the strength of the steady magnetic field and γ is a constant (the gyromagnetic ratio) (Stryer, 1981).

Depending on the chemical environment (bondings, etc.) the precise resonance frequency of any one ^{13}C atom will be shifted by a few parts per million (ppm) from that of the standard. This is the chemical shift. The formula for calculating the chemical shift δ, then, is

$$\delta = \frac{v_{sample} - v_{ref}}{v_{ref}} \cdot 10^6 = \frac{\Delta v_{sample}(Hz)}{v_{ref}(MHz)} \quad (10.2)$$

(adapted from Brevard and Grainger, 1981; Friebolin, 1993).

The nucleus of an atom is surrounded by electrons. The electron cloud shields the nucleus to some extent from the applied magnetic field. The amount of shielding depends on the nature of the electrons around the nucleus and is given by the formula

$$B_{eff} = B_0 - \sigma B_0 = (1 - \sigma) B_0 \quad (10.3)$$

where σ is the shielding constant, B_{eff} is the magnetic field at the nucleus and B_0 is the applied magnetic field, thus altering the chemical shift. Then, the chemical shift of a given atom varies as a function of its chemical environment, thus allowing the description of the chemical environment from the chemical shift.

For most studies, the same deuterated solvents are employed in both ^{13}C and proton NMR. A major advantage of ^{13}C spectroscopy over proton NMR is the larger chemical shift range that for most organic substances is ~200ppm in comparison with ~10ppm for proton NMR.

The chemical shift of carbon depends on the hybridization of carbon and its structural environment, and from high to low field is sp^3, sp and sp^2. This trend is observed in the typical olive oil ^{13}C spectrum. The chemical shift of methyl and methylene sp^3 carbon is in the range 10–90 ppm; the range of olefinic sp^2 carbons is 127–130 ppm; the carbonyl carbons appear in the range 170–175 ppm. The glycerol carbon shifts are found in the 60–70 ppm region (Vlahov, 1996). The high-resolution ^{13}C NMR spectra of olive oil, usually dissolved in $CDCl_3$, are readily obtained running 250–300 scans. Spectra consist of 40–45 signals. Most of the signals are assigned according to the chemical shift of standard compounds and literature data. A typical spectrum for olive oil is shown in Fig. 10.1. The principal assignments are given in Fig. 10.2.

10.3.2 Pyrolysis mass spectrometry

Pyrolysis mass spectrometry (PyMS) is a technique that (via Curie-point pyrolysis) thermally degrades a sample of interest at a known temperature in an inert atmosphere or a vacuum. It causes molecules to cleave at their weakest points to produce smaller, volatile fragments called pyrolysate (Irwin, 1982). The mass spectrometer can then be used to separate the

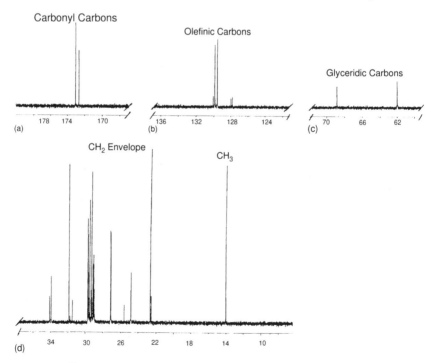

Figure 10.1 ^{13}C NMR spectrum of extra virgin olive oil obtained from the Dritta cultivar: (a) carbonyl region; (b) olefinic region; (c) glyceridic region; (d) CH_2 envelope and methyl region.

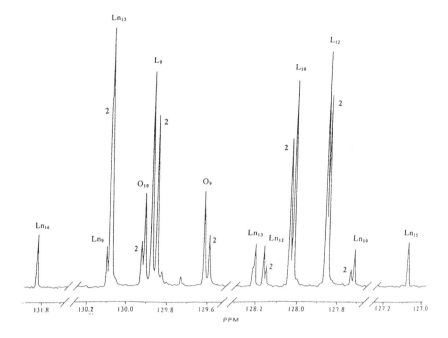

Figure 10.2 ^{13}C NMR spectrum of a seed oil: the expanded olefinic region. Ln = linolenic acid; L = linoleic acid; O = oleic acid.

components of the pyrolysate on the basis of their mass-to-charge ratio (m/z) to produce a pyrolysis mass spectrum (Meuzelaar, Haverkamp and Hileman, 1982), which can then be used as a 'chemical profile' or fingerprint of the sample material analysed. The spectrum obtained may then be used as an input to some analysis tool such as Neural Nets, PCA, PCR, PLS or other statistical or computational techniques (Michie, Spiegelhalter and Taylor, 1994; Weiss and Kulikowski, 1991), allowing the sample to be categorized. Much work has been done in this field already, including the use of bacterial strains with both standard back-propagation ANNs (Goodacre, Neal and Kell, 1994) and Kohonen ANNs (Goodacre et al., 1996), complex binary and tertiary mixtures (Goodacre, Neal and Kell, 1994), casamino acids and glycogen (Goodacre, 1993; Neal, 1994), microbial fermentations (Goodacre, 1994b; Goodacre and Kell, 1996; Goodacre et al., 1995) and the adulteration of foodstuffs such as olive oils (Goodacre, Kell and Bianchi, 1992, 1993) and orange juice (Goodacre, Hammond and Kell, 1997). It has also been successfully applied to other discrimination problems, by Aylott et al. (1994) with PCA and CVA for classification of some of the 500 brands of Scotch whisky, and by Kajioka and Tang (1984) to distinguish between six species of the family *Legionellaceae*. The latter led to the identification of a new strain of the bacterium.

330 LIPID ANALYSIS IN OILS AND FATS

In Curie-point pyrolysis the material is placed on an iron–nickel alloy foil which is heated to the Curie point of the foil (this is 530°C for 50:50 Fe–Ni foils). For a given type of foil, the Curie-point temperature is constant, therefore this type of pyrolysis is very reproducible. The foil holding the sample is rapidly heated to its Curie point by passing a radio-frequency current for 3 s (in the case of the Horizon 200-X instrument used at Aberystwyth) through a coil surrounding the foil. The foil takes around 0.5 s to reach this point; at this temperature the material on the foil is thermally

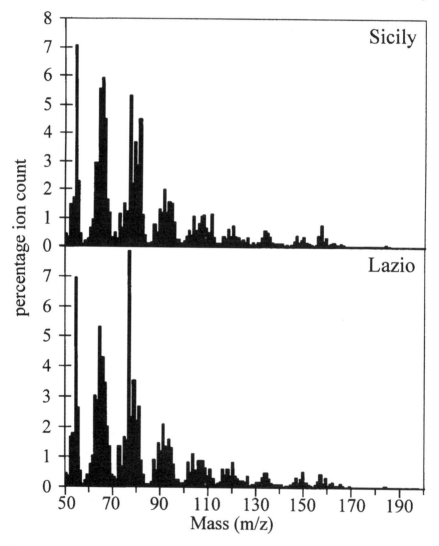

Figure 10.3 Two typical pyrolysis mass spectrometry spectra of extra virgin olive oils from Sicily and Lazio.

degraded into its pyrolysate. The pyrolysate is then separated into its components by the mass spectrometer part of the equipment, producing a pyrolysis mass spectrum (Fig. 10.3). Goodacre (1994a) and Goodacre and Kell (1996) give a very thorough analysis of the PyMS technique and equipment, along with a brief history. More detail is given in the books by Irwin (1982) and Meuzelaar, Haverkamp and Hileman (1982).

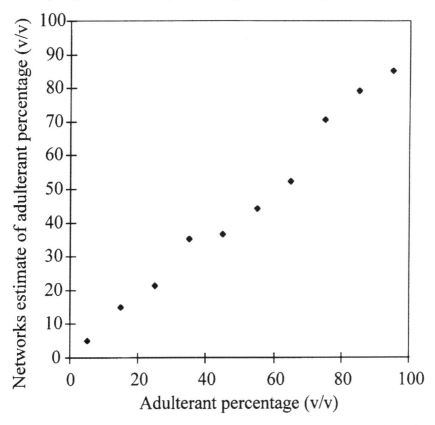

Figure 10.4 A typical prediction curve for adulteration of extra virgin olive oil. Adulteration series using from 0%–100% adulterant were prepared using husk oil, with samples every 5% [a total of 21 samples (in triplicate)]. 1.5 µl of sample was analysed by pyrolysis mass spectrometry and the spectra collected over the m/z range 51–200. Data were normalized as a percentage of total ion count to remove the most direct influence of sample size *per se*. Normalized spectra were sorted according to the level of adulteration, with triplicates being kept together. Half of these samples (every other one) were used to train a neural net, the remaining half being used as a test set. The network architecture is composed of an input layer of 150 nodes, one node for each mass within the spectrum of each oil, a hidden layer of 8 nodes and a single output node. Headroom was maintained at ±10% for each input. A sigmoid (logistic) transfer function was used, as the data relationship is suspected to be non-linear. The supervised learning algorithm was standard back propagation (that updates weights as each pattern is presented) with a learning rate of 0.2 and a momentum of 0.8. Patterns were presented randomly with noise of 0.05 added. All data going into and out of the net were scaled automatically (each column individually) by the software.

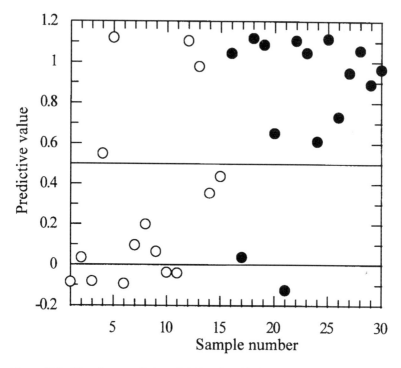

Figure 10.5 Neural net prediction of virgin olive oil adulterated at the 1% level by high oleic sunflower oil. (a) Prediction with variable selection, 60 samples, 30 adulterated by 1% vol./vol. high oleic sunflower oil and 30 unadulterated, were run through the pyrolysis mass spectrometer and the spectra collected. Data were normalized to the total ion count of the samples and principal components analysis (PCA) performed. The first factor from the PCA analysis was then used to select the variables (58) which were fed into an artificial neural net using a stochastic back-propagation algorithm; 15 adulterated and 15 unadulterated samples, encoded, respectively, as 1 and 0, were used to train the net (58 nodes on the input layer, 4 nodes in the hidden layer and a single output node), and 30 samples were used to test the model. The results for the unseen test set are shown, and were obtained after 29 000 epochs with an error on the training set of 0.01. The horizontal line at 0.5 is an aid to interpretation. All samples with a value close to 1 (i.e. those above the line) are the predicted adulterated samples (actual adulterated samples being filled circles); all those with a value close to 0 (i.e. those below the line) are the predicted unadulterated samples (actual unadulterated samples being open circles); 11 out of 15 (73%) unadulterated samples and 13 out of 15 (87%) adulterated samples being predicted correctly.

One of the great advantages of PyMS is that it is relatively cheap compared with other methods of analysis. Many samples can be run through the PyMS machine in a short time (typically less than 2 min each) at a cost of less than £1 sterling per sample.

The atomic masses up to 50 are discarded since they include very common compounds such as methane (CH_4, 16 amu), ammonia (NH_3, 17 amu), water (H_2O, 18 amu), methanol (CH_3OH, 32 amu) and hydrogen sulphide (H_2S, 34 amu), which are likely to be present in large quantities in any pyrolysate. Fragments with an *m/z* ratio of over 200 are rarely analytically important for bacterial discrimination so these are also discarded (Good-

CHEMOMETRIC METHODS 333

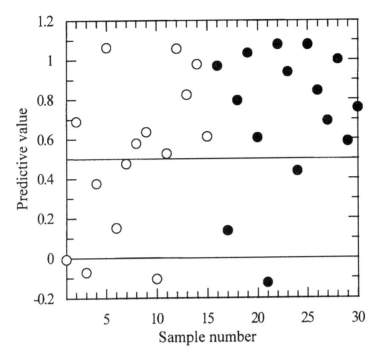

Figure 10.5(b) Prediction with no variable selection. The experimental rationale was as for part (a), except that the artificial neural network was run on all 150 variables, that is, no variable selection was used. Optimization in this case resulted in having 8 nodes in the hidden layer; the outputs were obtained after 10 000 epochs with an error on the calibration set of 0.01; 6 out of 15 (40%) unadulterated samples and 12 out of 15 (80%) adulterated samples were predicted correctly. It may be seen that very poor separation was achieved – the two data bases do not line up on the predictive value of 0 or 1 but form a loose cloud in the middle area around 0.5, indicating that the net was unable to separate the two groups clearly.

acre, 1994a). It is assumed that these fragments are also unimportant for other organic compounds, although no reference has been found to support or contradict this belief.

The exploitation of PyMS for assessing the adulteration of olive oil (Goodacre, Kell and Bianchi, 1992, 1993) has been continued by Bianchi, Giansante and Lazzari (1996) and by Salter *et al.* (1997) who have shown that the principle is generally and quantitatively applicable to a wide range of potential adulterants, including olive oil, hazelnut oil, husk oil, corn oil, peanut oil, maize oil, soya oil, sunflower oil, high oleic acid sunflower oil and grape stone, along with rapeseed/soya and palm/peanut/sunflower oil mixes (Fig. 10.4). Predictions may be improved upon in some cases by the use of variable selection, and in many cases adulterants may be detected when present at less than 2% (vol./vol. Figs 10.4–10.6). Bianchi *et al.* (1994b) suggest that adulteration by lower grades of olive oil may be detected by the presence of long chain esters.

334 LIPID ANALYSIS IN OILS AND FATS

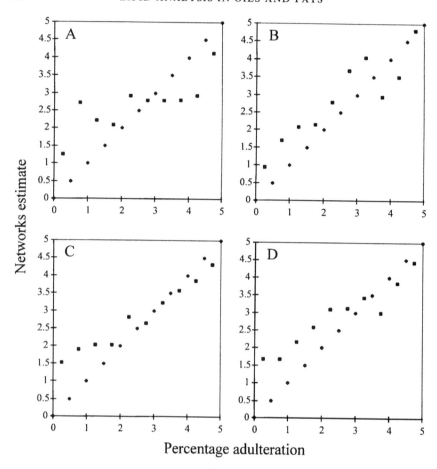

Networks estimate

Percentage adulteration

Figure 10.6 Adulteration series from 0%–100% were prepared using husk oil, with samples every 5% [a total of 21 samples (in triplicate)]; 1.5 µl of sample was analysed by pyrolysis mass spectrometry in triplicate over the m/z range 51–200. Data were normalized as a percentage of total ion count to remove the most direct influence of sample size *per se*. Normalized spectra were sorted according to level of adulteration, with triplicates being kept together. Half of these samples (every other one) were used to train a neural net, the other half being used as a test set. The network architecture is composed of an input layer corresponding to the number of variables used, one node for each mass used within the spectrum of each oil. The optimum number of nodes within the hidden layer varied, whilst a single output was used that represented the network's estimation of the adulteration as a numeric value. Headroom was maintained at ±10% (for each input) for all networks. A sigmoid (logistic) transfer function was used (unless stated otherwise) as the data relationship was suspected to be non-linear. The supervised learning algorithm was standard back propagation (that updates weights as each pattern is presented) with a learning rate of 0.2 and a momentum rate of 0.8. Patterns were presented randomly with noise of 0.05 added. All data going into and out of the net were scaled automatically (each column individually) by the software. (A) Prediction using no variable selection. All 150 masses are used, with 8 nodes in the hidden layer. The network was run over 3622 epochs to a training error of 0.025 and a test set RMSEP of 1.25. After this point overtraining occurred. (B) Mutual information was used as a method of variable selection. All 150 variables were assessed using mutual information (Battiti, 1994). The three variables that

10.4 Multivariate methods

For the purposes of this chapter, it is convenient to consider the complete set of measured quantities (e.g. the different m/z ratios of PyMS ion counts) for a given sample as being a single n-dimensional position vector, $x^T = (x_1\, x_2 \ldots x_n)$ with x_i being the measured value of quantity i. A set of m such vectors is represented as a matrix, $\mathbf{X}$, with rows, x_i^T. The value stored at row i, column j of $\mathbf{X}$ is denoted X_{ij}.

A scores matrix is one in which the columns contain the values of transformed variables. An example of a scores matrix is that containing principal component values. A loadings, or weightings, matrix is one by which a data matrix may be multiplied to obtain a scores matrix.

10.4.1 Principal components analysis

PCA was devised by Hotelling (1933) as an aid to the interpretation of academic test results. The reasoning behind PCA is that a set of test results often shows correlations between the different types of test. These results are presumably reflecting some 'mental factor'. Hotelling derived PCA as a method for extracting a set of uncorrelated variables as linear combinations of the original variables. The 'mental factors' are contained in the new data set. Hotelling named the new variables the 'components' and suggested that, if the components were arranged in decreasing order of variance, then the first few, the 'principal components', should be those which characterized the 'mental factors'. Later components would reflect less important effects, with some possibly reflecting measurement errors. Tests were suggested which would allow later components to be discarded as noise, so leaving a lower-dimensional data set, with the most easily interpreted effects appearing first. Figure 10.7 demonstrates how this can be useful. Even a three-dimensional cloud of points can be difficult to interpret if it is viewed from a poor vantage point. By rotating the cloud in a manner which is in some way

---

were clearly the most important were fed into an artificial neural network (ANN) having 4 nodes in the hidden layer. After some 363 483 epochs a training error of 0.082 and a test RMSEP of 1.1 was recorded. (C) The w value (text, Section 10.4.11) was used as a method of variable selection. All masses were ranked according to the w value and the best 50 selected as an input for an ANN that had 8 nodes in the hidden layer. After some 9357 epochs a training error of 0.05 and test RMSEP of 0.99 was achieved. (D) Principal components analysis was used as the method for variable selection. All masses were ranked according to the score of the first principal component, the best 50 variables being selected as the input for an ANN. The ANN consisted of an input layer of 50 nodes, a hidden layer of 8 nodes and a single output node. In this case a linear transfer function was found to be optimal, perhaps reflecting the linear nature of the variable selection. After some 233 epochs a training error of 0.1 and a test RMSEP of 1 14 for the illustrated prediction was achieved.

optimal, one can obtain a much clearer view of any structure within the data. PCA helps to achieve this aim.

Hotelling's derivation of PCA was made by assuming that the original n variables in $\mathbf{X}$ were normally distributed. The overall distribution of the n-dimensional vector set of samples in $\mathbf{X}$ is therefore a multivariate normal distribution and will appear ellipsoidal if plotted as a scatter plot in n-dimensional space. The method used was to derive a formula for the optimally fitting ellipsoid, using least squares methods for the optimization.

An alternative derivation may be used which gives the same results as Hotelling's but which does not require the assumption of normality of the variables in $\mathbf{X}$, as follows.

Suppose one has a cluster of n dimensional points, x_i. The requirement is to find a set of axes for the cluster such that when the x_i are transformed linearly to the new axes the transformed variables are uncorrelated, that is one needs to find n orthogonal unit vectors, p_i such that the $\mathbf{X}p_i$ and $\mathbf{X}p_j$ are uncorrelated for $i \neq j$. Assuming that the variables in $\mathbf{X}$ have been centred to have zero mean, and scaled to unit variance, one has $\mathbf{S} = \mathbf{X}^T\mathbf{X}$ as the correlation matrix of $\mathbf{X}$.

One must therefore find an orthonormal matrix, $\mathbf{P}$, such that $\mathbf{S}' = [\mathbf{XP}]^T[\mathbf{XP}]$ has non-zero values on the diagonal only, that is:

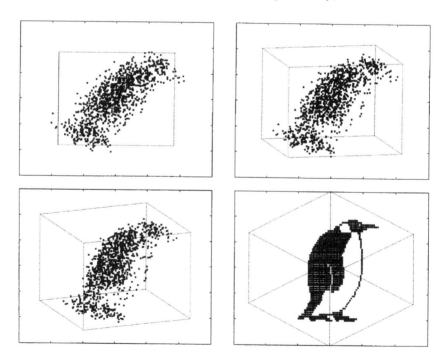

Figure 10.7 Example of how principal components analysis can help to extract useful information.

$$\mathbf{S}' = \begin{bmatrix} a_1 & 0 & 0 & \cdots & 0 & 0 \\ 0 & a_2 & 0 & \cdots & 0 & 0 \\ \vdots & \vdots & & \ddots & & \\ 0 & 0 & 0 & \cdots & 0 & a_n \end{bmatrix} = \text{diag}(a_1, \ldots, a_n), \quad (10.4)$$

$$\mathbf{S}' = [\mathbf{XP}]^T[\mathbf{XP}] = \mathbf{P}^T\mathbf{X}^T\mathbf{XP} = \mathbf{P}^T\mathbf{SP} = \text{diag}(a_1, \ldots, a_n). \quad (10.5)$$

$\mathbf{S}$ is a symmetric matrix [since correlation (X_i, X_j) = correlation (X_j, X_i)] of real values. Therefore, from the theory of eigenanalysis, it follows that $\mathbf{P}$ is the matrix whose columns are the orthogonal eigenvectors of $\mathbf{S}$ and that $a_1, \ldots, a_n$ are the corresponding eigenvalues. For a more detailed discussion of eigensystem theory see, for example, Morris (1982).

The problem of finding the components of $\mathbf{X}$ is therefore that of deriving the eigensystem of $\mathbf{S}$, the correlation matrix of $\mathbf{X}$.

Next, consider a component, T_i. The sample values of T_i, t_i are derived from $\mathbf{X}$ by $t_i = \mathbf{X}p_i$, where p_i is an eigenvector. Let λ_i be the corresponding eigenvalue. Then

$$\begin{aligned} \text{var}(T_i) &= t_i^T t_i \\ &= (\mathbf{X}p_i)^T(\mathbf{X}p_i) \\ &= p_i^T \mathbf{X}^T \mathbf{X} p_i \\ &= p_i^T \mathbf{S} p_i \\ &= p_i^T \lambda_i p_i \\ &= \lambda_i p_i^T p_i \\ &= \lambda_i \end{aligned} \quad (10.6)$$

So, to extract the components in decreasing order of variance, one need only extract the eigenvectors in decreasing order of eigenvalue.

Finally, consider

$$\sum_{i=1}^{n} \text{var}(T_i) = \sum_{i=1}^{n} \text{var}\left(\sum_{j=1}^{n} P_{ji} X_i\right) = \sum_{i=1}^{n}\sum_{j=1}^{n}\sum_{k=1}^{n} P_{ji} P_{ki} S_{jk} \quad (10.7)$$

However, $\mathbf{P}$ is orthonormal, so its columns (and rows) must be linearly independent and the dot product of any two columns (rows) must be 0 if they are not the same, or 1 if they are. That is,

$$\sum_{i=1}^{n} P_{ji} P_{ki} = \delta_{jk}, \quad \text{where } \delta_{jk} = 1 \text{ if } j = k, 0 \text{ otherwise}; \quad (10.8)$$

$$\sum_{i=1}^{n} \text{var}(T_i) = \sum_{k=1}^{n} \sum_{j=1}^{n} S_{jk} \delta_{jk}$$

$$= \sum_{i=1}^{n} S_{ii} \qquad (10.9)$$

$$= \sum_{i=1}^{n} \text{var}(X_i) \ .$$

So, the proportion of the variance in **X** explained by component i is therefore $\lambda_i \left[\sum_{i=1}^{n} \text{var}(X_i) \right]^{-1}$, or λ_i/n for X_i scaled to variance 1.

By using the above, one may then extract principal components (by calculating eigenvectors) in decreasing order of variance (by observing the corresponding eigenvalues) and know the proportion of the variance in **X** explained by any component (by observing the eigenvalue size).

Extracting principal components. The mathematical literature details a number of methods for determining the eigensystem of a matrix. In the case of principal component extraction, the matrix $X^T X$ is square and symmetric, as

$$(X^T X)_{ij} = \sum_{k=1}^{n} X_{ik}^T X_{kj} = \sum_{k=1}^{n} X_{jk}^T X_{ki} = (X^T X)_{ji}, \qquad (10.10)$$

which makes a number of methods available.

Jacobi's method. Jacobi's method allows all n eigenvectors and eigenvalues to be extracted at the same time. It works by applying a sequence of transformations to reduce $X^T X$ to diagonal form. The product of the transformation matrices is a matrix having the eigenvectors as columns, and the diagonal elements of the reduced matrix are the eigenvalues. The off-diagonal elements of the matrix are zeroed by applying a plane rotation matrix to each off-diagonal position (i, j). A full description of this method may be found in Gourlay and Watson (1973).

Unfortunately, Jacobi's method is computationally intensive, requiring of the order of $6n^3$ arithmetic operations to calculate the eigensystem.

The power method. The power method may be used to determine the largest eigenvalue and its corresponding eigenvector. The method uses the iterative scheme $z_k = A z_{k-1}, k = 1, 2, 3, \ldots$ where A is the $n \times n$ matrix and z_0 is initialized to a first estimate of the eigenvector. Then z_k tends to the eigenvector and $(z_k)_i/(z_{k-1})_i$ to the eigenvalue as $k \to \infty$ [where $(z)_i$ indicates element i of the vector, z]. Each iteration uses order n^2 operations, with the convergence rate being proportional to the ratio of the first two eigenvalues, ratios close to unity giving slow convergence. The method fails for largest eigenvalues of equal modulus, with the eigenvector oscillating in sign.

Subsequent eigenvalues can be found by transforming the matrix, **A**, such that the remaining eigensystem is retained, but the influence of z is removed. This process is known as deflation and may be achieved in a number of ways. Again, see Gourlay and Watson (1973) for further details.

Simultaneous iteration. Simultaneous iteration (Clint and Jennings, 1970) allows the simultaneous extraction of the k largest eigenvalues and their corresponding eigenvectors. The method works by reducing the problem to that of calculating the eigensystem of a reduced matrix of size $k \times k$. This eigensystem is used to improve an estimate of the eigenvectors of the larger matrix. The process is iterated until the required accuracy is obtained. If $k \ll n$, then a great saving in time may be obtained.

The non-linear iterative partial least squares method. The previous methods required the covariance matrix, $\mathbf{X}^T\mathbf{X}$ to be calculated prior to extraction of the eigensystem. The non-linear iterative partial least squares (NIPALS) algorithm (Wold, 1966, 1975) sidesteps this requirement by noting that component weights may be derived from **X** and the corresponding components by linear regression. The components are, by definition, obtained from **X** and the weights vector. Thus, an iterative process can be defined:

- make an initial estimate for the scores vector, t_0;
- repeat the following:
  - Regress **X** on t_{n-1} to obtain an estimate of the loadings vector, p_n:

$$p_n^T = (t_{n-1}^T t_{n-1})^{-1} t_{n-1}^T \mathbf{X};$$

  - normalize p_n to have unit length;
  - project the **X** matrix to form a new scores vector, $t = \mathbf{X}p_n$;

  until p_n converges.

This process generates the first principal component loadings and scores for **X**. Their effect may then be removed from **X** by projecting t back into the coordinate system used for **X** and subtracting:

$$\mathbf{X}' = \mathbf{X} - tp^T$$

The iteration can be repeated on $\mathbf{X}'$ to obtain the second and subsequent components. This algorithm is of order n^2 for each component extracted, depending on the convergence rate.

10.4.2 Predictive models

The methods discussed thus far allow the exploration of a multidimensional data set, and may allow clusters to be separated, but they do not result in predictive models.

To build a system capable of predicting characteristics of interest from measured values, we need to form a model which relates the measurements to the physical effect of interest. This model can then be applied to future measurements to predict the effect. In mathematical terms, the objective is to find the multidimensional function, f, such that $y = f(x)$, where x is a new multivariate reading and y is the effect to be predicted. For the present, we shall assume that f is a linear function, for simplicity. Artificial neural networks (ANNs), a non-linear method, will be discussed later.

In the linear case, then, we have y being a simple weighted sum of the variables x: $y = b_0 + b_1 x_1 + b_2 x_2 + \ldots + b_n x_n$. In vector notation, this may be expressed $y = x \cdot b + b_0$, b_0 being a constant offset. Finally, if we augment x with an entry, $x_0 = 1$, then b may be augmented with b_0 to give $y = x \cdot b$.

If we wish to predict a number of values of y, we may express this as a single matrix multiplication, $y = Xb$, where the x vectors have now been placed as rows of X.

For model formation, then, the problem is that of choosing the vector, b to give good modelling of the values of y. In the following discussions, let X_c and y_c be a set of measured calibration values for the variables x and y. A number of methods for choosing an estimate for b, $\hat{b}$ from X_c and y_c will now be considered.

10.4.3 Multiple linear regression

Multiple linear regression (MLR) provides the most 'obvious' course of action for estimating b, that of minimizing errors in the model. A relationship, $y_c = X_c b + e$ is assumed, with e containing all unmodelled variation in y_c. For an arbitrary value of b, e will be made up of two parts: the systematic error due to poor choice of b, and the random error of measurement in X_c and y_c. For MLR, the random measurement errors are assumed to have zero mean (i.e. the true value) and equal variance. Under this assumption, it is obvious that the best model we can choose is that which reduces the length of e to a minimum. In practice, it is simpler to minimize the squared length of e, or $(y_c - X_c b)^T (y_c - X_c b)$. This method is known as least squares regression and is familiar in the single dimensional case as finding the 'line of best fit'. The least squares estimator for b is given by $b = (X_c^T X_c)^{-1} X_c^T y_c$ (for a derivation, see Wonnacott and Wonnacott, 1981).

Note that this formula involves the inversion of $X_c^T X_c$, the covariance matrix of X. If this inverse does not exist, then $\hat{b}$ cannot be calculated by means of MLR. Singularity of $X_c^T X_c$ corresponds to the linear dependence of a subset of the X variables. So, if X_c contains such relationships, MLR cannot be applied. In practice, it is rare to have exact linear dependence, since the measurement errors will tend to preclude this. However, near linearity will tend to make the inverse numerically unstable and subject to large errors, so much the same effect occurs. Another formulation of the

MLR equation can be obtained by expressing it in terms of the matrix of eigenvectors, **P**, and the diagonal matrix of eigenvalues, diag(λ_i), as follows:

$$\mathbf{P}^T(\mathbf{X}^T\mathbf{X})\mathbf{P} = \text{diag}(\lambda_i), \qquad (10.11)$$

$$(\mathbf{X}^T\mathbf{X})^{-1} = \mathbf{P}\,\text{diag}\left(\frac{1}{\lambda_i}\right)\mathbf{P}^T \qquad (10.12)$$

$$\hat{\boldsymbol{b}} = \mathbf{P}\,\text{diag}\left(\frac{1}{\lambda_i}\right)\mathbf{P}^T\mathbf{X}^T\boldsymbol{y}. \qquad (10.13)$$

If **X** contains (almost) collinear variables, then λ_i will be (close to) zero for some i, giving an eigenvalue matrix containing large values which are susceptible to small alterations in the calibration data. In our application, collinearity of the **X** variables is to be expected, since harmonic responses at similar frequencies and amplitudes are likely to be similar.

MLR may be modified to reduce the problems of collinearity through the use of variable selection. Rather than calculating the model relating y_c to $\mathbf{X}_c$, we relate y_c to $\mathbf{X}'_c$, a matrix containing a subset of the original **X** variables. The subset is chosen to eliminate collinearity and to improve the model by removing irrelevant variables (a discussion of overfitting is given in Section 10.4.7). Three common methods are used for subset selection: Forward selection (FS), backward elimination (BE) and stepwise multiple linear regression (SMLR).

Forward selection. Here the model is built up by adding variables in **X**, one at a time, to the model. The variables are added by choosing those which improve the model most at each step according to some statistical test. Variables which are collinear with a variable already added will contribute little to the quality of a subsequent model. Likewise irrelevant variables contribute little. Such variables will not, therefore, be added. Variables are added until a 'stopping' criterion is reached, for example when the improvement afforded by addition of the next variable drops below a threshold value.

Backward elimination. Here, all the variables are used to form an initial model. Variables are then selected for deletion, once again by using a statistical test and threshold to determine when to stop. In the case of BE, exactly collinear variables must be eliminated before the initial model is formed.

Stepwise multiple linear regression. This is a modified form of forward selection. The model starts out including only one variable, and more variables are subsequently added. But at each stage a BE-style test is also applied. If a variable is added, but becomes less important as a result of subsequent additions, SMLR will allow its removal from the model.

A similar method, 'best of all subsets', regression considers models formed from all possible subsets of the **X** variables. For each subset the linear regression is formed and tested for its quality. The best of these is selected as the final regression set. Although this method is guaranteed to do at least as well as the stepwise methods, the problem of combinatorial explosion rears its head: for n variables in **X** the number of possible subsets is 2^n. Even for moderate values of n, the number of regressions required can be too large to be practical. Near-infra-red spectroscopic methods, for example, typically have hundreds of variables. Heuristics may be used to select a feasible subset of the original variables on which to run 'best of all subsets' regression, but one is still left with the problem of selecting this subset.

10.4.4 Ridge regression

Ridge regression (RR) (Hoerl and Kennard, 1970a, b) is an attempt to solve the problems of collinearity in MLR by modifying the regression equation. Rather than having $\hat{b} = (\mathbf{X}^T\mathbf{X})^{-1}\mathbf{X}^T y$, as for MLR, RR forms the family of regressors $\hat{b}^* = (\mathbf{X}^T\mathbf{X} + k)^{-1}\mathbf{X}^T y$, where $0 \leq k \leq 1$. In eigenvector and eigenvalue form, this is stated as $\hat{b}^* = \mathbf{P}\,\text{diag}\,(1/(\lambda_i + k))\mathbf{P}^T\mathbf{X}^T y$, where the λ_i and **P** are eigenvalues and eigenvectors of $\mathbf{X}^T\mathbf{X}$. By varying k we can prevent division by small values and so make the model more stable against perturbations of the calibration set. The new regressor, $\hat{b}^*$ will be biased, tending to centre on a value removed from the optimal **b**, but will have a smaller mean square error. For a well-chosen value of k, we can make $\hat{b}^*$ lie close to the true **b** and have better stability.

RR introduces the concept of the 'ridge trace'. This is a graph showing the weightings, $\hat{b}_i^*$ as k varies from 0 to 1. The ridge trace highlights those variables which are unstable, as these will show great variation over the range of the trace. These variables are then candidates for deletion. The ridge trace can be used to guide the choice of k. A value can be chosen where the individual traces have 'flattened' out.

10.4.5 Principal components regression

Principal component regression (PCR) combines PCA and MLR to give another method for removing the effects of collinearity (Massy, 1965). PCA generates components which are orthogonal and in decreasing order of eigenvalue (variance). Hence, to remove variables with small eigenvalues, one needs merely to select a threshold eigenvalue and discard all components below this threshold. This corresponds to selecting the first n components. A simple MLR model is then formed on the remaining components.

For modelling, then, we have $\hat{b} = (\mathbf{T}^T\mathbf{T})^{-1}\mathbf{T}^T y$, where $\mathbf{T} = \mathbf{X}\mathbf{P}$, **P** being the matrix of retained component weightings. For prediction, we must first transform the new **X** data by using **P**; $\hat{b}$ is then used to make the prediction. So, $\hat{y} = (\mathbf{XP})\hat{b}$. If we combine $\mathbf{P}$ and $\hat{b}$ to give a new $\hat{b}$, $\hat{b}_{PCR}$, we may predict values by using the standard prediction $\hat{y} = \mathbf{X}\hat{b}_{PCR}$, with $\hat{b}_{PCR} = \mathbf{P}\hat{b}$.

If no components are excluded, then we have

$$\begin{aligned}\hat{b}_{PCR} &= \mathbf{P}[(\mathbf{XP})^T(\mathbf{XP})]^{-1}(\mathbf{XP})^T y \\ &= \mathbf{P}(\mathbf{P}^T\mathbf{X}^T\mathbf{XP})^{-1}\mathbf{P}^T\mathbf{X}^T y \\ &= \mathbf{PP}^T(\mathbf{X}^T\mathbf{X})^{-1}\mathbf{PP}^T\mathbf{X}^T y \\ &= (\mathbf{X}^T\mathbf{X})^{-1}\mathbf{PP}^T\mathbf{X}^T y \\ &= (\mathbf{X}^T\mathbf{X})^{-1}\mathbf{X}^T y \\ &= \hat{b}_{MLR}\end{aligned} \qquad (10.14)$$

Hence, PCR and MLR provide exactly the same estimator when no principal components are excluded.

PCR resembles RR when considered from an eigenvalue perspective. Where RR uses the gambit of offsetting eigenvalues to prevent them being small, PCR merely deletes them. RR gives a biased estimation through application of the offset, whereas the PCR estimator is unbiased. A good coverage of PCA and PCR applications is given by Jolliffe (1986).

Although PCR allows us to delete noise effects, we are still left with the possibility of irrelevant systematic effects being retained. Jolliffe (1982) and Davies (1995) both note the substantial inertia amongst users of PCR to the application of selection criteria to the generated principal components.

10.4.6 Latent variables

Latent variables are a useful concept when dealing with high dimensional sets containing collinearity. Consider, for PCA, we have $\mathbf{T} = \mathbf{XP}$, where $\mathbf{P}$ is the eigenvector matrix and $\mathbf{T}$ is the component scores matrix. We have already seen that only a few of the components may be significant so define $\mathbf{P}'$ to contain only the first few eigenvectors. Given $\mathbf{T}$ and $\mathbf{P}'$, we can estimate $\mathbf{X}$ by $\hat{\mathbf{X}} = \mathbf{P}^T\mathbf{T}$. Viewing the model in this form, we have a small number of variables (the components) which are responsible for the $\mathbf{X}$ observations. These underlying variables are termed 'latent variables'.

The latent variables may be considered as causal effects for our observations $\mathbf{X}$. We wish to model the data $\mathbf{Y}$ from our observations. Hence it makes sense to form the model for $\mathbf{Y}$ on the causes rather than the effects. Given the data in $\mathbf{X}$ we can estimate the causes and from these form the model.

Figure 10.8 shows the relationship between traditional methods such as MLR and latent-variable-based methods. Using the traditional approach, the effects of any t_i are spread throughout the variables in $\mathbf{X}$. We have to select one of these to represent each t_i. The value of x_i selected is unlikely

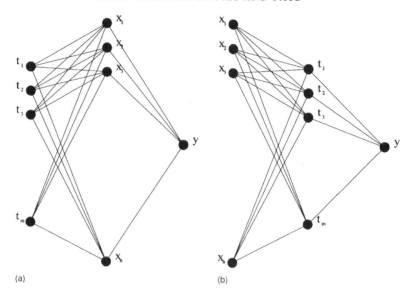

Figure 10.8 Causal relationships in modelling: (a) traditional multiple linear regression; (b) latent-variable-based modelling.

to represent the variations of t_i alone, so the selection process is likely to introduce irrelevant effects which are inseparable from the relevant. By choosing a single value, x_i, to represent t_i, we are also making the model susceptible to noise. An error in measuring the single x_i will strongly affect the prediction, since we are relying entirely on x_i for the contribution of t_i.

Now consider the latent variable method. In this case, we are using all the x_i to model each t_i. An error in measuring a single x_i will not cause a major error in t_i, since we are (weighted) averaging over a number of x_i. At the second stage, we are predicting y by using a number of independent variables, so the effects of irrelevant latent variables can be completely removed.

Latent variable methods have another advantage when the number of variables in **X** is high. For simple MLR, we have to invert $(\mathbf{X}^T\mathbf{X})$, which, as noted, means that the columns of **X** must be linearly independent. If **X** contains more columns than rows, this is not possible. Hence, for simple MLR we must have more samples than variables for the model to be stable. For spectroscopic applications, where hundreds of variables can be present, this is often not feasible. In these cases, then, we are forced to take fewer samples and immediately discard a number of variables, losing information in the process. The use of latent variables avoids this. If m samples of n variables are taken, with $m < n$, only m latent variables will have non-zero variance. All of the original variability in the data can therefore be retained for investigation.

10.4.7 Validation

When forming a model from calibration data, we encounter a problem. We are calibrating $\mathbf{X}_c$ against y_c. Since we have no extra information, the only method is to form a model which explains the y_c readings in terms of the X_c readings. It is possible to form a model which succeeds in relating X_c to $\hat{y}_c$, but has poor predictive power.

As an example, consider the following extreme model:

$$\hat{y} = \begin{cases} y_{ci} & \text{if } x = x_{ci}, \text{for some } i; \\ 0 & \text{otherwise} \end{cases} \quad (10.15)$$

This model will predict $\hat{y}_c$ perfectly from $\mathbf{X}_c$ for any calibration set, but is unlikely to supply correct answers in the future. This situation, called overfitting, can occur very easily. Without reference to an external set of test data, we have no way of knowing whether we are losing predictive power. The model can be improved to the point where, rather than modelling real effects, it is modelling noise. At this point the data have been overfitted.

Conversely, if we are too cautious and stop improving the model too soon, we will underfit the calibration data by not taking relevant effects into consideration. In this case, when we come to using the model for prediction, the unconsidered effects may change and push the model away from correct answers.

So, how can we tell the difference between an underfitted, good, or overfitted model when we have no external reference points? This problem is the basis of model validation.

In order to assess the utility of a model, we need to obtain a value which gives a measure of its ability to predict future values. A commonly used measure is the mean square error of prediction (MSEP) and its root (RMSEP) (Allen, 1971). This is simply the mean squared difference between the predicted values of Y and the true values of Y for a given model. Obviously, better models will yield better MSEP values, so the aim of model formation is to minimize the MSEP.

When forming a model, we wish to minimize the influence of random variations in the data and maximize the influence of the underlying causal effects. It is therefore important to use many (X, Y) pairs for training, as the standard error of a sample set is inversely proportional to the number of samples. Conversely, for estimation of the MSEP we need many (X, Y) test pairs, for exactly the same reasons. When forming a model, we usually have a limited number of (X, Y) pairs to use. There is therefore a trade-off between using samples for model formation and MSEP estimation. A number of methods have been suggested for addressing this problem and these will now be discussed.

Self-prediction. It may be tempting to use self-prediction to estimate the MSEP for a model. In self-prediction the whole calibration set is used to

346   LIPID ANALYSIS IN OILS AND FATS

form the model. The MSEP is then estimated with the same calibration set. This method makes good use of the available observations because as many as possible are used for both modelling and validation. However, this is not a sensible method to use in general. Consider our trivial model above. It predicts every value in the calibration set perfectly and therefore has an estimated MSEP of zero for self-prediction validation. The true MSEP will be much greater and the model is in fact useless in all but contrived circumstances.

Self-prediction does, however, have the advantage that its MSEP estimate gives a lower bound on the true MSEP of the model. It is often instructive, when using latent variable models, to view a graph of the MSEP for self-prediction versus the number of latent variables. The 'true' MSEP graph will usually tend to show a local minimum as we move from underfitting to overfitting. In the case of self-prediction, the MSEP graph often shows

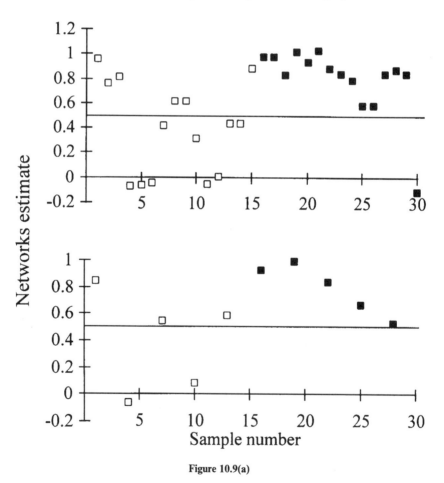

Figure 10.9(a)

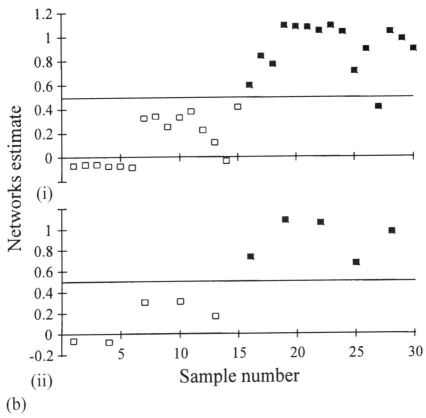

Figure 10.9 Prediction of the region of origin of extra virgin olive oils by means of artificial neural networks (ANNs). (a) ANNs trained on randomly split data (the output was taken after 27 000 epochs, with a training error of 0.009): (i) all samples in triplicate form; (ii) averaged output for each triplicate form. (b) ANNs trained on data split by means of the Duplex partitioning method (the output was taken after 8000 epochs, with a training error of 0.0196): (i) all samples in triplicate form; (ii) averaged output for each triplicate form. In each training method a three-layered ANN (150 input nodes, 6 hidden nodes and a single output node) was trained with 33 objects (15 oil samples from Lazio, coded 0, and 18 samples from Sicily, coded 1) and interrogated periodically by using 30 test objects [15 Lazio (□) and 15 Sicily (■)] previously unseen by the training net. The horizontal line corresponds to a network estimate of 0.5.

similar reductions in MSEP until the 'true' MSEP local minimum, tending then to trail off slowly to zero as the number of latent variables increases. A sudden change of slope can often be seen in the self-prediction MSEP graph, this being useful in estimating the number of latent variables to use.

So, if there is no alternative to self-prediction, it may be used with great care to make an attempt to select the number of factors, but its use is in general 'dangerous' and to be avoided.

Training and test sets. If more data can be obtained easily, then the validity of a model can be tested by using a new set of observations to estimate the

MSEP directly. The calibration set is used merely for forming the model, with the new data being used for validation.

Unfortunately this method depends on the availability of these extra observations. The aim of multivariate calibration is often to allow prediction of values of Y which are difficult or expensive to measure. When this is the case, time or expense can preclude the acquisition of large amounts of test data. In fact it is often the case that a single set of observations is presented with which to calibrate and validate the model. This complicates matters somewhat.

The simplistic approach in this situation is to partition the set of observations into two sets and to use one for calibration and the other for validation. Each set must be chosen to contain a representative spread of values of X and Y, otherwise the model will be formed or validated from a subset of the available range, leading to erroneous predictions or estimations of validity. Snee (1977) suggests a number of methods by which a suitable partition of the data may be achieved. Figure 10.9 shows the effect of different partitioning regimes.

The trade-off between calibration and prediction set sizes mentioned above is especially important in this situation. A model based on only half the observations is quite likely to contain an unrepresentative set of calibration objects. Increasing the number of observations for calibration will reduce the chances of choosing a poor calibration set while reducing the quality of the MSEP estimate.

Full cross-validation. Ideally, what is needed is a hybrid of self-prediction and training and test set validation. We need to use all observations in both model formation and validation without encountering the problems of self-prediction. Full cross-validation (Geisser, 1975; Stone, 1974) attempts to do just this. The term 'cross-validation' is often applied to the partitioning form of training and test set validation and it is from this that full cross validation was developed (from here on we will use the term 'cross-validation' to refer to full cross-validation).

When the observations are partitioned into two equal sets we can form the model from either set and validate using the other. An improved idea is to form two models, one from each set, and in each case use the remaining set for validation. We then have two estimates of MSEP and two models to try. If the partition splits the data well then we would expect these models to place similar significance on each variable and thereby have similar properties and MSEPs. A model formed with use of the full data set should also have a similar structure and similar (or better) MSEP.

We can therefore use the following steps to estimate the MSEP of a model where the full data set is used:

1. partition the observations into two sets, S_1 and S_2;
2. form a model from S_1, use S_2 to estimate its MSEP (MSEP1);
3. form a model from S_2, use S_1 to estimate its MSEP (MSEP2);
4. form a model from the full set of observations and average MSEP1 and MSEP2 to estimate its MSEP.

The problem of choosing a 'good' partition still remains, but we now have a model based on all the observations, and an MSEP also based on all observations, but no self-prediction.

Consideration of the above steps leads to a method for reducing the partition-selection problem. We need to increase the proportion of observations used in the individual models. Instead of partitioning into two sets, the set can be split into n sets of approximately equal size. For each set, S_i, a model is formed from all observations not in S_i. The MSEP for this submodel is then estimated by using S_i as the test set. All n estimated MSEP values are then averaged to give an estimate of the MSEP for the full model. Again, we have the full observation set being used to create and validate the model, but in this case the larger calibration partition size reduces the chance of selecting a 'bad' spread of observations. For $n = 2$, this procedure equates to that detailed above. The limiting case occurs when each S_i contains a single observation. This is called 'leave-one-out' validation. This limit should give a good MSEP estimate as each submodel differs from the full model only in the effect of the single omitted observation. The choice of n is a trade-off between modelling time for n models and MSEP estimate quality.

Cross-validation is a good method for estimating the optimal number of components in latent variable methods because in these cases we expect the model to move from underfitted to optimal to overfitted through addition of components. A graph of MSEP versus number of components will therefore show a minimum at the point where the model is optimally fit (Wold, 1978).

Leverage correction. Leverage is a concept applied to observations used in a calibration model. It is a value between $1/n$ and 1, where n is the number of observations. The leverage of an object indicates its importance to the structure of the model. Observations having high leverage are important to the model in that they contribute greatly to its structure.

Consider the case when a model is fitted increasingly closely to its calibration set. Any observations which contain high noise levels (i.e. outliers) will start to gain in leverage as they skew the model to fit them. For such an overfitted model, then, there will be an increasing number of observations having high leverage.

This consideration gives rise to the idea of using leverage to estimate the validity of a model. Indeed it has been shown (Martens and Næs, 1989) that the residuals (and hence MSEP) for cross-validation, $\hat{f}_{i(\mathrm{CV})}$, can be estimated from the leverage of observations in an MLR model by means of the relations

$$\hat{f}_{i(\text{CV})} = \frac{\hat{f}_i}{1 - h_i}, \tag{10.16}$$

$$h_i = \frac{1}{n} + \hat{t}_i^{\text{T}}(\hat{\mathbf{T}}^{\text{T}}\hat{\mathbf{T}})^{-1}\hat{t}_i, \tag{10.17}$$

where h_i is the leverage of observation i and $\hat{f}_i$ its Y residual. This scheme works as expected in that if an observation is an outlier and the model is forced to fit it then its leverage h_t will be close to unity and hence the estimated cross-validation residual large. The MSEP estimate,

$$\frac{1}{n} \sum_{i=1}^{n} f_i^2 \, (\text{CV}),$$

will therefore also be large. Notice also that a high leverage value for an observation in a non-overfitted model is an indicator that that observation may be a significantly bad outlier and is worthy of investigation.

For cross-validation the $\hat{t}_i$ will be different for each model. But, as noted, for similar models, these $\hat{t}_i$ should be similar. Hence the residuals, $\hat{f}_i$ should also be similar for each submodel used. We may therefore simply use the residuals for the full calibration model and avoid the multiple calibration of cross-validation.

The process for leverage correction validation is therefore:

1. model on full calibration set and retain the Y residuals;
2. calculate the leverage for each observation from the model scores;
3. estimate the MSEP from residuals and leverages.

For calibration methods other than MLR, the cross-validation relation [equation (10.16)] does not hold, but Martens and Næs (1989) have suggested adjusted leverage-correction MSEP, $E_{\text{MSEP}}^{\text{LC}}$, estimators for PCR and PLSR:

$$E_{\text{MSEP}}^{\text{LC}} = \frac{1}{F_{\text{df}}} \sum \hat{f}_i^2 \left[1 - \sum_{a=1}^{A} \left(\frac{\hat{t}_{ia}^2}{\hat{t}_a^{\text{T}}\hat{t}_a} \right) \right]^{-1}, \tag{10.18}$$

where A is the number of factors, $\hat{t}_a$ the PCA (or PLS) scores vector for factor a and $\hat{f}_i$ the residuals; F_{df} is the number of degrees of freedom. Martens and Næs recommend that a good value for F_{df} is $n - 1 - A$, this value reducing the underestimation of MSEP that is inherent in leverage correction.

10.4.8 Partial least squares regression

Consider the NIPALS algorithm for PCA discussed earlier (Section 10.4.1). It was stated that the method uses the fact that the PCA loadings are regression coefficients for scores, and vice versa; that is, for $\mathbf{X}$, an $r \times c$

data matrix, with $p(c \times 1)$ and $t(r \times 1)$ being the corresponding loadings and scores vectors for the first component. We have $t = \mathbf{X}p$ by definition. This may be written in two ways to give rise to regression equations:

$$\mathbf{X} = tp^T, \tag{10.19}$$

or

$$\mathbf{X}^T = pt^T. \tag{10.20}$$

Regressing $\mathbf{X}$ on t in the first equation gives $\hat{p}^T = (t^T t)^{-1} t^T \mathbf{X}^T$, and regression of $\mathbf{X}$ on p in the second equation gives $\hat{t}^T = (p^T p)^{-1} p^T \mathbf{X}$. Simplifying this and normalizing so that $\hat{p}$ is of unit length gives the PCA NIPALS algorithm. Lyttkens (1966) showed that this algorithm was convergent to the required values for p and t.

Equations (10.19) and (10.20) represent a system of three variables where each is linearly dependent on the other two. The NIPALS method was soon shown to be applicable to many such systems (e.g. Lyttkens, 1973).

Wold (1975) developed a methodology for designing NIPALS models for multivariate regression. The basis of a model is an 'arrow diagram'. The variables are split into blocks, each of which is considered to be related to a latent variable. Each variable, both latent and observed, is represented as a box (or circle). Arrows are placed between the boxes, these indicating causal relationships. Hence arrows will lie between a latent variable and its block of observed variables. If the observed variable is considered to be causal to the latent variable then the arrow points towards the latent variable. If the variations of the latent variable are seen to explain the observed variations then the arrow points in the opposite direction. Interrelationships between latent variables may also be indicated by arrows. Wold's method allows the arrow diagram to be decomposed into a set of regressions which, when iterated, will provide the latent variable loadings. A number of methods are defined for forming the regression at this level, termed modes A, B and C. Wold (1980) terms mode A as being a regression where the 'outgoing' variables are explained by the latent variable, mode B being a regression where the incoming variables explain the latent variable. Mode C is the case where both A and B are used in a model.

An example should clarify matters. Figure 10.10 shows a path model for PCR. The latent variable (first principal component) is caused by the X variations and it is this which is assumed to cause the Y variations. Hence the X arrows point into the latent variable and the Y arrows point outward. The latent variable is therefore a weighted sum of the X_i. Let $\hat{p}$ and $\hat{q}$ be the X and Y weight vectors, $\mathbf{X}$ and $\mathbf{Y}$ the X and Y data matrices, and t the latent variable vector. Then we have:

$$t = \mathbf{X}p, \quad \text{normalized to length 1}; \tag{10.21}$$

which, by rearranging, gives

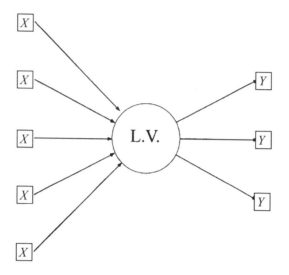

Figure 10.10 Path model diagram for principal components regression. L.V. = latent variable; X and Y are observed variables. The direction of the arrows indicates the causal relationship; for instance, variables X are causal to the latent variable, and variations in the latent variable explain variations in Y.

$$\mathbf{X} = tp^{\mathrm{T}}; \qquad (10.22)$$

$$\mathbf{Y} = tq; \qquad (10.23)$$

and, by MLR,

$$\hat{p}^{\mathrm{T}} = (t^{\mathrm{T}}t)^{-1}t^{\mathrm{T}}\mathbf{X} \qquad (10.24)$$

and

$$\hat{q} = (t^{\mathrm{T}}t)^{-1}t^{\mathrm{T}}\mathbf{Y}. \qquad (10.25)$$

So, by initializing $\hat{p}$ to a non-zero value and iterating over

$$\hat{t} = \mathbf{X}p, \quad \text{normalized}, \qquad (10.26)$$

and equations (10.24) and (10.25) until convergence of $\hat{p}$, we can obtain a PCR model. In fact, $\hat{q}$ is not used inside the loop, so it may be calculated after $\hat{p}$ has converged. With $\hat{q}$ removed from the loop we can see that the iterative part is equivalent to the NIPALS PCA algorithm, with the final $\hat{q}$ regression forming the PCR. [A discussion of the case where there are several interacting variables with associated blocks is given by Wold (1982); Bookstein (1980) gives an insight into the geometric interpretation of NIPALS methodology, now commonly called PLS.]

The method we shall refer to as PLS is one of incorporating the Y data into the latent variable modelling step by bringing an extra projection into the iterative sequence, as follows:

$$\hat{w}^T = (t^T t)^{-1} t^T X, \qquad (10.27)$$

$$\hat{t} = X\hat{w}, \qquad (10.28)$$

$$\hat{q}^T = (t^T t)^{-1} t^T Y, \qquad (10.29)$$

$$\hat{t} = Y\hat{q}. \qquad (10.30)$$

To put it in a visual manner, the location of $\hat{t}$ gets pulled first towards the X variables then towards the Y variables each time round the iteration. In the case of PCR, the only alteration to $\hat{t}$ is to pull it towards the X variables, so we end up converging to principal components. The pull of the Y variables tends to bring convergence such that the latent variables chosen are of relevance to both X and Y. Both sets of loadings ($\hat{w}$ and $\hat{q}$) are required for prediction in the case of PLS.

A simplified PLS model exists for cases when only a single Y variable is to be modelled. PLS 1 is non-iterative and can therefore be used to generate a result quickly in these cases (details are given in Martens and Næs, 1989).

PLS has come to the fore as an important modelling method because of the improved latent variables it generates. Since these variables are extracted in decreasing relevance to Y, the problem of variable selection is bypassed to a significant degree. Each PLS factor is extracted and its effect subtracted from the remaining X (and Y) variation before the next is obtained. The PLS factors obtained in this manner are always generated in decreasing order of importance to the overall model and the problem of selection is reduced to that of the number to use. This problem is handled easily by standard validation methods. De Jong (1993a) has shown that for any number of latent variables, PLS will fit at least as well as PCR.

Standard spectroscopic methods have tended to benefit from the use of PLS for gaining useful insights into data (Bhandare *et al.*, 1993; Haaland and Thomas, 1988a, b). In this discipline it is usual to obtain intensities at a large number of wavelengths. The wavelengths are considered as variables and the intensities their values. There is often a high degree of collinearity within such data sets and the phenomenon of interest is often distributed throughout a large number of the measured wavelengths. MLR-based methods are therefore of limited use, as a large number of wavelengths would have to be dropped to escape the collinearity problem, and this would tend also to lose the interesting phenomena.

10.4.9 Artificial neural networks

The X and Y matrices of the statistical models are analogous to the training inputs and outputs of a neural network used to create the model, and the test inputs and outputs used to predict values. Cheng and Titterington

(1994) take a look at neural networks from a statistical point of view. Indeed, Sarle (1994) suggests that neural networks are no more than nonlinear regression and discriminant models that can be implemented by means of standard statistical software; Ripley (1994) agrees with this statement.

The human brain is a very complex organ. It contains around 10^{10} neurons (that is, the basic processing, or nerve, cell). Neurons consist of two types (Beale and Jackson, 1990):

- interneuron cells, with input and output locally (over a distance up to 100 μm);
- output cells, connecting to different regions of the brain, to muscles, or connecting from sensory organs (e.g. the eye) into the brain.

If enough active inputs are received to an individual neuron at any one time, that neuron is activated (it 'fires'), otherwise it remains inactive. This is analogous to the McCulloch–Pitts model of the neuron, proposed as long ago as 1943. Their model of neural activity is described in detail by Aleksander (1989).

Thus the brain is massively parallel, any one processing job being shared between many neurons. The consequence of this is that any one single neuron is not generally very important; if one neuron were to malfunction for some reason, it would be relatively unlikely to affect other neurons significantly. This kind of processing, spread over many processing units, is known as 'distributed processing', and is widely considered to be fairly tolerant of errors (to have 'fault tolerance').

The traditional computer has one (or maybe two or a few more) processors. The consequences of this are clear: in addition to the inferior processing power, if the only processor should malfunction then the whole system would be unviable. As the term 'neural network' implies, this field of computing was originally aimed towards modelling networks of real neurons in the brain (Hertz, Krogh and Palmer, 1991).

Neural networks, which naturally lend themselves to parallel processing (although they are often presently run, by necessity of available equipment, on single-processor computers), clearly overcome this problem of possible failure by allowing processing to continue in the event of a failure of one of the constituent neurons.

Traditionally, neural networks are seen as a branch of artificial intelligence (AI), although some AI academics disagree over this. It surely could be said that it is one of the most unquestionably AI areas in computing science. In most other areas of AI, learning takes place in a fully understood and predictable way using explicitly represented knowledge (Luger and Stubblefield, 1989), whereas for neural networks the way in which the net learns is not known, nor is it predictable or repeatable, because of the random nature of the network initialization (discussed below).

Neural networks can learn to recognize patterns within sets of data (which may be an encoded image, a spectrum or any encoded set of related data). Fault tolerance helps in pattern recognition, allowing the net to cope with differences between input examples it receives, placing greatest importance only on the parts of the input data which are important for distinguishing the data it is learning.

So what is meant by 'learning'? A good definition is given by Judd (1990, p. 3), who states that 'Learning is the capacity of a system to absorb information from its environment without requiring some external intelligent agent to program it'. By 'program' it can be understood that Judd is referring to more direct methods of information input, such as that used in an expert system. Judd goes on to say that, unfortunately, all learning algorithms so far reported are unacceptably slow for large networks. Given the size of the human brain it is no great wonder that attempts to simulate this sort of complexity have not succeeded! Nonetheless, the principle of neurons as elements of processing is not lost on reduction to a smaller sized network, such as those modelled by the various packages available on desktop computers. A tutorial review of neural networks, with particular reference to multivariate PyMS applications, is given by Goodacre, Neal and Kell (1996).

With standard back propagation, the most common type of neural network, there are a number of input nodes (equal to the number of inputs), each connected to every node of a hidden layer, which are in turn each connected to the output node(s) (Fig. 10.11). Each node in the input layer brings into the network the value of one independent variable. The hidden layer nodes (called 'hidden' because they are hidden from the outside world) do most of the work (Smith, 1993). Each output node passes a single dependent variable out of the network.

In neural net jargon, the neuron is known as a 'perceptron' (Rosenblatt, 1958). The learning rule for these 'multilayer perceptrons' is called the back-propagation rule. This is usually ascribed to Werbos in his thesis of 1974 (Werbos, 1993), but was popularized by Rumelhart and McClelland (1986) as recently as 1986, since when there has been a revival in interest in neural networks.

The network is initialized with random weights on the connections between the perceptrons, and the input is applied to the input nodes. By using a training set of data and comparing the output from the net with the desired (known) output it is possible to calculate new values for the weights to increase the accuracy by decreasing the error between known and actual values. Each of these iterations is known as an 'epoch'. An error function is defined to represent the difference between the network's output and the desired output. The aim therefore is to reduce the value of this function as far as possible. The back-propagation rule does this by calculating the value of the error function for one particular input and propagating the error back from one layer to the previous one. Thus each connection has its weights

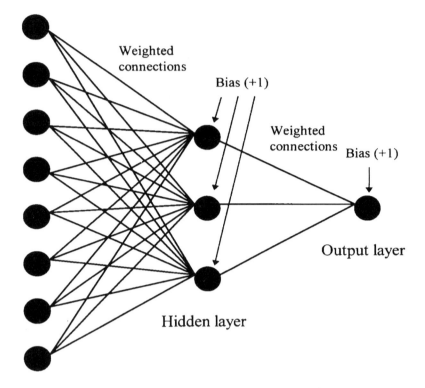

Figure 10.11 Structure of a standard back-propagation 8–3–1 neural network.

adjusted. The mathematics involved in this procedure are often badly presented, but some of the most understandable explanations are by Beale and Jackson (1990) and Bishop (1995).

One problem frequently encountered with neural networks is that of overtraining. This occurs when the net trains so closely to the training set that any other data will not be recognized. In this case, when the net is tested, it will be found that examples which were in the training set will give highly accurate results, and other examples may be wildly inaccurate.

Take, for example, the training and test data shown in Fig. 10.12. Overtraining results in the fit shown with a dashed line, where the points in the test set are poorly fitted to the line. The ideal line is shown by a solid line.

According to Hecht-Nielsen (1989, p. 116), the exact origin of this problem has still not been fully elucidated, but it seems to be related to the manner in which the afflicted networks form their mapping approximations. When testing, it is normal to use the training set as well as a test set which has not been seen by the network during training – this helps to show up any over-

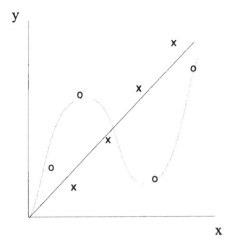

Figure 10.12 Graph showing how a network may be overtrained to the data in the training set. ○ = training set; × = test set; ――― = desired train; ----- = overtrained.

training as well as showing how well the network has learnt what it has seen and how good at generalization (dealing with previously unseen data) it is.

It is clear, then, that neural networks, being good at dealing with large numbers of inputs, lend themselves to the analysis of spectra such as those obtained by the PyMS technique and NMR. PyMS, for example, produces a spectrum of 150 elements, therefore the network used will have 150 input nodes. Goodacre, Kell and Bianchi (1992, 1993) have used neural networks for the detection of adulteration in olive oil from PyMS data. Francelin, Gomide and Lanças (1993) compare three different types of neural network for the classification of vegetable oils, including olive oil, after gas chromatography analysis.

10.4.10 Chemometrics

The use of PLS and related methods in a wide range of sciences has led to the emergence of a new discipline, that of chemometrics. This is devoted to the study of pragmatic multivariate methods in sciences, and the literature has grown rapidly over recent years (e.g. Brereton, 1992; Brown *et al.*, 1996; Ramos *et al.*, 1986). With the increasing use of PLS has come the realization that it is not well understood in terms of its statistical properties. The method was designed to avoid making assumptions about structure within the data except those that are built into the path model. No assumption of normality is placed on the variables, for instance.

Frank and Friedman (1993) and Helland (1988) have studied the more commonly used chemometric regression methods (namely PLS and PCR) from a statistical viewpoint. Their studies highlight the general similarity between these methods and that of ridge regression. This tends to confirm the proposition that there is a continuum of regression methods extending from MLR to PCR, with PLS lying within this continuum. Stone and Brooks (1990) derived the continuum regression (CR) algorithm, for which there are two controlled parameters: the number of factors (termed ω) and a position indicator, α. When $\alpha = 0, \frac{1}{2}, 1$, CR is equivalent to MLR, PLS and PCR, respectively. The optimal values for α and ω are estimated using cross-validation. Malpass *et al.* (1994) have investigated the continuum regression methodology and have simplified the selection of optimal α values (1993). Seasholtz and Kowalski (1993) have highlighted the importance of choosing parsimonious models. That is to say that if two models are formed then the one which requires fewer parameters for its description is more likely to provide good future predictions.

On the algorithmic front, a number of new PLS methods have been proposed, for example those of Lindgren, Geladi and Wold (1993) and de Jong (1993b). Extensions have been proposed to improve the situation when non-linear data are encountered. The methods discussed so far assume that the observed and latent variables are linearly related. When non-linearity is encountered, the methods tend to fail to make a good model. This phenomenon shows up in the requirement of large numbers of factors in model formation, and in curvature in plots of the residuals after modelling. Taavitsainen and Korhonen (1992) have shown how PLS can be extended to incorporate certain types of non-linearity at the latent variable stage, and Oman, Næs and Zube (1993) have used a PCR-based method augmented with products of the components for similar purposes.

10.4.11 Variable selection

Variable selection is a technique whereby those variables which are considered to be most important in the creation of a model are used, whereas others are discarded. There are many arguments for selecting variables and a number of ways of selecting them, some of which are described in this chapter. To date, comparatively few researchers have used variable selection methods, most concentrating on improving prediction by using all the variables.

Why select variables? The rationale behind using latent variable methods is the exclusion of effects that contribute only noise. Latent variable methods assume that there are underlying effects which are expressed to varying degrees in the measured variables. If, however, some of the measured variables do not reflect any relevant underlying effect then there is no reason to include them in the modelling process. Indeed, since latent variable methods rarely assign zero relevance to any given variable, one may expect such irrelevant variables to introduce noise despite the application of such methods.

The principle of parsimony (de Noord, 1994; Flury and Riedwyl, 1988; Seasholtz and Kowalski, 1993) states that if a simple model (that is, one with relatively few parameters or variables) fits the data then it should be preferred to a model that involves redundant parameters. A parsimonious model is likely to be better at prediction of new data and to be more robust against the effects of noise (de Noord, 1994). Despite this, the use of variable selection is still rare in chromatography and spectroscopy (Brereton and Elbergali, 1994). Note that the terms 'variable selection' and 'variable reduction' are used by different researchers to mean essentially the same thing.

Chatfield (1995) warns of the dangers of selecting variables in such a way as to enable some sort of model to be made from pure noise. This is a risk with some methods such as genetic algorithms (Bangalore et al., 1996; Broadhurst et al., 1997; Horchner and Kalivas, 1995; Jouan-Rimbaud et al., 1995; Kubinyi, 1994a, b, 1996); if these methods are used, thorough validation of the results is necessary to avoid this problem.

Some variable selection techniques for classification. It can easily be shown that certain variables in a data set not only contribute little to the model but actually detract from the optimum model. Let us take a simple, imaginary, case to demonstrate this. If we have a set of data describing two different varieties of olive oil, Leccino and Frantoio, and only three variables, we can look at the data and see which of the variables is most valuable for discriminating between the two varieties (Table 10.3).

If we take the standard deviation (StDev) of the Leccino and Frantoio oils for variables 1, 2, and 3 and then calculate the average of these, we have a value which represents the 'inner variance' or 'reproducibility'. The higher

Table 10.3 Finding the most valuable variables for discrimination between two olive oil varieties, Leccino (Le) and Frantoio (Fr). StDev = standard deviation; w is defined in text in equation (10.31) and indicates the characteristicity of a variable.

Variety	Variable		
	1	2	3
Le 1	3.4	5.4	6.4
Le 2	3.2	2.8	4.5
Le 3	3.5	8.6	5.9
Fr 1	3.0	3.9	5.1
Fr 2	3.2	7.5	3.5
Fr 3	3.1	5.5	4.9
StDev Le	0.152	2.905	0.984
StDev Fr	0.1	1.803	0.871
Average StDev	0.126	2.354	0.928
StDev All	0.186	2.162	1.027
W	0.678	1.088	0.903

this value, the higher the inner variance and the lower the reproducibility (Eshuis, Kistemaker and Meuzelaar, 1977). By taking the standard deviation over all the samples, we calculate a value which represents the 'outer variance' or 'specificity'. The higher this value the greater the outer variance and the greater the specificity. The ratio between inner variance and outer variance represents the 'characteristicity'. [The terminology 'characteristicity', 'reproducibility' and 'specificity' has been adopted from Eshuis, Kistemaker and Meuzelaar (1977).] So, by dividing our value for the average of the standard deviations by the standard deviation of the whole we get a value w which is an indication of the characteristicity of the variable (for varieties, var. 1, ... var. n):

$$w = \frac{\text{average [St Dev (var. 1), St Dev (var. 2), ..., St Dev (var. } n)]}{\text{St Dev (all samples)}}. \quad (10.31)$$

Now, if w has a value greater than 1 then the inner variance is greater than the outer variance; therefore this variable is a hindrance to correct discrimination and so it should definitely be discarded.

The PCA scores plots of the data shown in Table 10.3 demonstrate the effect of selecting variables. Taking all three variables [Fig. 10.13(a)] it is apparent that discrimination is possible, but not easy. Eliminating the worst variable (with $w > 1$) makes discrimination easier [Fig. 10.13(b)]. If the best variable is removed, discrimination is impossible [Fig.10.13(c)]

When looking at data sets with many varieties, where one variety may contain more samples than another, it could be desirable to weight w in favour of varieties with a greater representation. In this case, we can use:

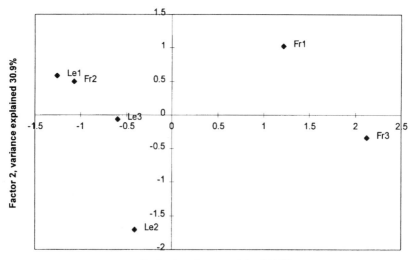

Figure 10.13(a)

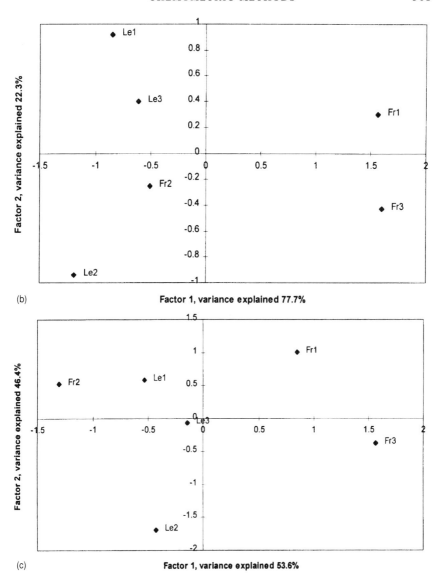

Figure 10.13 Principal components analysis scores plots: (a) using all three example variables (first two principal components; discriminations of the varieties, particularly sample Le2, would be very difficult); (b) using the best two variables, unweighted w selected [equation (10.31) in text] (discrimination of the varieties is now possible using only the first principal component); (c) discarding the best variable, unweighted w selected (linear discrimination of the varieties would not appear to be possible from this chart). Details of the variables are given in Table 10.3.

$$w = \frac{[\text{StDev}(\text{var.}1) \times N(\text{var.}1)] + [\text{StDev}(\text{var.}2) \times N(\text{var.}2)] + \ldots + [\text{StDev}(\text{var.}n) \times N(\text{var.}n)]}{N(\text{total}) \times \text{StDev}(\text{all samples})}$$

(10.32)

Where N (var. n) is the number of samples for variety n and N (total) is the total number of samples.

Another method of variable selection for use in classification problems involves use of the Fisher ratio, whereby a value is calculated for each variable according to the following formulae.

Calculation of the between-group variation, V_b:

$$V_b = \sum_{i=1}^{g} n_i(\bar{y}_i - \bar{y})^2. \qquad (10.33)$$

Calculation of the within-group variation, V_w:

$$V_w = \sum_{i=1}^{g} \sum_{j=1}^{n_i} (y_{ij} - \bar{y}_i)^2. \qquad (10.34)$$

Calculation of the Fisher coefficient, F:

$$F = \left(\frac{1}{g-1} V_b\right)\left(\frac{1}{n-g} V_w\right)^{-1}, \qquad (10.35)$$

where g is the number of groups;
n_i is the number of elements in group i;
$\bar{y}_i$ is the mean value of group i;
$\bar{y}$ is the total mean;
y_{ij} is the value of object j in group i.

The three selection methods are henceforth referred to as weighted w, unweighted w, and Fisher.

It may often be found that factors other than that searched for (e.g. olive oil variety) may be having some influence on the data (for example, time of harvesting or region of origin). Although variables containing data so affected may have a value for w of less than 1, they may still be having a great influence on the model by causing oils of, say, a similar harvesting date, to cluster together, or at least to be pulled away from their hoped-for varietal clustering, in a PCA scores plot. In order to eliminate this effect and to increase the parsimony of the model, we would wish to discard many variables with a value of w less than 1. We cannot easily know which variables are affected most in this way, and the optimal model complexity is not fixed for a given problem; it is data set dependent (de Noord, 1994). One solution is to start from a minimum number of variables (two or three), selected with a low threshold value of weighted or unweighted w, and work upwards towards $w = 1$ to see at what point the best model is reached.

It could be expected, when trying to identify varieties, that the optimum model will be achieved at or near a value for w which selects variables that contribute to the discrimination of varieties but does not select any that also contribute a large amount to the discrimination of other factors (such as region or harvesting date). In practice it is found that the ideal threshold ranges from that which selects only the best three or four variables right up to $w = 1$, depending on the data and the desired factor (whether variety, region, etc.). Undoubtedly, this is at least in part arising from the principle of parsimony.

10.4.12 Exploitation of multivariate spectroscopies in the identification of the geographical origin of olive oils

For the next olive oil harvest (1996/97 season) the Italian oil producers will provide the consumers with DOC (*Denominazione di Origine Controllata*) or CBO (Certified Brands of Origin) virgin olive oils. [The so-called DOC (Controlled Denomination of Origin) classification of extra virgin olive oil is being introduced in Italy according to law 169/1992.] Similar provisions can be found in EU regulations 2081/1992 and 2082/1992, governing the DOP [*Denominazione di Origine Proteggètta* (Protected Denomination of Origin)] for agricultural food products. These highly priced oils will be obtained from either a single (or a high percentage content from a single) variety or from several varieties growing in a specified region. Clearly, to enforce the legal situation suitable methods for determining the geographical origins of an extra virgin olive oil will be essential.

The methods used to attain the correct identification of olive oils will also have to take into account the potential large variability arising from variety, location and environmental differences in the compositional characteristics of 'pure' virgin olive oils; thus the availability of reliable methods for authentication of the geographical origin of the oils will be crucial. As discussed above, one possible method is Curie–point PyMS (Section 10.3.2) combined with a powerful multivariate or chemometric analysis technique such as ANNs (Section 10.4.9) (Fig. 10.14).

The use of variable selection has been shown to improve the prediction of the variety and region of origin of olive oils considerably, using both ^{13}C NMR (Shaw *et al.*, 1996, 1997) and PyMS data. The results that are shown here were produced with use of PLS and PCR software written in-house by Jones, in conjunction with Microsoft Excel 5 macros written by Shaw.

The NMR data used consisted of five varieties; four Italian: Coratina (14), Dritta (12), I-77 (16) and Moraiolo (16) and one Israeli (8). The samples were run in duplicate to ensure reproducibility. In such circumstances it is important to ensure that the duplicates are kept together, not split between test and training sets. The Dritta oil, however, was all from one sample and so was unavoidably split between the test and training sets. The regions used were from the same data, being a mixture of varieties from Abruzzo (12),

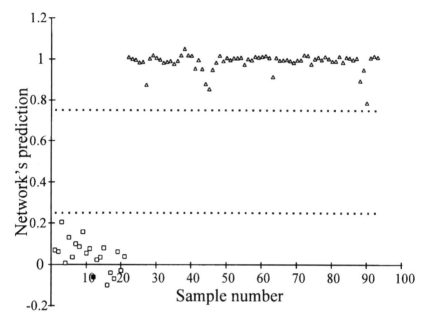

Figure 10.14 Prediction of the region of origin for extra virgin olive oils by means of artificial neural networks (ANNs). Raw data were separated into training and test sets by means of the Duplex partitioning method. A three-layered ANN [150 input nodes (±10% headroom), 8 hidden nodes and a single output node] was trained with 183 objects (39 Abruzzo, coded 0, and 144 Sardinia, coded 1) and interrogated periodically using 93 separate objects [21 Abruzzo (□) and 72 Sardinia (△)] previously unseen by the training net. The output after 110 000 epochs is shown above, with a training error of 0.009 root mean square. The lines represent network estimates of 0.25 and 0.75. Squares below the 0.25 line are deemed as being Abruzzo, and triangles above the 0.75 line as Sardinia. In this strict test all 93 are correctly identified.

Puglia (14), Toscana (12), Israel (8) and the 12 measurements of the Dritta sample, which was also from Abruzzo.

Figure 10.15(a) shows how only the best six or seven variables allow a 100% prediction of variety I-77, whereas if all variables are used only around 85% can be achieved. Figure 10.15(b) shows that similar success can be achieved with regions, again with less than half the variables; region Toscana has proved to be easier to predict than other regions, which is fortuitous as it is these oils that are the most highly prized by connoisseurs. For these two examples, one Y variable was used, a 1 being used to represent the variety being predicted, and a 0 all other varieties.

The remaining examples are predictions of five varieties simultaneously, rather than just one. The output matrix, **Y**, was therefore encoded by using five variables of 1s and 0s, using a 1 to represent each variety in the appropriate column.

MLR performs significantly worse than PLS or PCR on the data used; as has been pointed out by Martens and Næs (1989), the MLR predictor has seriously deficient performance where there is collinearity in the X data.

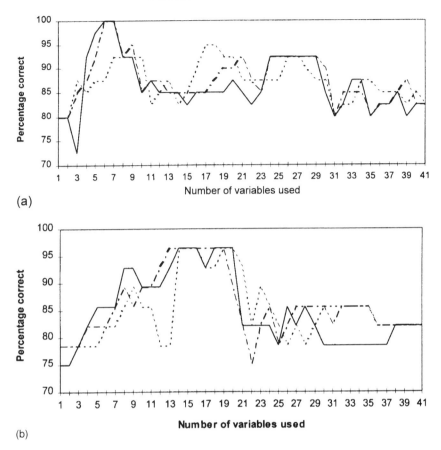

Figure 10.15 PLS 1 (partial least squares) prediction of: (a) olive oil variety I–77 from ^{13}C NMR data containing five varieties (all oils were correctly predicted with Fisher and weighted w selection using the best six or seven variables); (b) olive oil region of origin Toscana (Tuscany) from ^{13}C NMR data containing four regions (at best, all but one are correctly predicted). Selection procedures: ——— = Fisher; – – – – = weighted w; - - - - = unweighted w. Selection procedures are described in text, Section 10.4.11, equations (10.31)–(10.35).

Variable selection, however, greatly improves the prediction obtained by MLR, as can be seen in Fig. 10.16. Without variable selection, these results would be worse than a random guess!

The same data, using PLS 2 rather than MLR, gives a better prediction (Fig. 10.17), using weighted and unweighted w selection methods. Although the best prediction using weighted w (91.6%, all but three) is achieved with all but one of the variables, the inclusion of that one variable significantly reducing the prediction, to 75% (all but eight).

Since we know to what most of the carbon signals correspond, it is possible to draw some conclusions on the chemical significance of the order of variable selection.

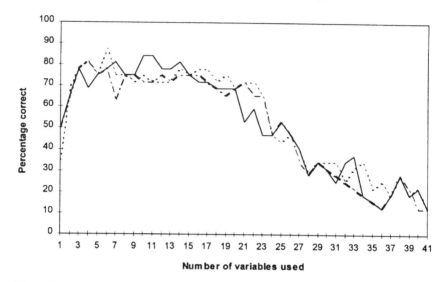

Figure 10.16 Multiple linear regression prediction of olive oil variety from ^{13}C NMR data containing five varieties (at best, all but four predictions are correct with selection by unweighted w). Selection procedures: ——— = Fisher; – - – - = weighted w; - - - - = unweighted w. Selection procedures are described in text, Section 10.4.11, equations (10.31)–(10.35).

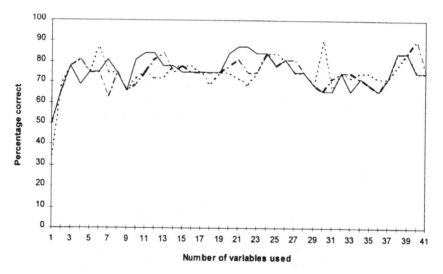

Figure 10.17 PLS 2 (partial least squares) prediction of olive oil variety from ^{13}C NMR data containing five varietes (at best, all but three predictions were correct with use of weighted and unweighted w). Selection procedures: ——— = Fisher; – - – - = weighted w; - - - - = unweighted w. Selection procedures are described in text, Section 10.4.11, equations (10.31)–(10.35).

For region discrimination the most important variables are found in the aliphatic region of the spectrum, from both oleic and saturated chains [Fig. 10.18(a)]. The fatty acids corresponding to the most significant 10 variables are identified. All the assignments are either for the α or the α+β positions of the glycerol backbone. Since there are two α positions for each β position it seems likely that it is the strength of the signal (strong signals probably contain less noise) which causes those variables corresponding to the α position to be selected above those corresponding to the β position.

For variety discrimination, the carbonyl region is much more prominent, but again oleic acid predominates [Fig. 10.18(b)]. The most significant finding here is that position of oleic acid (α or β) is not indicated by any of the 10 most significant signals in the aliphatic region, but is indicated by CS_2 and CS_4 in the carbonyl region, as is the position of linoleic with CS_5. Therefore, it would appear that variety discrimination is aided by the knowledge of the position of monounsaturated fatty acids on the glycerol backbone, whereas this is not so important for region discrimination. It follows, then, that one of the main distinguishing features of olive oil varieties is the position of the monounsaturated fatty acids on the glycerol backbone, whereas for regions the main factor is the relative proportions of fatty acids in the oil. It is unfortunate that the assignment of CS_22 is unknown, as this is important for both variety and region discrimination.

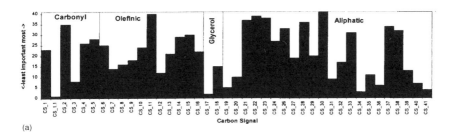

(a)

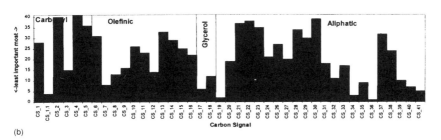

(b)

Figure 10.18 Relative significance of carbon signals (CS) for (a) region discrimination; (b) variety discrimination. Assignments are α+β except where indicated. O = oleic; S = saturated; L = linoleic.

10.5 Concluding remarks and future prospects

For reasons of space we have been unable to cover all of the major chromatographic and spectroscopic methods which might be applied to olive oils and other lipids or other chemometric methods which might be used to turn the complex, multivariate data so obtained into information. In particular, we recognize the immense potential of vibrational spectroscopies (Kemsley, Appleton and Wilson, 1994; Winson *et al.*, 1997a, b) such as Fourier transform infra-red, near infra-red and Raman (Section 10.2.4). Although the chemometric methods we have described can make excellent predictive models the non-linear ones in particular do not easily lend themselves to explaining how they do it. This is known as the assignment problem. By contrast, rule-based methods such as classification and regression trees (Breiman *et al.*, 1984) and fuzzy multivariate rule induction (Alsberg *et al.*, 1997) can provide models which are much easier to interpret according to the IF...THEN rules which they can produce. The production of simpler rules and models is contingent on the extraction of appropriate features in a pre-processing step; to this end we have found a peak parameter representation to be of value (Alsberg, Winson and Kell, 1997), and less familiar statistical and multivariate methods that may prove to be of particular use in the extraction of relevant information from such spectra include wavelet analysis and regression, Fourier regression, Fourier deconvolution and B-spline splitting (Alsberg, Woodward and Kell, 1997). Finally, genetic programming methods (Koza, 1992, 1994a, b), although computationally intensive, have the potential to search the high-dimensional space with high efficiency (Edmonds, Burkhardt and Adjei, 1995) to find the models which best relate the spectroscopic data available to the biological or chemical properties of interest (Gilbert *et al.*, 1997).

Acknowledgements

We thank the UK BBSRC, the Higher Education Funding Council for Wales and the Ministry of Agriculture, Fisheries and Foods for financial support. Thanks are also due to the Italian Ministry of Agriculture.

References

Afifi, A. A. and Clark, V. (1996) *Computer-aided Multivariate Analysis*, 3rd edn, Chapman & Hall, London.

Akitt, J. W. (1983) *NMR and Chemistry: An Introduction to the Fourier Transform–Multinuclear Era*, 2nd edn, Chapham & Hall, London.

Aldridge, W. N. (1992) The Toxic Oil Syndrome (TOS, 1981): from the disease towards a toxological understanding of its chemical aetiology and mechanism. *Toxicology Letters*, **64–65**, 59–70.

Aleksander, I. (1989) *Neural Computing Architectures: the Design of Brain-like Machines*, MIT Press, Cambridge, MA.

Allen, D. M. (1971) Mean square error of prediction as a criterion for selecting variables. *Technometrics*, **13**, 469–75.
Alsberg, B. K., Winson, M. K. and Kell, D. B. (1997) Improving interpretation of multivariate and rule induction models by using a spectral peak parameter representation. *Chemometrics and Intelligent Laboratory Systems*, **36**, 95–109.
Alsberg, B. K., Woodward, A. M. and Kell, D. B. (1997) An introduction to wavelet transforms for chemometricians: a time-frequency approach. *Chemometrics and Intelligent Laboratory Systems*, **37**, 215–39.
Alsberg, B. K., Goodacre, R., Rowland, J. J. and Kell, D. B. (1997) Classification of pyrolysis mass spectra by fuzzy multivariate rule induction; comparison with regression, K-nearest neighbour, neural and decision-tree methods. *Analytica Chimica Acta*, in press.
Anon. (1994) Per l'olio di oliva fine delle frodi? *L'Informatore Agrario*, **18**, 41.
Aparicio, R. and Alonso, V. (1994) Characterization of virgin olive oils by SEXIA expert system. *Progress in Lipid Research*, **33**, 29–38.
Aparicio, R. and Morales, M. T. (1995) Sensory wheels – a statistical technique for comparing QDA panels – application to virgin olive oil. *Journal of the Science of Food and Agriculture*, **67**, 247–57.
Aparicio, R., Alonso, V. and Morales, M. T. (1994). Detailed and exhaustive study of the authentication of European virgin olive oils by SEXIA expert-system. *Grasas y Aceites*, **45**, 241–52.
Aparicio, R., Alonso, V. and Morales, M. T. (1996) Developments in olive oil authentication, in *Food Authenticity '96*, Norwich.
Aparicio, R., Ferreiro, L. and Alonso, V. (1994) Effect of climate on the chemical-composition of virgin olive oil. *Analytica Chimica Acta*, **292**, 235–41.
Aparicio, R., Gutierrez, F. and Morales, J. R. (1992) Relationship between flavor descriptors and overall grading of analytical panels for virgin olive oil. *Journal of the Science of Food and Agriculture*, **58**, 555–62.
Aparicio, R., Navarro, M. S. and Ferreiro, M. S. (1991) Definite influence of the extraction methods on the chemical composition of virgin olive oil. *Grasas y Aceites*, **42**, 356–62.
Armanino, C., Leardi, R. and Lanteri, S. (1989) Chemometric analysis of Tuscan olive oils. *Chemometrics and Intelligent Laboratory Systems*, **5**, 343–54.
Aylott, R. I., Clyne, A. H., Fox, A. P. and Walker, D. A. (1994). Analytical strategies to confirm Scotch whisky authenticity. *Analyst*, **119**, 1741–6.
Baeten, V., Meurens, M., Morales, M. T. and Aparicio, R. (1996) Detection of virgin olive oil adulteration by Fourier transform Raman spectroscopy. *Journal of Agricultural and Food Chem* **44**, 2225–30.
Bangalore, A. S., Shaffer, R. E., Small, G. W. and Arnold, M. (1996) Genetic algorithm-based method for selecting wavelengths and model size for use with partial least squares regression. Application to near infrared spectroscopy. *Analytical Chemistry*, **68**, 4200–12.
Baroni, M., Clementi, S., Cruciani, G. *et al.* (1992) Predictive ability of regression-models. 2. Selection of the best predictive PLS model. *Journal of Chemometrics*, **6**, 347–56.
Battiti, R. (1994) Using mutual information for selecting features in supervised neural net learning. *IEEE Transactions Neural Networks*, **5**, 537–50.
Beale, R. and Jackson, T. (1990) *Neural Computing: An Introduction*, IOP, Bristol.
Belton, P. S. (1995) NMR in context. *Annual Reports on NMR Spectroscopy*, **31**, 3–18.
Bhandare, P., Mendelson, Y., Peura, R. A. *et al.* (1993) Multivariate determination of glucose in whole blood using partial least-squares and artificial neural networks based on mid-infrared spectroscopy. *Applied Spectroscopy*, **47**, 1214–21.
Bianchi, G., Giansante, L. and Lazzari, M. (1996) Analisi per la tutela di genuinità, origine geografica e varietale degli oli vegetali. *L'Informatore Agrario*, **20**, 45–8.
Bianchi, G., Angerosa, F., Camera, L. *et al.* (1993) Stable carbon isotope ratios (13C/12C) of olive oil components. *Journal of Agricultural and Food Chemistry*, **41**, 1936–40.
Bianchi, G., Gussoni, M., Limiroli, R. *et al.* (1994a) NMR and chemical studies of the morphologically different parts of the olive fruit (*Olea Europaea* L.). *Acta Horticulturae*, **356**, 260–3.
Bianchi, G., Tava, A., Vlahov, G. and Pozzi, N. (1994b) Chemical-structure of long-chain esters from sansa olive oil. *Journal Of the American Oil Chemists' Society*, **71**, 365–9.
Bishop, C. M. (1995) *Neural Networks for Pattern Recognition*, Clarendon Press, Oxford.

Bookstein, F. L. (1980) Data analysis by partial least squares, in *Evaluation of Econometric Models* (eds J. Kmenta and J. B. Ramsey), Academic Press, London, pp. 75–90.
Bosaeus, I., Belfrage, L., Lindgren, C. and Andersson, H. (1992) Olive oil instead of butter increases net cholesterol excretion from the smallbowel. *European Journal of Clinical Nutrition*, **46**, 111–15.
Boschelle, O., Giomo, A., Conte, L. and Lercker, G. (1994) Caratterizzazione della cultivar di olivo del Golfo di Trieste mediante metodi chemiometrici applicati ai dati chimica-fisici. *La Rivista Italiana delle Sostanze Grasse*, **71**, 57–65.
Boskou, D. (1996) *Olive oil: chemistry and technology*. AOCS Press, Champaign, IL.
Breiman, L., Friedman, J. H., Olshen, R. A. and Stone, C. J. (1984) *Classification and Regression Trees*, Wadsworth International, Belmont, CA.
Brekke, T., Barth, T., Kvalheim, O. M. and Sletten, E. (1990) Multivariate analysis of carbon-13 nuclear magnetic resonance spectra. Identification and quantification of average structures in petroleum distillates. *Analytical Chemistry*, **62**, 56–61.
Brereton, R. G. (1992) *Multivariate Pattern Recognition in Chemometrics*, Elsevier, Amsterdam.
Brereton, R. G. (1995) Deconvolution of mixtures by factor-analysis. *Analyst*, **120**, 2313–36.
Brereton, R. G. and Elbergali, A. K. (1994) Use of double windowing, variable selection, variable ranking and resolvability indices in window factor analysis. *Journal of Chemometrics*, **8**, 423–37.
Brevard, C. and Grainger, P. (1981) *Handbook of High Resolution NMR*, John Wiley, New York.
Broadhurst, D., Goodacre, R., Jones, A. et al. (1997) Genetic algorithms as a method for variable selection in MLR and PLS regression, with applications to pyrolysis mass spectrometry. Submitted for publication in *Analytica Chimica Acta*.
Brown, P. J. (1993) *Measurement, Regression, and Calibration*, Oxford Science Publications, Oxford.
Brown, S. D., Sum, S. T., Despagne, F. and Lavine, B. K. (1996) Chemometrics. *Analytical Chemistry*, **68**, R21–R61.
Campbell, I. D. and Dwek, R. A. (1984) *Biological Spectroscopy*, Benjamin Cummings, London.
Chatfield, C. (1995) Model uncertainty, data mining and statistical inference. *Journal of the Royal Statistical Society*, **158**, 419–66.
Cheng, B. and Titterington, D. M. (1994) Neural networks: a review from a statistical perspective. *Statistical Science*, **9**, 2–54.
Clint, M. and Jennings, A. (1970) The evaluation of eigenvectors of real symmetric matrices by simultaneous iteration. *The Computer Journal*, **13**, 76–80.
Collins, E. J. T. (1993) Food adulteration and food safety in Britain in the 19th and 20th centuries. *Food Policy*, (April), 95–109.
Cruciani, G. and Watson, K. A. (1994) Comparative molecular-field analysis using grid force-field and GOLPE variable selection methods in a study of inhibitors of glycogen phosphorylase B. *Journal of Medicinal Chemistry*, **37**, 2589–601.
Davies, A. M. C. (1995) The better way of doing principal component regression. *Spectroscopy Europe*, **7**, 36–8.
Defalguerolles, A. and Jmel, S. (1993) Variable selection criteria – based on specific Gaussian graphical models in principal components analysis. *Canadian Journal of Statistics – Revue Canadienne de Statistique*, **21**, 239–56.
de Jong, S. (1993a) PLS fits closer than PCR. *Journal of Chemometrics*, **7**, 551–7.
de Jong, S. (1993b) SIMPLS: an alternative approach to partial least squares regression. *Chemometrics and Intelligent Laboratory Systems*, **18**, 251–63.
de Noord, O. E. (1994) The influence of data preprocessing on the robustness and parsimony of multivariate calibration models. *Chemometrics and Intelligent Laboratory Systems*, **23**, 65–70.
EC (1991) Commission Regulation (EEC) no. 2568/91 of 11 July 1991 on the characteristics of olive oil and olive-residue oil and on the relevant methods of analysis. *Official Journal of the European Communities* **L248**, 1–83.
Edmonds, A. N., Burkhardt, D. and Adjei, O. (1995) Genetic programming of fuzzy logic production rules, in *IEEE International Conference on Evolutionary Computation*, vols 1–2, IEEE, Perth, pp. 765–70.
Eshuis, W., Kistemaker, P. G. and Meuzelaar, H. L. C. (1977) Some numerical aspects of reproducibility and specificity, in *Analytical Pyrolysis* (eds C. E. R. Jones and C. A. Cramers), Elsevier, Amsterdam, pp. 151–6.

Firestone, D., Carson, K. L. and Reina, R. J. (1988) Update on control of olive oil adulteration and misbranding in the United-States. *Journal of the American Oil Chemists Society*, **65**, 788–92.

Firestone, D. and Reina, R. J. (1987) Update on control of olive oil adulteration in the United-States. *Journal of the American Oil Chemists' Society*, **64**, 682–82.

Firestone, D., Summers, J. L., Reina, R. J. and Adams, W. S. (1985) Detection of adulterated and misbranded olive oil products. *Journal of the American Oil Chemists' Society*, **62**, 1558–62.

Flury, B. and Riedwyl, H. (1988) *Multivariate Statistics: A Practical Approach*, Chapman & Hall, London.

Forina, M. and Tiscornia, E. (1982) Pattern recognition methods in the prediction of Italian olive oil origin by their fatty acid content. *Annali di Chimica*, **72**, 143–55.

Francelin, R. A., Gomide, F. A. C. and Lanças, F. M. (1993) Use of artificial neural networks for the classification of vegetable oils after GC analysis. *Chromatographia*, **35**, 160–6.

Frank, I. E. and Friedman, J. H. (1993) A statistical view of some chemometrics regression tools. *Technometrics*, **35**, 109–35.

Fraser, G. E. (1994) Diet and coronary heart disease: beyond dietary fats and low-density-lipoprotein cholesterol. *American Journal of Clinical Nutrition*, **59**, S1117–23.

Friebolin, H. (1993) *Basic One- and Two-dimensional NMR Spectroscopy*, 2nd edn. VCH, Weinheim.

Galli, C., Petroni, A. and Visioli, F. (1994) Natural antioxidants, with special reference to those in olive oil, and cell protection. *European Journal of Pharmaceutical Sciences*, **2**, 67–8.

Garcia, M. V. A. and López, R. A. (1993) Characterization of European virgin olive oils using fatty-acids. *Grasas y Aceites*, **44**, 18–24.

Garcia, J. M., Gutiérrez, F., Castellano, J. M. *et al.* (1996) Influence of storage temperature on fruit ripening and olive oil quality. *Journal of Agricultural and Food Chemistry*, **44**, 264–7.

Geisser, S. (1975) The predictive sample reuse method with applications. *Journal of the American Statistical Association*, **70**, 320–8.

Gigliotti, C., Daghetta, A. and Sidoli, A. (1994) Caratterizzazione geografica e merceologica di oli di oliva mediante valutazione della composizone trigliceridica per HPLC. *La Rivista Italiana delle Sostanze Grasse*, **71**, 51–6.

Gilbert, R. J., Goodacre, R., Woodward, A. M. and Kell, D. B. (1997) Genetic programming, a novel method for the quantitative analysis of pyrolysis mass spectral data. *Analytical Chemistry*, in press.

Goodacre, R. (1994a) Characterization and quantification of microbial systems using pyrolysis mass spectrometry: introducing neural networks to analytical pyrolysis. *Microbiology Europe*, **2**, 16–22.

Goodacre, R. (1994b) Characterization and quantification of microbial systems using pyrolysis mass spectrometry: introducing neural networks to analytical pyrolysis. *Microbiology Europe*, **2**, 16–22.

Goodacre, R. and Kell, D. B. (1996) Pyrolysis mass spectrometry and its applications in biotechnology. *Current Opinion in Biotechnology*, **7**, 20–8.

Goodacre, R., Hammond, D. and Kell, D. B. (1997) Quantitative analysis of the adulteration of orange juice with sucrose using pyrolysis mass spectrometry and chemometrics. *J. Anal. Appl. Pyrol*, **40/41**, 135–58.

Goodacre, R., Howell, S. A., Noble, W. C. and Neal, M. J. (1996) Sub-species discrimination, using pyrolysis mass spectrometry and self-organising neural networks, of *Propionibacterium acnes* isolates from normal human skin. *Zentralblatt für Bakteriologie*, **284**, 501–15.

Goodacre, R., Kell, D. B. and Bianchi, G. (1992) Neural networks and olive oil. *Nature*, **359**, 594–594.

Goodacre, R., Kell, D. B. and Bianchi, G. (1993) Rapid assessment of the adulteration of virgin olive oils by other seed oils using pyrolysis mass-spectrometry and artificial neural networks. *Journal of the Science of Food and Agriculture*, **63**, 297–307.

Goodacre, R., Neal, M. J. and Kell, D. B. (1994) Rapid and quantitative analysis of the pyrolysis mass spectra of complex binary and tertiary mixtures using multivariate calibration and artificial neural networks. *Analytical Chemistry*, **66**, 1070–85.

Goodacre, R., Neal, M. J. and Kell, D. B. (1996) Quantitative analysis of multivariate data using artificial neural networks: a tutorial review and applications to the deconvolution of pyrolysis mass spectra. *Zentralblatt für Bakteriologie*, **284**, 516–39.

Goodacre, R., Neal, M. J., Kell, D. B. et al. (1994) Rapid identification using pyrolysis mass spectrometry and artificial neural networks of *Propionibacterium acnes* isolated from dogs. *Journal of Applied Bacteriology*, **76**, 124–34.

Goodacre, R., Trew, S., Wrigley-Jones, C. et al. (1995) Rapid and quantitative analysis of metabolites in fermentor broths using pyrolysis mass spectrometry with supervised learning: application to the screening of *Penicillium chryosgenum* fermentations for the overproduction of penicillins. *Analytica Chimica Acta*, **313**, 25–43.

Gourlay, A. R. and Watson, G. A. (1973) *Computational Methods for Matrix Eigenproblems*. John Wiley, Chichester, Sussex.

Grob, K., Biedermann, M., Bronz, M. and Schmid, J. P. (1994a) Recognition of mild deodorization of edible oils by the loss of volatile components. *Zeitschrift für Lebensmittel-Untersuchung und -Forschung*, **199**, 191–4.

Grob, K., Giuffré, A. M., Leuzzi, U. and Mincione, B. (1994b) Recognition of adulterated oils by direct analysis of the minor components. *Fat Science Technology*, **96**, 286–90.

Guinda, A., Lanzón, A. and Albi, T. (1996) Differences in hydrocarbons of virgin olive oils obtained from several olive varieties. *Journal of Agricultural and Food Chemistry*, **44**, 1723–6.

Gussoni, M., Greco, F., Consonni, R. et al. (1993) Application of NMR microscopy to the histochemistry study of olives (*Olea Europaea* L.). *Magnetic Resonance Imaging*, **11**, 259–68.

Gussow, J. D. (1995) Mediterranean diets: are they environmentally responsible? *American Journal of Clinical Nutrition*, **61** (supplement), 1383S–9S.

Haaland, D. M. and Thomas, E. V. (1988a). Partial least squares methods for spectral analyses. 1. Relation to other quantitative calibration methods and the extraction of qualitative information. *Analytical Chemistry*, **60**, 1193–202.

Haaland, D. M. and Thomas, E. V. (1988b) Partial least squares methods for spectral analyses. 2. Application to simulated and glass spectral data. *Analytical Chemistry*, **60**, 1202–8.

Harris, R. K. (1986) *Nuclear Magnetic Resonance Spectroscopy*. Longman Scientific and Technical, Harlow, Essex.

Haumann, B. F. (1996) Olive oil: Mediterranean product. *Inform*, **7**, 890–903.

Hazen, K. H., Arnold, M. A. and Small, G. W. (1994) Temperature-insensitive near-infrared spectroscopic measurement of glucose in aqueous solutions. *Applied Spectroscopy*, **48**, 477–83.

Hecht-Nielsen, R. (1989) *Neurocomputing*, Addison-Wesley, Reading, MA.

Heikka, R., Minkkinen, P. and Taavitsainen, V. M. (1994) Comparison of variable selection and regression methods in multivariate calibration of a process analyzer. *Process Control and Quality*, **6**, 47–54.

Helland, I. S. (1988) On the structure of partial least squares regression. *Communications on Statistical Simulations*, **17**, 581–607.

Hertz, J., Krogh, A. and Palmer, R. G. (1991) *Introduction to the Theory of Neural Computation*. Addison-Wesley, Redwood City, CA.

Hoerl, A. E. and Kennard, R. W. (1970a) Ridge regression: biased estimation for nonorthogonal problems. *Technometrics*, **12**, 55–67.

Hoerl, A. E. and Kennard, R. W. (1970b) Ridge regression: application to nonorthogonal problems. *Technometrics*, **12**, 69–82.

Horchner, U. and Kalivas, J. H. (1995) Further investigation on a comparative-study of simulated annealing and genetic algorithm for wavelength selection. *Analytica Chimica Acta*, **311**, 1–13.

Hotelling, H. (1933) Analysis of a complex of statistical variables into principal components. *Journal of Educational Psychology*, **24**, 417–41, 498–520.

IFR (1994) *Annual Report 1994*, Institute of Food Research.

Irwin, W. J. (1982) *Analytical Pyrolysis: A Comprehensive Guide*, Marcel Dekker, New York.

Ismail, A. A., van de Voort, F. R., Emo, G. and Sedman, J. (1993) Rapid quantitative determination of free fatty acids in fats and oils by Fourier transform infrared spectroscopy. *Journal of the American Oil Chemists' Society*, **70**, 335–41.

Jolliffe, I. T. (1982) A note on the use of principal components in regression. *Applied Statistics*, **31**, 300–3.

Jolliffe, I. T. (1986) *Principal Component Analysis*, Springer, Berlin.

Jouan-Rimbaud, D., Massart, D. L., Leardi, R. and de Noord, O. E. (1995) Genetic algorithms as a tool for wavelength selection in multivariate calibration. *Analytical Chemistry*, **67**, 4295–301.

Judd, J. S. (1990) *Neural Network Design and the Complexity of Learning*. MIT Press, Cambridge, MA.

Kafatos, A. and Comas, G. (1991) Biological effects of olive oil on human health, in *Olive Oil* (ed. A. K. Kiritsakis), Americal Oil Chemists' Society, Champaign, IL, pp. 157–81.

Kajioka, R. and Tang, P. W. (1984) Curie-point mass spectrometry of *Legionella* species. *Journal of Applied and Analytical Pyrolysis*, **6**, 59–68.

Kemp, W. (1986) *NMR in Chemistry: A Multinuclear Introduction*, Macmillan Education, London.

Kemsley, E. K., Appleton, G. P. and Wilson, R. H. (1994) Quantitative-analysis of emulsions using attenuated total reflectance (ATR). *Spectrochimica Acta Part a – Molecular Spectroscopy*, **50**, 1235–42.

Kiritsakis, A. K. (1984) Effect of selected storage conditions and packaging materials on olive oil quality. *Journal of the American Oil Chemists' Society*, **61**, 1868–70.

Kiritsakis, A. K. (1991) *Olive oil*, AOCS, Champaign, IL.

Kiritsakis, A. and Dugan, L. R. (1985) Studies in photoxidation of olive oil. *Journal of the American Oil Chemists' Society*, **62**, 892–6.

Kiritsakis, A. and Markakis, P. (1991) Olive oil analysis, in *Essential Oils and Waxes* (eds. H. E. Linskens and J. F. Jackson), Springer, Berlin, pp. 1–20.

Koza, J. R. (1992) *Genetic Programming: On The Programming of Computers by Means of Natural Selection*. MIT Press, Cambridge, MA.

Koza, J. R. (1994a) Genetic programming as a means for programming computers by natural selection. *Statistics and Computing*, **4**, 87–112.

Koza, J. R. (1994b) *Genetic Programming II: Automatic Discovery of Reusable Programs*. MIT Press, Cambridge, MA.

Kubinyi, H. (1994a) Variable selection in QSAR studies. 1. An evolutionary algorithm. *Quantitative Structure–Activity Relationships*, **13**, 285–94.

Kubinyi, H. (1994b) Variable selection in QSAR studies. 2. A highly efficient combination of systematic search and evolution. *Quantitative Structure–Activity Relationships*, **13**, 393–401.

Kubinyi, H. (1996) Evolutionary variable selection in regression and PLS analyses. *Journal of Chemometrics*, **10**, 119–33.

Kvalheim, O. M., Aksnes, D. W., Brekke, T. *et al.* (1985) Crude oil characterization and correlation by principal component analysis of ^{13}C nuclear magnetic resonance spectra. *Analytical Chemistry*, **57**, 2858–64.

Lai, Y. W., Kemsley, E. K. and Wilson, R. H. (1994) Potential of Fourier transform-infrared spectroscopy for the authentication of vegetable-oils. *Journal of Agricultural and Food Chemistry*, **42**, 1154–9.

Lai, Y. W., Kemsley, E. K. and Wilson, R. H. (1995) Quantitative-analysis of potential adulterants of extra virgin olive oil using infrared-spectroscopy. *Food Chemistry*, **53**, 95–8.

Li-Chan, E. (1994) Developments in the detection of adulteration of olive oil. *Trends in Food Science and Technology*, **5**, 3–11.

Lindgren, F., Geladi, P., Berglund, A. *et al.* (1995) Interactive variable selection (IVS) for PLS2. Chemical applications. *Journal of Chemometrics*, **9**, 331–42.

Lindgren, F., Geladi, P. and Wold, S. (1993) The Kernel algorithm for PLS. *Journal of Chemometrics*, **7**, 45–59.

Linos, A., Kaklamanis, E., Kontomerkos, A. *et al.* (1991) The effect of olive oil and fish consumption on rheumatoid-arthritis – a case control study. *Scandinavian Journal of Rheumatology*, **20**, 419–26.

Luger, G. F. and Stubblefield, W. A. (1989) *Artificial Intelligence and the Design of Expert Systems*, Benjamin Cummings, Redwood City, CA.

Lyon, D. H. and Watson, M. P. (1994) Sensory profiling – a method for describing the sensory characteristics of virgin olive oil. *Grasas y Aceites*, **45**, 20–5.

Lyttkens, E. (1966) On the fix-point property of Wold's iterative estimation method for principal components, in *Multivariate Analysis* (ed. K. R. Krishnaiah), Academic Press, New York, pp. 335–50.

Lyttkens, E. (1973) The fix-point method for estimating interdependent systems with the underlying model specification. *Journal of the Royal Statistical Society, Series A*, **135**, 353–94.

MAFF (1995) *Manual of Nutrition: Reference Book 342*, 10th edn, Ministry of Agriculture, Fisheries and Food, The Stationery Office, London.

Malpass, J. A., Salt, D. W., Ford, M. G. *et al.*, (1994) Continuum regression: a new algorithm for the prediction of biological activity, in *Advanced Computer-Assisted Techniques in Drug Discovery* (ed. H. van der Waterbeemd), VCH, Weinheim, pp. 163–89.

Malpass, J. A., Salt, D. W., Wynn, E. W. *et al.* (1995) Prediction of biological activity using continuum regression, in *Trends in QSAR and Molecular Modelling 92* (ed. C. G. Wermuth), ESCOM, pp. 314–16.
Mark, H. (1991) *Principles and Practice of Spectrosocpic Calibration*, John Wiley, New York.
Martens, H. and Næs, T. (1989) *Multivariate Calibration*, John Wiley, Chichester, Sussex.
Martin-Moreno, J. M., Willett, W. C., Gorgojo, L. *et al.* (1994) Dietry fat, olive oil intake and breast cancer risk. *International Journal of Cancer*, **58**, 774–80.
Massy, W. F. (1965) Principal components regression in exploratory statistical research. *Journal of the American Statistical Association*, **60**, 234–56.
Meuzelaar, H. L. C., Haverkamp, J. and Hileman, F. D. (1982) *Pyrolysis Mass Spectrometry of Recent and Fossil Biomaterials*, Elsevier, Amsterdam.
Michie, D., Spiegelhalter, D. J. and Taylor, C. C. (1994) Machine learning: neural and statistical classification, in *Ellis Horwood Series in Artificial Intelligence* (ed. J. Campbell), Ellis Horwood, Chichester, Sussex.
Morales, M. T., Alonso, M. V., Rios, J. J. and Aparicio, R. (1995) Virgin olive oil aroma – relationship between volatile compounds and sensory attributes by chemometrics. *Journal of Agricultural and Food Chemistry*, **43**, 2925–31.
Morris, A. O. (1982) *Linear Algebra – An Introduction*, Van Nostrand Reinhold, London.
Mottram, R. F. (1979) *Human Nutrition*, 3rd edn, Edward Arnold, London.
Murphy, D. J. (1995) New oils for old. *Chemistry in Britain*, **31**, 300–2.
Norinder, U. (1996) Single and domain mode-variable selection in 3D QSAR applications. *Journal of Chemometrics*, **10**, 95–105.
Oman, S. D., Næs, T. and Zube, A. (1993) Detecting and adjusting for nonlinearities in calibration of near-infrared data using principal components. *Journal of Chemometrics*, **7**, 195–212.
Ozaki, Y., Cho, R., Ikegaya, K. *et al.*, (1992) Potential of near-infrared Fourier transform Raman spectroscopy in food analysis. *Applied Spectroscopy*, **46**, 1503–7.
Peri, C. and Rastelli, C. (1994) Implications for the future and recommendations for modifications to current regulations concerning virgin olive oil. *Grasas y Aceites*, **45**, 60–1.
Perrin, J.-L. (1992) Les composés mineurs et les antioxygènes naturels de l'olive et de son huile. *Revue Française des Corps Gras*, **39**, 25–32.
Rade, D., Strucelj, D., Mokrovcak, Z. and Hrboka, Z. (1995) Influence of olive storage and processing on some characteristics of olive oil. *Prehrambeno-Tehnoloska I Biotehnoloska Revija*, **33**, 119–22.
Ramos, L. S., Beebe, K. R., Carey, W. P. *et al.* (1986) Chemometrics. *Analytical Chemistry*, **58**, 294R–315R.
Ranalli, A. and Martinelli, N. (1994) Extraction of the oil from the olive pastes by biological and not conventional industrial technics. *Industrie Alimentari*, **33**, 1073–83.
Ripley, B. D. (1994) Neural networks and related methods for classification. *Journal of the Royal Statistical Society, Series B – Methodological*, **356**, 409–37.
Rosenblatt, F. (1958) The perceptron: a probabilistic model for information storage and organization in the brain. *Psychological Review*, **65**, 386–408.
Rumelhart, D. E. and McClelland, J. L. (1986) *Parallel Distributed Processing. Experiments in the Microstructure of Cognition*, MIT, Cambridge, MA.
Sadeghi-Jorabchi, H., Hendra, P. J., Wilson, R. H. and Belton, P. S. (1990) Determination of the total unsaturation in oils and margarines by Fourier-transform Raman spectroscopy. *Journal of the American Oil Chemists' Society*, **67**, 483–6.
Sadeghi-Jorabchi, H., Wilson, R. H., Belton, P. S. *et al.* (1991) Quantitative analysis of oils and fats by Fourier-transform Raman spectroscopy. *Spectrochimica Acta A*, **47**, 1449–58.
Salter, G. J., Lazzari, M., Giansante, L. *et al.* (1997) Determination of the geographical origin of Italian extra virgin olive oil using pyrolysis mass spectrometry and artificial neural networks. *Journal of Analytical Applied Pyrolysis*, **40/41**, 159–70.
Salunkhe, D. K., Chavan, J. K., Adsule, R. N. and Kadam, S. S. (1991) *World Oilseeds: Chemistry, Technology and Utilization*. Van Nostrand Reinhold, New York.
Sarle, W. S. (1994) Neural networks and statistical models, in *Nineteenth Annual SAS Users Group International Conference*.
Sato, T. (1994) Application of principal-component analysis on near-infrared spectroscopic data of vegetable-oils for their classification. *Journal of the American Oil Chemists' Society*, **71**, 293–8.

Schwaiger, I. and Vojir, F. (1994) Anwendung Multivariater Statistischer Verfahren zur Überprüfung der Authentizität von Speiseölen. *Deutsche Lebensmittel-Rundschau*, **90**, 143–6.
Seasholtz, M. B. and Kowalski, B. (1993) The parsimony principle applied to multivariate calibration. *Analytica Chimica Acta*, **277**, 165–77.
Segre, A. L., Mannina, L., Barone, P. and Sacchi, R. (1996) Quality and geographical origin of virgin olive oil as determined by high-field ^{1}H-NMR. *Bruker Report*, **143**, 27–8.
Shahidi, F. (1990) *Canola and Rapeseed: Production, Chemistry, Nutrition and Processing Technology*, Van Nostrand Reinhold, New York.
Shaw, A. D., di Camillo, A., Vlahov, G. *et al.* (1996) Discrimination of different olive oils using ^{13}C NMR and variable reduction, in *Food Authenticity '96*, Norwich.
Shaw, A. D., di Camillo, A., Vlahov, G. *et al.* (1997) Discrimination of the variety and region of origin of extra virgin olive oils using ^{13}C NMR and multivariate calibration with variable reduction. *Analytica Chimica Acta*, in press.
Shepherd, J. and Packard, C. J. (1992) Atherosclerosis in perspective: the pathophysiology of human cholesterol metabolism, in *Human Nutrition: A Continuing Debate* (eds M. Eastwood, C. Edwards and D. Parry), Chapman & Hall, London, pp. 33–50.
Simpkins, W. and Harrison, M. (1995a) The state of the art in authenticity testing. *Trends in Food Science and Technology*, **6**, 321–8.
Simpkins, W. and Harrison, M. (1995b) The state of the art in authenticity testing. *Trends in Food Science and Technology*, **6**, 321–8.
Smith, M. (1993) *Neural Networks for Statistical Modeling*, Van Nostrand Reinhold, New York.
Snee, R. D. (1977) Validation of regression models: methods and examples. *Technometrics*, **19**, 415–28.
Sreerama, N. and Woody, R. W. (1994) Protein secondary structure from circular dichroism spectroscopy: combining variable selection principle and cluster analysis with neural network, ridge regression and self-consistent methods. *Journal of Molecular Biology*, **242**, 497–507.
Stone, M. (1974) Cross-validatory choice and assessment of statistical predictions. *Journal of the Royal Statistical Society*, **36**, 111–33.
Stone, M. and Brooks, R. J. (1990) Continuum regression: cross-validated sequentially constructed prediction embracing ordinary least squares, partial least squares and principal components regression. *Journal of the Royal Statistical Society, Series B*, **52**, 237–69.
Stryer, L. (1981) *Biochemistry*, 2nd edn, Freeman, San Francisco, CA.
Taavitsainen, V.-M. and Korhonen, P. (1992) Nonlinear data analysis with latent variables. *Chemometrics and Intelligent Laboratory Systems*, **14**, 185–94.
Trichopoulou, A. (1995) Olive oil and breast-cancer. *Cancer Causes and Control*, **6**, 475–6.
Trichopoulou, A., Gnardellis, C., Katsouyanni, K. *et al.* (1995a) Consumption of olive oil and specific food groups in relation to breast-cancer risk in Greece – response. *Journal of the National Cancer Institute*, **87**, 1022.
Trichopoulou, A., Katsouyanni, K., Stuver, S. *et al.* (1995b) Consumption of olive oil and specific food groups in relation to breast-cancer risk in Greece. *Journal of the National Cancer Institute*, **87**, 110–16.
Trichopoulou, A., Kouris-Blazos, A., Vassilakou, T. *et al.* (1995c) Diet and survival of elderly Greeks: a link to the past. *American Journal of Clinical Nutrition*, **61**(supplement), 1346s–50s.
Tsimidou, M. (1995) The use of HPLC in the quality control of virgin olive oil. *Chromatography and Analysis*, (Aug/Sept), 5–7.
Tsimidou, M. and Karakostas, K. X. (1993) Geographical classification of Greek virgin olive oil by nonparametric multivariate evaluation of fatty-acid composition. *Journal of the Science of Food and Agriculture*, **62**, 253–7.
van de Voort, F. R. (1994) FTIR spectroscopy in edible oil analysis. *INFORM*, **5**, 1038–42.
van de Voort, F. R., Ismail, A. A. and Sedman, J. (1995) A rapid, automated method for the determination of *cis* and *trans* content of fats and oils by Fourier-transform infrared-spectroscopy. *Journal of the American Oil Chemists' Society*, **72**, 873–80.
van de Voort, F. R., Ismail, A. A., Sedman, J. *et al.* (1994a) The determination of peroxide value by Fourier transform infrared spectroscopy. *Journal of the American Oil Chemists' Society*, **71**, 921–6.

van de Voort, F. R., Ismail, A. A., Sedman, J. and Emo, G. (1994b) Monitoring the oxidation of edible oils by Fourier transform infrared spectroscopy. *Journal of the American Oil Chemists' Society*, **71**, 243–53.

Visioli, F. and Galli, C. (1994) Oleuropein protects low density lipoprotein from oxidation. *Life Sciences*, **55**, 1965–71.

Visioli, F. and Galli, C. (1995) Natural antioxidants and prevention of coronary heart-disease – the potential role of olive oil and its minor constituents. *Nutrition Metabolism and Cardiovascular Diseases*, **5**, 306–14.

Visioli, F., Vinceri, F. F. and Galli, C. (1995) Waste-waters from olive oil production are rich in natural antioxidants. *Experientia*, **51**, 32–4.

Vlahov, G. (1996) Improved quantitative C-13 nuclear magnetic resonance criteria for determination of grades of virgin olive oils. The normal ranges for diglycerides in olive oil. *Journal of the American Oil Chemists' Society*, **73**, 1201–3.

Vlahov, G. and Angelo, C. S. (1996) The structure of triglycerides of monovarietal olive oils: a ^{13}C-NMR comparative study. *Fett/Lipid*, **98**, 203–5.

Wehrli, F. W., Marchand, A. P. and Wehrli, S. (1988) *Interpretation of Carbon-13 NMR Spectra*, 2nd edn, John Wiley, Chichester, Sussex.

Weiss, S. H. and Kulikowski, C. A. (1991) *Computer Systems that Learn: Classification and Prediction Methods from Statistics, Neural Networks, Machine Learning, and Expert Systems*, Morgan Kaufmann, San Mateo, CA.

Werbos, P. J. (1993) *The Roots of Back-propagation: From Ordered Derivatives to Neural Networks and Political Forecasting*, John Wiley, Chichester, Sussex.

Williams, D. A. R. (1986) *Nuclear Magnetic Resonance Spectroscopy*, John Wiley, Chichester, Sussex.

Williams, K. P. J., Pitt, G. D., Batchelder, D. N. and Kip, B. J. (1994) Confocal Raman microspectroscopy using a stigmatic spectrograph and CCD detector. *Applied Spectroscopy*, **48**, 232–5.

Williams, K. P. J., Pitt, G. D., Smith, B. J. E. and Whitley, A. (1994) Use of a rapid scanning stigmatic raman imaging spectrograph in the industrial environment. *Raman Spectroscopy*, **25**, 131–8.

Winson, M. K., Goodacre, R., Timmins, É. et al. (1997a) Diffuse reflectance absorbance spectroscopy taking in chemometrics (DRASTIC). A hyperspectral FT-IR-based approach to rapid screening for metabolite overproduction. *Analytica Chimica Acta*, in press.

Winson, M. K., Todd, M., Rudd, B. A. M. et al. (1997b) A DRASTIC (diffuse reflectance absorbance spectroscopy taking in chemometrics) approach for the rapid analysis of microbial fermentation products: quantification of aristeromycin and neplanocin A in *Streptomyces citricolor* broths., in press.

Wold, H. (1966) Estimation of principal components and related models by iterative least squares, in *Multivariate Analysis* (ed. K. R. Krishnaiah), Academic Press, New York, pp. 391–420.

Wold, H. (1975) Soft modelling by latent variables: the non-linear iterative partial least squares (NIPALS) approach, in *Perspectives in Probability and Statistics, Papers in Honour of M. S. Bartlett* (ed. J. Gani), Academic Press, London, pp. 117–42.

Wold, H. (1980) Model construction and evaluation when theoretical knowledge is scarce (theory and application of partial least squares), in *Evaluation of econometric models* (eds J. Kmenta and J. B. Ramsey), Academic Press, London, pp. 47–74.

Wold, H. (1982) Soft modeling: the basic design and some extensions, in *Systems under Indirect Observation: Causality, Structure, Prediction. Part II* (eds K. G. Jöreskog and H. Wold), North Holland, Amsterdam, pp. 1–53.

Wold, S. (1978) Cross validatory estimation of the number of components in factor and principal components models. *Technometrics*, **20**, 397–405.

Wonnacott, T. H. and Wonnacott, R. J. (1981) *Regression: A Second Course in Statistics*, John Wiley, Chichester, Sussex.

Yoder, C. H. and Schaeffer, C. D. J. (1987) *Introduction to Multinuclear NMR*, Benjamin/Cummings, Menlo Park, CA.

Zamora, R., Navarro, J. L. and Hidalgo, F. J. (1994) Identification and classification of olive oils by high-resolution C-13 nuclear magnetic resonance. *Journal of the American Oil Chemists' Society*, **71**, 361–4.

Zupan, J. and Gasteiger, J. (1993) *Neural Networks for Chemists: An Introduction*, VCH, Weinheim.

Index

Acetic acid 68, 69.
Acetonitrile 41.
Acetylcholinesterase (A'ChE) 228.
Acrolein 77.
Acylcarnitine 186.
Acyl glycerides 22, 202.
Acyl sitosteryl glycoside 116, 118.
Acyl transferase 198.
Adduct 313.
Adenosine triphosphate 224.
Adulteration 322, 325.
Aging process 204.
Alga (*Chlorella minutissima*) 310.
Algorithms 326, 358.
O-Alkyl phospholipids 122.
Alpine currant (*Ribes alpinum*) 204.
Alumina 2.
Aluminium chloride
Ammonia CI/MS 207, 231.
Anaerobic bacteria 67.
Angular momentum 326.
p-Anisidine Value 79.
Antigen 229.
Antioxidants 320.
Apple juice 283.
Apricot kernel oil 94.
Arachidic acid 324.
Arachidonic acid 100, 187, 189, 191, 204, 209, 227, 301, 319.
Arachis oil (peanut or ground nut) 274.
Argentation chromatography 41.
Artificial Neural Networks (ANN) 318, 324, 340, 347, 353, 364.
Atherogenic 321.
Atheroma 194, 226.
Atherosclerotic lesions 321.
Atmospheric Pressure Ionisation 294, 301.
Authenticity 265, 266, 325.
Autoradiography 13,
Autosampler 64.
Autoxidation 186, 193, 204, 223, 251.

Azelaoyl group 225.
Avenasterol 278.
2,2'-Azobis-2, 4-dimethylvaleronitrile (AMUN) 224, 225.
2,2'-Azobis (2-amidinopropane)dihydrochloride (AAPH) 225.

Babassu kernel oil 266, 267, 274.
Bacillus acidocaldarius 136.
Bacillus spp. 221, 222.
Back flushing 257, 259.
Backward Elimination 341.
Band broadening 38.
Basophils 196.
Beef (cooked) 24.
Beef heart 211, 220, 222.
Benzene 83.
Benzoate derivative 196.
per-O-Benzoyl 1-monomycoloylglycerol 196.
Biller-Biemann enhancement 200.
Binding Energy 251.
Biomembranes 291.
Bismuth subnitrate 14.
Blackcurrant (*Ribes nigrum*) 201, 204.
Blackcurrant oil 109, 110.
Blood 191, 307.
Blood lipids 22, 210.
Bombaceae 147.
Bombax munguba 149.
Borage (*Borago officinalis*) 204.
Borage oil 100, 102.
Boric acid TLC. 6.
Breast cancer 321.
Boron trifluoride-methanol 144, 168, 173.
Bovine brain 212, 217, 221, 225, 232, 234, 296, 304, 307.
Bovine heart phospholipids 127.
Bovine liver 216, 219.
Bovine lung surfactant 122, 127.
Bovine milk fat 201, 296, 299, 304, 307.

Brassica sterol 278.
Brownlee silica gel 215.
Butanal 77.
9-(2'-But-cis-1-enyl-cyclopentenyl) nonanoate 165.
9-(2'-But-cis-1-enyl-cyclopentyl) nonanoic 155, 158.
9-(2'-But-trans-1-enyl-cyclopentyl) nonanoic 155, 158.
Butter 82, 198, 199, 200, 201.
9-(2'-Butyl-cyclopent-cis-3'-enyl) nonanoate 159.
9-(2'-Butyl-cyclopentyl) nonanoate 160.
tert-Butyldimethyl silyl esters 187, 196, 197.
tert-Butyl hydroperoxides 205.
Butyric acid 67, 68, 69.

CAD-MIKE (mass analysed ion kinetic energy) 153.
C-pathway 269.
C-pathway 270.
Calibration 90.
Calidris pusilla 26.
Caloncoba echinata 141.
Camelina sativa 51.
Campesterol 278.
Capacity factor 38.
Capillary column 38, 39.
Caproic acid 68, 69.
Carbohydrates 253.
Carbon dioxide 35, 269.
Carbowax 75, 172.
Carcinogenesis 204.
Cardiolipin 220, 221.
Carotene 320.
Castor oil 44, 322.
Ceramide 230, 231, 232, 233, 291, 296, 299, 307, 308.
Cerebrosides 231.
Cerebroside sulphate 302, 310.
Certified brands of origin (CBO) 363.
Characteristicity 360.
Chaulmoogric acid 137, 139, 141.
Cheese 67, 70.
Chemometric methods 317, 357.
Chicken 232.
Chicken meat 79, 255, 263.
Chiral phase 197, 198, 220.
Chloroform 83.
Chlorophyll 320.
Chlostridia 67.

Cholera toxin 13.
Cholesterol 14, 18, 22, 93, 96, 97, 192, 194, 277, 278, 321.
Cholesterol hydroperoxide 225.
Cholesterol sulphate 192.
Cholesterol triol 193.
Cholesteryl arachidonate 193.
Cholesteryl linoleate 194.
[H] Choline 226.
Chromarods 19, 20.
Chromatography 250.
Circular development 10.
Clausius-Clapeyron equation 61.
Cloud berry (*Rubus chaemaemorus*) 204.
Co source 251.
Coal 27.
Coal tar pitch 27.
Cocoa butter 18.
Cocoa butter substitutes 272.
Coconut 266, 267, 278, 281, 282.
Cod liver
Codex alimentarius 265, 270.
Coenzyme A 224.
Collisionally activated dissociation 145.
Comamonas testosteroni 238.
Confection industry 24.
Coomassie blue 13.
Copper (II) sulphate 23.
Coratina variety 363.
Core aldehydes 193, 205, 206, 226.
Corn oil 198, 322, 33.
Cottonseed 147, 149, 150, 199, 213, 265, 274, 278, 281, 284, 285, 286.
Covariance matrix 340.
Cows 201.
CP-Sil 84 172, 174.
Crassulacean pathway 270.
Creosote 26.
Cryofocusing 66.
Cryogenic focusing 65.
Cupric acetate 13.
Curie point pyrolysis 328, 330, 363.
Cyclic Fatty acids 136 seq., 157.
Cyclic Fatty acid monomers 152, 154.
Cyclobutanones 252, 253, 257.
Cyclohexenyl acids 157.
Cyclohexenyl esters 164.
12-Cyclohexyldodec-cis-9-enoate 159.
11-Cyclohexylundecanoic acid 137.
11-Cycloheptylundecanoic acid 137.

15-Cyclopent-2-enyl-pentadecanoate 140.
15-Cyclopent-2-enyl-pentadec-9-enoate 140.
11-Cyclopent-2-enyl-undecanoate 140.
13-Cyclopent-2-enyl-tridecanoate 140.
13-Cyclopent-2-enyl-tridec-4-enoate 140, 143.
Cyclopropane acids 136.
Cyclopropene acids 136.
Cysteine 254.

Dansylhydrazine 260, 261.
DB 5 147.
Decadienal 73, 74, 75, 76, 77.
Decanoic acid 69.
Declustering Potential 311.
Denominazione di Origine Controllata DOC 363.
Denominazione di Origine Proteggetta DOP 363.
Densitometric methods 16.
Desulfobulbus spp 144.
per Deuterated ethanol 198.
Diacylglycerides 99, 184, 185, 195, 299, 305, 306, 323.
Diatoms 25.
2, 7'Dichlorofluorescein 12, 14.
Diethanolamine 226.
Digalactodiacylglycerides 114, 115, 116, 117, 296.
Digalactosyldiacylglycerols 46, 53.
Diglycerides 6, 11, 197, 198, 202, 207, 208, 304.
Dihydromalvic acid 144.
Dihydrosterculic acid 144, 145, 146, 150.
Dihydroxyacetone 253.
Dimethyl 2-carbomethoxy-3-propyl-glutarate 168.
1, 3-Di (9, 10-methyleneoctadecanoyl)-glycerol 145.
Dimethyl nonanedioate 168.
Dimethyloxazoline derivatives DMOX 139, 140, 143, 145, 149, 151, 154, 155, 164, 166, 167, 170, 171.
Dimethyl sulphide 84.
Dinitrophenylhydrazone derivatives 193.
Dinitrophenyl urethane derivatives 196, 197, 198, 220.

Dinoflagellates 25.
2, 3-Dinor-5, 6-dihydro-8-isoprostaglandin 224.
Dioleoylglycerols 198.
Diphosphatidyl glycerol 215.
Direct probe MS 182.
Distearoylphosphatidylglycerol 123.
Docosahexaenoic acid 93, 100, 104, 118, 175, 186, 224.
Dogfish liver 5.
Dolichol phosphates 222.
Dritta variety 363.
Druppe 318.

EDTA 6.
Egg lecithin 18, 96, 129, 130.
Egg lipids 3, 117, 118.
Egg yolk 206, 214, 225, 229, 232, 296, 297.
Eicosanoids 189, 191, 291, 298.
Eicosanolactones 191.
Eicosapentaenoic acid 175.
Eicosenoic acid. 319.
Eigen values 337, 341.
Eigen vectors 338.
Electron Impact MS 183.
Electrospray 181, 294, 306.
-Eleostearic acid 3.
Endothelial cells 211.
Enrichment factor 280.
Enterobacter agglomerans 237.
Enterobacteria 136.
Epidioxides 205.
Epoxy acids 6, 185, 205.
Epoxy alcohols 204.
12, 13-Epoxy-octadecenoate 223.
Erucic acid 272.
Erythrocytes 228, 233, 299, 301.
Erythrocyte plasma membrane 210, 214, 307, 309.
Escherichia coli 221, 311.
Ethane 60.
Ethanolamine glycerophospho-lipids 212.
Eukaryotes 228.
Euonymus oil 107, 108.
Euphoria longana 145.
Evaporative Light Scattering Detector ELSD 45, 262, 301.
Evening primrose (*Oenothera biennis*) 204.

Fast Atom Bombardment FAB 17, 181, 192, 223, 230.
Fatty acids 184.
Federation of Oils, Fats and Seeds Association FOSFA 266.
Ferns 310.
Field desorption 185.
Fingerprint 329.
Fisher ratio 362.
Fishmeal (menhaden) 24.
Fish oil 42, 49, 90, 95, 100, 222, 284.
Flagellates 25.
Flame ionisation detectors 17, 20.
Flame thermionic ionisation detector 23.
Flow injection 209, 217.
Fluorescent biological probe 208.
Fluorescent labelling 264.
Folch extraction 173, 312.
Food 250.
Forward selection 341.
Fourier Transform Infra red (FTIR) 155, 323, 368.
Fossil fuels 27.
Free Fatty Acid Phase (FFAP) 67.
French fries 173, 174.
Frying 136.
Full Cross-Validation 348.
Fungal Mycelium 219, 301, 310.

Galactosyl ceramide 293, 302, 310, 311.
Gamma rays 251.
Gangliosides 13, 230, 234, 235.
Gastric mucosa 233.
Gastrointestinal tract 233.
GC-FTIR 154.
Genetically modified soyabean oil 202.
Germination 250.
Glioma cell 218, 303.
Glucose 229.
Glucosylated phosphatidyl ethanolamine 215.
Glycerobenzoate derivatives 299.
Glycolipids 15, 24, 109, 300, 301.
Glycosphingolipids 2, 18, 233, 234, 235, 307.
Glycosylated aminophospholipids 229.
Glycosylated glycerophosphates 227.
Glycosylated phosphoinositol 228, 298.
Golay equation 39.
Gorlic acid 137, 138, 141, 142.

Gradient 47.
Gradient development 9.
Grapeseed oil 274.
Grignard reaction 198.
Groundnut oil (peanut) 267, 272, 276, 282, 284, 285, 286.

HPLC-UV 296, 299.
HPTLC 3.
Halphen test 147.
Hatch slack 270.
Haemophilus spp. 237.
Hazel nut oil 33.
Head space GC 59.
Heart mitochondria 226.
Hepatic perisinusoidal stellate cell line 188.
Heptadecene 256, 257.
Heptadienal 77, 81.
Heptanal 77, 81.
Heptanoic acid 69.
Heptanone 86.
Heptenal 71, 77.
Herring silage 23.
Hexaenoic acid 13.
Hexane 83.
Hexanal 60, 71, 72, 76, 79.
1-O-Hexadecyl-2-acetyl-sn-Glycero-3-phosphocholine 211.
Hexenal 76, 77, 80.
7-(2'-Hexyl-cyclopent-cis-4'-enyl)heptanoate 159.
7-(2'-Hexyl-cyclopentyl)heptanoate 171, 185.
High density lipoprotein HDLP 321.
High Oleic Sunflower oil 153, 159, 160, 168, 169, 175, 333.
High performance liquid chromatography 290, 296, 324.
Homonuclear coupling 89.
Human depot fat 112, 113.
Husk oil 33, 334.
Hydnocarpic acid 137, 139, 141.
Hydnocarpus spp. 136.
Hydnocarpus anthelmintica 140, 141.
Hydrazine 147.
Hydrogen sulphide 84, 332.
Hydrogenated milk fat 3.
Hydrogenation 154, 190, 205.
Hydroperoxides 186, 188, 190, 203, 204, 205, 210, 223, 225, 252.
Hydroxyacids 6, 186, 188, 191.

7-Hydroxycholesterol 192, 195.
Hydroxyeicosatetraenoic acid. 224.
Hydroxyheptenal 187.
3-Hydroxymyristate 238.
16-Hydroxypalmitic acid 191.
2-Hydroxysterculic acid 149, 151.
6-Hydroxy-2, 3-undecadienal 187.
Hyperglycaemia 229.

Iatroscan 1, 17, 19.
Infra red 148.
Interleukine 304.
Immunochemical 226.
Injector 73.
Iodine 12, 15, 23.
Ion trap 194.
Integration 132.
Internal standard 62.
Irradiation 251.
Isobutane 199, 207.
Isobutyric acid 68, 69, 79.
Iso caproic acid 68, 69.
Isoprostenoids 220.
Isoprostane 189.
Isoprostane phospholipids 223.
Isovaleric acid 68, 69, 79.
Iteration 338, 339.

Jacobi's method 338.

Karanja 19.
Karanjin 19.
Karanjone 19.
Ketene 208.
7-Ketocholesterol 193, 194, 195.
Kieselguhr 2.
Kokum butter 18.

Lactobacillic acid 137, 144.
Lactobacillus spp. 136, 222.
Lactosamine 236.
Lard 203, 322.
Laser 46.
Latent variables 343, 344.
Lauric acid 277.
Lecithin 18.
Legionellaceae 329.
Leukocytes 213, 224.
Leukotrienes 187, 188.
Leverage correction 349.
Lignoceric acid 319.
Limnanthes alba (Meadowform) 192.

Lingon berry (*Vaccinium vitisidaea*) 201.
Linola 90.
Linoleic acid 11, 152, 163, 171, 185, 191, 287, 319, 367.
Linolenic acid 11, 138, 152, 157, 161, 163, 166, 171, 175, 185, 201, 203, 272, 319.
-Linolenic acid 42, 201, 203.
Linseed oil 101, 153, 157, 158, 162, 166, 168.
Lipase 267.
Lipid A 236.
Lipolysis 223.
Lipo oligosaccharide (LOS) 237.
Liposomes 117, 121, 126.
Lipoxygenase 186, 189, 191, 195.
Liquid chromatography-mass spectrometry LC-MS 295.
Litchi chinensis 144, 145.
Liver 198.
Liverworts 310.
Longan 145.
Loop injection 290, 300.
Low density Lipoprotein (LDL) 195, 206, 321.
Low erucic rapeseed 266, 275, 280, 281.
Lysophosphatic acid 120.
Lysophosphatidyl choline 119, 120, 121.
Lysophosphatidyl ethanolamine 120.
Lysophosphatidyl inositol 216.
Lysophospholipids 24.

Maize germ 267, 274, 281, 282, 283, 284, 286, 288, 333.
Malondialdehyde 253.
Malvaceae 136, 147.
Malvalic acid 137, 147, 149, 150.
Mammalian cell culture 202.
Margarine 321.
Marine lipids 3.
Marker substance 255.
Mass spectrometry EI 145.
Mass spectrometry CI 145.
Mass spectrometry Fast atom bombardment (FAB) 145.
Mast cells 196, 209.
Matrix 335, 337.
McLafferty rearrangement 185.
Meadowfoam 191.
Methane 332.
Methanethiol 84, 147.

Methanol 332.
Menhaden 5.
Mediterranean basin 318.
Mediterranean diet 320.
Meningococcal 237.
Methoxime derivatives 184, 190.
Methoxy ether derivatives 148.
Methyl butyrate 81.
12-(2'-methyl-cyclopentyl)dodec-cis-9-enoate 159.
11-(2'-methyl-cyclohexyl)undec-cis-9-enoate 159.
11-(2'-methyl-cyclohexyl)undec-trans-9-enoate 159.
9, 10-methylenehexadecanoic acid 141.
7, 8-methylenehexadecanoic acid 144, 145.
8, 9-methyleneheptadecanoic acid 144.
11, 12-methyleneoctadecanoic acid 141.
9, 10-methyleneoctadecanoic acid 144, 145.
Methyl heptadecanoate 174.
Methyl linoleate 170, 190.
Methyl linolenate 170, 190.
Methyl malvalate 148, 149.
Methyl oleate 170, 190.
Methyl phenyl silicone 199.
Methyl silicones 256.
Methyl sterculate 148, 149.
Microalgae 25.
Micropacked columns 40.
Milk 232.
Milk fat 67, 201.
Milk powder 80, 81.
Minnows 26.
Mobile phase 39, 40.
Modifiers 301.
Molybdophosphoric acid 13.
Momordica charantia 3.
Monoacylglycerides 99, 184, 195, 197.
Monogalactosyldiacylglycerides 46, 298.
Monoglycerides 24, 93, 202.
Monooleoylglycerol 198.
Moraiolo variety 363.
Mosses 310.
Moving belt interface 207.
Mucosa 233.
Multiple development 7.
MHE Multiple headspace extraction 60.
Multiple linear regression 340, 341, 344, 366.

Multiple reaction mode 191.
Multivariate analysis 317, 335, 351.
Multivariate spectroscopy 363.
Mungaba oil 149, 152.
Mustard seed 274.
Myocardial ischaemia 186.
Mycolic acid 136, 300.

-Naphthol 15.
Natural waters 23.
R(+)Naphthylethylamine polymer 197.
Near infra red spectroscopy 52, 325.
Nebulizer 183, 198, 294.
Neisseria sp. 237.
Negative Ion Mode 181, 198.
Neural nets 329, 354.
Neurons 354.
Neutral lipids 8.
Nicotinate derivatives 145, 184.
Ninhydrin 14.
Nitrogen dioxide 190.
3-Nitrobenzoyl alcohol 226.
3-Nitrophenyl alcohol 209.
Nocardia asteroides 300.
Nonadienal 72.
Nonanal 72, 77, 81.
Nonanoic acid 69.
Non-specific reagents 13.
4-(2'-nonyl-cyclopentyl)butanoate 159.
3-(2'-nonyl-cyclohexyl)propanoate 160, 171.
Nuclear Magnetic Resonance spectroscopy 87, 318, 326.
Nuclear Overhouser effect 89.
Nuclear spin quantum number 326.
Nucleosil 5SA 162, 263.

Pachira aquatica 149.
Packed column 39.
Palmitate 141.
Palmitic acid 287, 288, 319, 324.
Palmitoleic acid 319, 324.
sn-1-Palmitoyl 2-stearoylglycerol 196.
Palm kernel oil 45, 198, 266, 267, 274, 277, 278, 280, 281, 282, 284, 285.
Palm mid fraction 272, 273.
Palm oil 45, 199, 265, 266, 267, 270, 271, 272, 273, 278, 281, 284, 285, 286.
Palm olein 265, 272, 273, 275, 277, 284, 285, 286.
Palm stearin 273, 275.
Pancreatic islets 233.

Parsimony 359.
Partial least squares regression 318, 339, 350, 366.
Peanut oil 171, 175, 333.
Pee Dee Belemnite 270.
Pentadecane 256.
Pentaenoic acid 13.
Pentafluorobenzyl derivatives 153, 157, 162.
Pentafluoro benzyl oxime 187, 204, 211.
Pentanal 76, 77, 78.
Pentane 60, 71, 77, 80, 256.
Pentanol 76, 77.
Pentanone 81.
Pentyl furan 77.
Peptostreptococcus anaerobius 68.
Perception 355.
Perchlorethylene 82.
Peroxide value 79.
Peroxides 204.
Peroxidation 206, 223.
Petroleum 27.
Petroleum ether 266.
Phase diagram 36.
Phenacyl bromide 156, 157, 162.
Phenol 319.
Phorbol esters 225.
Phosphatidic acid 119, 124, 125, 215, 291.
Phosphatidyl choline 14, 17, 46, 93, 117, 120, 121, 126, 127, 132, 133, 175, 190, 207, 297, 300.
Phosphatidyl ethanolamine 6, 14, 117, 118, 120, 128, 212, 292, 300.
Phosphatidyl glycerol 18, 188, 206, 226, 230.
Phosphatidyl inositol 6, 120, 215, 218, 292, 300, 302.
Phosphatidyl serine 6, 14, 120, 128, 215, 216, 292, 302.
Phosphoceramide 229.
Phospholipids 8, 18, 24, 109, 117, 122, 206, 290, 291, 296, 301, 303, 309, 310.
Phospholipase 18, 206, 226, 230.
Phosphoric acid 18.
Photooxidation 224.
Photosynthetic fixation 269.
Picolinyl ester 139, 140, 141, 142, 145, 146, 154, 164, 165, 170, 171, 184.
Pig depot fat 109, 112.
Planck's constant 326.

Plasma human 193, 195, 216, 225, 234, 304.
Plasma rabbit 193.
Plasma spray 240, 294.
Plasmalogens 122, 208, 211, 213, 218, 299.
Plasmanyl choline 210.
Plasmanyl inositol 228.
Plasmenyl choline 210.
Plasmenyl ethanolamine 214, 304.
Polar lipids 290, 291.
Polyetherether ketone (PEEK) 312.
Polyisoprenyl phosphates 222, 223.
Polynucleotides 181.
Polypeptides 181.
Polyunsaturated glycerides 202.
Polyvinyl alcohol 292.
Polyvinylidene difluoride membrane 234.
Pongamia glabra 19.
Pongaglabrone 19.
Pongamol 19.
Positive ion mode 181.
Potassium dichromate 13.
Potato strips 168.
Power method 338.
Principal component analysis 318, 328, 330, 331.
Principalk component regression 318, 342, 352.
Process monitoring 18.
9-(2'-Prop-trans-1-enyl-cyclohex-cis-4-enyl)nonanoate 155, 158.
9-(2'-Prop-cis-1-enyl-cyclohex-cis-4-enyl) nonanoate 158, 165.
9-(2'-Propyl-cyclohex-cis-4-enyl)non-cis-8-enoate 158, 159.
9-(2'-Propyl-cyclohex-cis-4-enyl)non-trans-8-enoate 158, 166.
9-(2'-Propylcyclohexyl) nonanoate 160.
10-(2'-Propylcyclopentenyl)dec-cis-9-enoate 158, 165.
10-(2'-Propylcyclopentenyl)dec-trans-9-enoate 158.
10-(2'-Propylcyclopentyl)decanoate 171, 185.
Prostaglandins 189.
Prostanoids 189.
PGF 189.
Protein 254.
Protonated molecular ion 202, 203.
Protozoa 228, 310.

Pseudomolecular ion 217, 235.
Pseudomonas fluorescens 296, 299, 300.
Punicic acid 3.
Purge and trap 80.
Purity 265.
Purslane 298.
Pyrolysis mass spectrometry 318, 328, 330, 331.

Quadrupole mass spectrometry 182, 230.
Quantitation 16.
Quartz silica 19.
Quattro nuclei probe 88.

Rabbit 222.
Rabbit lung lavage fluid 209.
Raffination 320.
Raman Spectroscopy 325, 368.
Rancimat 77.
Randomized samples 203.
Rapeseed 5, 265, 266, 267, 275, 277, 278, 284, 286, 288, 322.
Raspberry (*Rubus idaeus*) 201.
Rat air pouch 191.
Rat brain 212, 217, 218, 231, 296, 299.
Rat heart 175.
Rat liver microsomes 224.
Rats 198.
Red Blood cells 217, 223, 224, 229.
Redfish 5.
Regioisomeric forms 203.
Repeatability 129.
Reproducibility 129.
Resorcinal 15.
Response factors 199, 203.
Retro-Diels Alder 164.
Reversed phase 5.
Reverse phase HPLC 202, 301.
Rhizobium leguminosarum 238.
Rheumatoid arthritis 321.
Rice bran 19.
Riprostil 189.
Robustness 131.
Ruminants 67.

Safflower oil 199, 267, 275, 284, 285, 286.
Saliva 211.
Salmon (Atlantic) 24.
Salmonella minnesota 237.
Salmon oil 106, 107.

Sampling 63.
Sapindaceae 147.
Sarcolemma (canine) 208, 212.
Schiff base 229.
Scotch whisky 329.
Sea scallop 5, 25.
Seal oil 107.
Self prediction 359.
Sensory wheels 323.
Sesame seed 275, 284, 286.
Shelf life 250.
Shigella flexneri 237.
Shrimp 25.
Sialic acid 234.
Signalling systems 291.
Silar 5CP 140, 267.
Silica gel 2, 3, 16.
Silver nitrate TLC 6, 18.
Silver ion HPLC 139, 154.
Single cell oils 200.
Single development 7.
Singlet oxygen 189.
Sitostanol 192.
Sitosterol 278.
Sitosteryl glycoside 116, 117.
Slip melting point 270, 273.
Snails 9.
Sodium adducts 201, 210, 217, 218.
Sodium Borohydride 205.
Sodium cyanoborohydride 229.
Sodium hypochlorite 15.
Soft ionization mass spectrometry 181, 294.
Soil 144.
Solid phase extraction 172.
Solvent systems 7.
Soyabean 73, 219, 265, 267, 272, 277, 280, 281, 284, 285, 299.
Soyabean lecithin 120.
Soyabean oil 175, 333.
Soyabean phospholipids 16, 122, 223.
Soya protein 283.
Sphingenine 219, 231, 307.
Sphingomyelin 132, 133, 210, 230, 291, 292, 299, 304, 308, 309.
Sphingosine 230, 304.
Squalene 325.
Stable isotope dilution 194, 195.
Stable carbon isotope ratio (SCIR) 269, 283.
Standard deviation 359.
Stationary phases 39, 144, 147.

Statistical methods 318.
Stearic acid 319, 324.
Stearin 271.
sn-1-Stearoyl-2-palmitoyl-glycerol 196.
1-Stearoyl-2-arachidonyl sn-3-glycerol 196, 197.
1-Stearoyl-2-oleoyl sn-3 phospho-serine 216.
Sterculia foetida 149, 152.
Sterculiaceae 136, 147.
Sterculic acid 137, 147, 149, 150, 151.
Steroids 18.
Sterols 13, 192.
Steryl esters 6, 9.
-Stigmastenol 278.
Stigmasterol 278.
Sulphatides 233.
Sunflower oil 43, 153, 167, 168, 169, 170, 174, 267, 275, 277, 278, 280, 281, 282, 284, 285, 286, 322, 333.
Supelcowax 10 148.
Supercritical fluid extraction 34.
Switching valve 257.

Tandem mass spectrometry 290.
n-Tetradecane 10, 11.
Tetradecene 256.
Tetrahydrofuran 301.
Tetrahydropyranyl derivatives 221.
Tetralinoleoyl GPG 221.
Tetramethylsilane 327.
Thermospray 181, 290, 294.
Thiobarbituric acid 79.
Thioglycerol 226.
Thrombin 304.
Thromboxanes 189, 191.
Tiliaceae 147.
TLC 1, 261, 300.
TNT 25.
Tocopherols 269, 277, 281.
Tocotrienols 269, 281.
Torpedo marmorata (Flounder) 299.
TPH Total Petroleum Hydrocarbon 22.
Triacylglycerides 35, 99, 198.

Trichosanthus anguina 3.
Triethanolamine 213, 216.
Triethylamine 187, 222.
Triglycerides 10, 12, 23, 42, 46, 51.
Trilinolenoyl glycerol 205.
Trinitrobenzene sulphonic acid 217.
Trimethylsilyl derivatives 184.
Triolein 300.
Tripalmitin 98.
Triphenylphosphine 186.
Triphenylphosphonium valerylamide derivative 190.
Triponosoma brucei 229.
Triponosoma cruzi 229.
Tropical grasses 270.
Tubular TLC 17.
Tumour 236.
Two dimensional TLC 9.

Ultrasphere-ODS 297.
Urea clathration 154, 171, 175.
Urine 189, 223.

Validation 129, 132, 345.
Van Deempter 47.
Vanillin 283.
Vapour pressure 61.
Variables 336.
Variance 338, 360.
Vectors 336.
Very Low Density Lipoprotein VLDL 198.
Vinyl methyl ether adducts 145.
Virgin olive oil 319, 331.

Water 12.
Waxes 319.
Wilkinson's catalyst 156.
Windows 295.

Xanthomas 194, 226.
X-rays 251.

Zinzadze reagent 14.